Science
and the
Making
of the
Modern
World

JOHN MARKS

Science
and the
Making
of the
Modern
World

HEINEMANN
EDUCATIONAL

Heinemann Educational
A Division of Heinemann Publishers (Oxford) Ltd
Halley Court, Jordan Hill, Oxford OX2 8EJ

OXFORD   LONDON   EDINBURGH
MADRID   ATHENS   BOLOGNA   PARIS
MELBOURNE   SYDNEY   AUCKLAND   SINGAPORE   TOKYO
IBADAN   NAIROBI   HARARE   GABORONE
PORTSMOUTH (NH) USA

First published 1983
94   95   96   97   15   14   13   12   11   10   9

British Library Cataloguing in Publication Data

Marks, John
    Science and the making of the modern world.
    1. Science—History
    I. Title
    509   Q125

ISBN 0 435 54781 X

FOR

Finola, Kevin, Rosalind and Cordelia

Filmset by Eta Services (Typesetters) Ltd, Beccles, Suffolk

Printed and bound in Great Britain by
Athenaeum Press Ltd, Gateshead, Tyne & Wear.

FOREWORD by Mary Hesse

The impact on society of science as a world-wide institution is perhaps the most striking feature of the late 20th century. The influence of science is felt through its applications, both benign and malign; through its experimental method, which can be both discouraging to dogma and an invitation to irresponsible exploitation; and through its effects in the world of ideas, where it can almost be said, in the anthropologist's sense, to constitute the modern mythology. For its own health such a dominant social institution needs critique and self-reflection. It is salutary, therefore, that since World War 2 there has been a proliferation of studies in the history, philosophy and sociology of science, and that courses based on these studies have been developed in schools, colleges and universities. Such courses attempt to introduce science students to the wider significance of their sometimes narrow specialisms, and also to encourage reflection on the place of science in history, philosophy and society among students of all disciplines.

The group of subjects that go under titles such as history, philosophy and social studies of science can hardly yet be said to be a unified discipline. There is a variety of approaches, ranging from the more traditional historical and philosophical, to investigations of the contemporary use of science and technology and the social and political dilemmas arising therefrom. Such studies inevitably raise political issues, not only in respect of particular scientific applications, but also in respect of the critical approach to science as such. Is the history, philosophy and sociology of science to underwrite the mood of 'anti-science' that appeared in some educational contexts in the '60s and '70s? Or is it to support unquestioningly the scientific orientation of modern western capitalism or of modern socialist societies? Or is it, in the enlightened tradition of liberal education, to attempt to stand back from immediate conflicts in order to reflect upon their origins, their varying arguments and social embodiments, and to encourage as far as possible objective debate? If it is to perform the latter, and surely most desirable, task, it is necessary for there to be not only a good grounding in the various sciences, but also an academic core which draws upon established standards in the relevant disciplines including history, philosophy and sociology. Only against such a background can current issues be responsibly discussed.

In this book John Marks has attempted just such a balanced and reflective approach. He has succeeded admirably in setting out what the educated scientist or scientific layperson needs to know to bring an informed and critical mind to the contemporary place and influence of science in society. He and his collaborators have ranged widely over the sciences, and he has broken new ground in a book of this sort by a comparative study of science in a variety of ideologically different 20th-century societies. The book should provide indispensable reading for introductory courses in history, philosophy and social studies of science in schools and institutions of higher education.

PREFACE

... that revolution ... outshines everything since the rise of Christianity and reduces the Renaissance and Reformation to the rank of mere episodes ... it changed the character of men's habitual mental operations ... while transforming the whole diagram of the physical universe and the very texture of human life itself ...
... [it is] the real origin both of the modern world and the modern mentality.
Professor Sir Herbert Butterfield in *The Origins of Modern Science, 1300–1800*.

The revolution in question is not the French Revolution of 1789 or the Russian Revolution of 1917. It is not a political revolution at all. It is the scientific revolution of the 16th and 17th centuries associated with names like Copernicus, Galileo and Newton. The very word 'revolution' in its modern sense comes from the title of Copernicus' classic book on astronomy *On the Revolution of the Heavenly Spheres* which was published in 1543. All Copernicus meant by it was 'rotation'. And it was because of the enormity of the changes which his work heralded—changes summarised so graphically by Professor Butterfield—that revolution has come to mean something like 'to turn the world upside down'.

Since 1949, when Professor Butterfield wrote those words, science has become even more a part of our everyday lives. Today's child takes television, space travel, transistors and antibiotics for granted, all in their infancy in 1949. Tomorrow's will probably do the same with silicon chip microprocessors, organ transplants and some of the products of the new biotechnology.

This book aims to give an account of how science has developed and how it has come to play such a central part in our lives. And it tries to provide students with some of the background knowledge they need if they are to make an informed and intelligent contribution to contemporary debates on the interactions between science, technology and society. The book is based on a course which is taken by all science undergraduates at the Polytechnic of North London. It could therefore be useful for staff and students on similar courses in other polytechnics, universities and colleges of higher and further education. Moreover, since it does not assume much knowledge of science, it could serve as an introduction to the natural sciences for students of the social sciences and also for a wide range of sixth-formers in their general studies.

CONTENT

The book deals with four main topics and the interactions between them:

(a) the historical development of scientific knowledge;
(b) the methods used in establishing this knowledge;
(c) the scientific community and how it works;

(d) how science has developed in different societies and how these societies have made use of science.

Parts 1–5 give an historical overview up to about the end of the 19th century. The rest of the book deals with the 20th century, starting in Part 6 with a descriptive review of some major advances in scientific knowledge. Chapter 7.1 outlines some general ideas about how science and technology develop and how their development is related to political and social systems of different kinds. The rest of the book attempts to test these ideas in two ways—by comparing the overall growth of science and technology in different kinds of society; and by studying a number of specific scientific and technological issues, looking at these in different kinds of society where possible.

Throughout, the main focus is on the development of scientific knowledge which, despite its limitations, is the most substantial and universal body of interlocking theoretical and empirical knowledge yet achieved. That is why science pervades so much of our lives in the 20th century and why it is important for us all to learn more about it.

But why treat science historically? Why not just give an overview of science today? And why compare science in different societies? There are many reasons and these are some of the most important.

Firstly we can try to show how scientific knowledge must always be tentative by describing the ways in which ideas, once firmly held, have changed—for example, how Newton's ideas were both superseded and incorporated in the work of Einstein or how the idea that species were permanent and fixed has been modified by Darwin's theory of evolution.

Next we can describe how, as science grew, specialist sciences developed—how physics and chemistry developed from natural philosophy and how biology and geology emerged from natural history. And we can also show the reunification of many of these specialisms in some of the great scientific discoveries of this century—for example, in the way that knowledge of atomic structure is now central to a proper understanding of much of physics and chemistry, or how knowledge of the structure of DNA, established largely by physical and chemical techniques, has revolutionised our understanding of biology.

Thirdly, we can indicate *both* how science depends on the climate of thought prevailing at any particular time *and* how much of science is cumulative and universally valid.

Finally, by studying science at many different times and in many different societies, we have a better chance of understanding just what is special about science, what conditions are needed if it is to flourish and how we can control potential abuses of its powers.

The approach adopted in this book differs from that of many books on science and society. Many of these concentrate on case-studies of problem areas such as environmental pollution or the development of nuclear power. This approach is less satisfactory because most students—and others—do not have sufficient basic knowledge of science and how it works to put such case-studies properly into context. More fundamentally, those who adopt the case-study approach often restrict themselves to studying these problems in only one kind of society—that is in liberal societies like our own—and make little or no effort to present evidence from other kinds of society. This is both intellectually dishonest and misleading since only by comparative analysis can it be seen whether certain problems are unique, specific to a particular kind of society, or universal, and, if they are not unique, whether the problems are worse in one kind of society rather than in another.

This book includes some case-studies, but it deals with them *after* the historical and comparative material has been presented. It is hoped that students will find this more intellectually satisfactory and that it will enable them to make a more balanced assessment of controversial issues of science policy.

* * *

POSTSCRIPT FOR THE ACADEMICS

Many sources have been used in writing this book both from the sciences themselves and from the history, sociology and philosophy of science. Some of these sources are acknowledged in the references and the guide to further reading (page 499). However, these lists are by no means complete since, if they were, they would be too long to be of very much use to students. The historical sections of the book attempt to combine the advantages of both internalist and externalist approaches while retaining a strong emphasis on the internal history of the development of scientific knowledge. The sociological sections are largely empirical and do not deal directly with approaches which speak of 'the social construction of scientific knowledge'. The philosophical sections emphasise the old-fashioned virtues of induction and empirical falsifiability and do not deal with topics such as the writings of 'critical theorists' of the Frankfurt school or with those who believe that, methodologically speaking, 'anything goes'.

These omissions are deliberate. The approach adopted here is designed to give students insight into the nature and development of scientific knowledge and to provide them with a coherent intellectual framework within which they can appraise current trends in the history, sociology and philosophy of science.

If any philosophical approach is favoured, it is the network model of science as recently emphasised in the work of Mary Hesse.[1]

And the theoretical approach to social questions which I have found most helpful is that of F. A. von Hayek,[2] most particularly his distinction between constructive and evolutionary rationalism and his concept of a spontaneous order.

REFERENCES

1 Hesse, M. B., *The Structure of Scientific Inference*, (London: Macmillan, 1974).
2 von Hayek, F. A., *Law, Legislation and Liberty* (3 vols.), (London: Routledge and Kegan Paul, 1973, 1976, 1979).

CONTENTS

FOREWORD *by Mary Hesse*   v

PREFACE   vi

PART ONE—INTRODUCTION
The importance of science in the making of the modern world   I

PART TWO
The origins of modern science

   2.1   Classical Greece and Greek astronomy   9
   2.2   The Copernican revolution   18

PART THREE
The 17th-century scientific revolution

   3.1   Galileo and his conflict with the Catholic Church   32
   3.2   Developments in the experimental tradition: new
         instruments and medical knowledge   44
   3.3   Descartes, the mechanical philosophy and the rise of
         mathematics   58
   3.4   Newton and his *Principia*   65
   3.5   Communication between scientists: the origins of the
         scientific community   78

PART FOUR
The development of the scientific revolution in the 18th and 19th centuries

   A WIDER CONSEQUENCES IN THE 18th CENTURY
   4.1   Newton and the Enlightenment   87
   4.2   The origins of the social sciences *by Caroline Cox*   103
   4.3   The American and French revolutions: ideas and
         consequences   117
   4.4   The early Industrial Revolution   129

   B THE GROWTH IN SCIENTIFIC KNOWLEDGE
   4.5   The origins of modern chemistry   144
   4.6   The making of geology   156

4.7   Physics in the early 19th century   165
4.8   Biology and Darwin's theory of evolution   174

C ACCELERATION IN THE 19th CENTURY
4.9   Science and the Industrial Revolution in the 19th century   186
4.10 The rise of the social sciences *by Caroline Cox*   196
4.11 Science, industrialisation and the world-wide expansion of
      European influence   212

PART FIVE
Science and technology in traditional China   222

PART SIX
The growth of science in the 20th century

6.1   Science and the 20th century   240
6.2   Electromagnetic waves and relativity: Maxwell and Einstein   251
6.3   Einstein's work and some of its social implications   263
6.4   Genetics and molecular biology   276
6.5   Developments in the medical sciences *by Caroline Cox*   294
6.6   The unification of chemistry and physics: Mendeleev, Bohr
      and Schrödinger   311
6.7   The origins and development of statistics and computing   324
6.8   The physical evolution of the universe: the modern world
      picture   338

PART SEVEN
**The philosophy of science, political systems and the
development of technology in the 20th century**

7.1   The philosophy of science, the scientific community and
      political systems   357
7.2   Science and National Socialism in Nazi Germany   368
7.3   Science, technology and government in liberal capitalist
      societies   378
7.4   Science and the Russian Revolution   389
7.5   Science and technology in Japan since the Meiji restoration   408
7.6   Science and technology in China before and after 1949   418

PART EIGHT
Case-studies of specific issues

8.1 Science, technology and economic development in the
third world *by Peter Mould*   429
8.2 Nuclear power and energy resources   441
8.3 Semiconductor physics and the computer revolution   453
8.4 Some applications of biological knowledge   464
8.5 Malthus revisited? The population explosion and the limits
to growth debate   471
8.6 The natural and the social sciences: similarities and
differences *by Caroline Cox*   481

PART NINE—CONCLUSION
Science and the making of the modern world   490

GUIDE TO FURTHER READING   499

INDEX   502

ACKNOWLEDGEMENTS

Many people have helped me in planning and writing this book and in developing the lecture course on which it is based. I would particularly like to thank Professor Mary Hesse (Professor of Philosophy of Science, University of Cambridge), who, in her role as external examiner for the course, has given generously of her time and has been a source of much advice and support which was especially valuable when the course was in its formative stages.

I am also deeply indebted to Lady Caroline Cox (Director, Nursing Education Research Unit, Chelsea College, University of London) who first introduced me to the sociology of science and who, over many years, has been an invaluable mentor, stimulus and critic. Without her wise counsel and constant encouragement, this book would never have been completed. In addition, she has contributed four chapters on the social and medical sciences which are based on lectures given to my students over the last four years.

I am also grateful to Peter Mould (Director, Federal Agricultural Coordinating Unit, Ibadan, Nigeria) for contributing a chapter on economic development in the third world which was also based on lectures given to my students.

Professor Hesse, Lady Cox and Professor Donald Cardwell (Professor of the History of Science and Technology, University of Manchester Institute of Science and Technology) have devoted much valuable time to reading a very bulky typescript; and their helpful comments have greatly improved the text and have enabled me to correct a number of errors and misinterpretations. The errors or defects which remain are, of course, my responsibility.

Mr J. P. Owen did sterling work on many of the illustrations and in chasing a number of elusive references, while Mrs Phyllis Domone and Mrs Bonnie Woon continually performed miracles in transforming—rapidly and efficiently—many pages of handwritten drafts into a clear and neat typescript.

Finally, I would like to thank my students at the Polytechnic of North London who helped to shape this book by their response to my early tentative attempts to discuss the complex interaction between science, technology and society.

Acknowledgement is due to the following for kindly supplying photographs and for permission to publish them:

American Institute of Physics 7.2.1; Ann Ronan Picture Library 2.2.8, 3.1.1, 3.1.3, 3.1.4, 3.2.3, 3.3.3, 3.4.4, 3.5.2, 4.1.2, 4.4.11, 4.9.6; Professor S Arnott 6.1.7; Argonne National Laboratory/Chicago Historical Society 6.3.3; The Associated Press Ltd 6.3.2; Barnabys Picture Library 4.2.5; BBC Hulton Picture Library 4.9.4, 7.4.4; Biophoto Associates 6.1.3, 6.4.7; The British Library 2.2.5, 2.2.7, 3.2.10, 3.2.11, 3.5.3, 4.6.1; The British Museum 4.4.2, 4.4.14; Courtesy of the Archives, California Institute of Technology 6.4.3; Gene Cox 6.1.2; Crown Copyright/COI 6.7.9; Derby Art Gallery 4.1.4; Direction des Archives de France 4.3.3; Ferranti Electronics Limited 6.7.8; Ferranti plc 6.7.1; The Fotomas Index 2.2.2, 2.2.3, 3.1.5, 3.2.5, 3.2.12, 4.1.7, 4.1.8, 4.1.9, 4.1.10, 4.1.11, 4.8.3; The Hebrew University of Jerusalem 6.2.4, 6.3.1; The Institution of Electrical Engineers (S C Mss 52) 4.9.7; Keystone Press Agency 6.3.4, 7.2.2, 7.4.3, 7.4.5 (photograph of Sakharov); The late Dr Joule: lent to Science Museum, London 4.7.2; Los Alamos National Laboratory 6.1.1; Manchester Central Library 4.4.5, 4.4.6; The Mansell Collection 3.1.2, 3.3.1, 4.1.3, 4.3.1, 4.3.2, 4.3.5, 4.5.1, 4.5.4, 4.8.2, 4.9.5, 4.9.8, 6.7.3; Mary Evans Picture Library 1.1; The Master and Fellows, Magdalene College, Cambridge 4.1.1; Dr Z A Medvedev 7.4.5 (photograph of Vavilov); Michael Holford 4.4.7; Mullard Radio Astronomy Observatory, Cambridge 6.8.5; NASA 6.1.4, 6.1.5, 6.8.1; National Portrait Gallery 3.4.1, 4.4.4; © 1948 by The New York Times Company 8.3.1; Novosti Press Agency 6.6.1; Courtesy of The Open University 4.6.4; Palomar Observatory 1.2, 6.8.2, 6.8.4; Popperfoto 6.3.5; Royal Astronomical Society (Monthly Notices of the RAS, vol. 144, no. 1, 1969. Dr G G Pooley) 6.8.6; By courtesy of the Royal Institution 4.7.3, 4.7.4; Science Museum, London 3.2.2, 3.2.4 (Crown copyright), 3.2.6, 3.2.7, 4.1.5, 4.4.8, 4.4.10, 4.4.12 (Crown copyright), 4.5.2 (Crown copyright), 4.5.3, 6.7.5, 6.7.6 (Crown copyright), 6.7.7 (Crown copyright); Smithsonian Institution (Photo. no. 61480-A) 2.2.1; Society for Cultural Relations with the USSR 7.4.1; Southeby Parke Bernet & Co. 5.13; Wellcome Institute Library, London 3.2.9, 4.10.4, 4.10.5, 6.5.2, 6.5.3, 6.5.4, 6.5.5, 6.5.6, 6.5.7; Reprinted by permission John Wiley and Sons Inc. from *Astronomy: Fundamentals and Frontiers* by R Jastrow and M H Thompson © 1972 by R Jastrow 6.8.3; The William Salt Library 4.10.2.

Quotations on pages 79, 81, 147, 163, 328 from *Dictionary of Scientific Biography*, edited by C. C. Gillispie, vols I, VIII, IX, XIV, copyright © 1970, 1973, 1974, 1976, American Council of Learned Societies, reprinted with the permission of Charles Scribner's Sons.

Quotations on pages 97, 147, 157, 161 from Charles Coulston Gillispie, *The Edge of Objectivity: An Essay in the History of Scientific Ideas*, copyright © 1960 by Princeton University Press, excerpts pp. 174, 206–7, 282, 284, 293 reprinted by permission of Princeton University Press.

The importance of science in the making of the modern world

Modern science is often taken for granted in many discussions of present and future events. It is assumed to be there—to be available—to be used or misused by us in shaping our societies and our lives.

One of the main purposes of this book will be to try to persuade you not to take it for granted; to learn a little about how it arose and how it develops; and to appreciate how extraordinary, how recent and, in some ways, how precarious a thing it is.

It is vital for us to try to understand how modern science works because the rise of science has had a much greater influence on recent world history than any other single factor. This influence has been both far-reaching and largely unplanned, and it goes far beyond the mere accumulation of knowledge, enormously important and fascinating as that is.

There are three main areas where the direct and indirect influence of science has dramatically altered the type of society in which we live.

Firstly, it was influential in the origins of the Industrial Revolution and in its subsequent development. The Industrial Revolution started in Britain in the later part of the 18th century. Its key components were cotton, iron and steam—the increasing mechanisation of cotton spinning and weaving, the manufacture of iron on a much larger scale than ever before, and, perhaps most important, the development by James Watt and others of a new source of power, the steam engine. Separately and in combination, these technical developments led to the emergence of the first factories and to the growth of a totally new form of transport, the railways. Starting in Britain in a relatively small way, industrialisation has now spread nearly all over the world. And it has probably changed people's lives more in the last 200 years than in the whole of the rest of recorded history—the only possibly comparable development being the emergence of agriculture around 12,000 years ago.

Secondly, certain ideas which have been fostered by modern science—for example, the idea of progress and of people's potential ability to control both nature and their own destiny—have significantly influenced three great political revolutions of the last 200 years: the American, the French and the Russian revolutions. The American Revolution of 1776 with its resounding Declaration of Independence and the French Revolution of 1789 with its famous slogan of 'Liberty, Equality and Fraternity', have inspired the political institutions of many countries, while the Russian Revolution of 1917 has probably, during this century, influenced the political systems of even more countries. Whatever you think of these revolutions and their consequences, they have undoubtedly shaped the modern world, and continue to do so.

Finally, science has had an enormous influence on man's idea of his place in the universe and of his relationship to other forms of life. From being the inhabitant of a body which was at the centre of the universe, and around which the rest of the

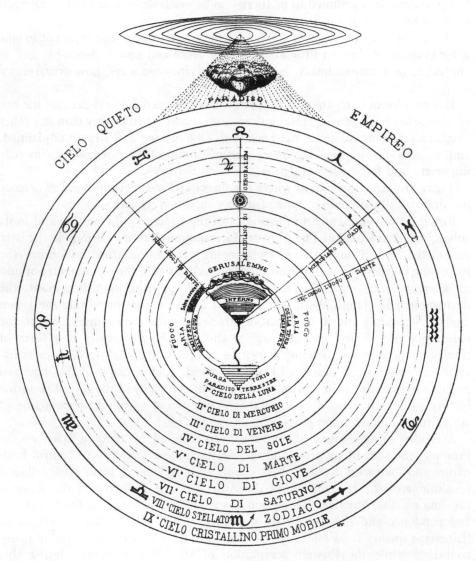

Fig. 1.1 The medieval view of the universe as portrayed in Dante's *Paradiso*.

universe revolved, man is now seen as living on a tiny planet of a minor, common star in one of millions of galaxies in a vast universe.

Fig. 1.1 is an illustration from *Paradiso*, the third book of the *Divine Comedy* by the 14th-century Italian poet Dante. It shows the medieval European view of the universe with the earth firmly at the centre, the planets and stars revolving about it and Paradise beyond.

Compare that diagram with a modern photograph (Fig. 1.2) of a rotating spiral galaxy, taken using the 200-inch telescope at Mount Palomar in California. The Milky Way is a galaxy like this and our sun is situated not at its centre but somewhere near the edge along one of the rotating spiral arms. And our galaxy is just one of many others which are now known to exist.

So man's physical habitat, the earth, is not so special as it was once thought to be. Nor does man now occupy the same special place as he was once thought to do in the vast diversity of living things. Charles Darwin's ideas on the evolution of life and the Descent of Man, the title of one of his best-known books, have firmly established man as being closely related to all other forms of life. These changes have profoundly affected man's view of his purpose in life and have greatly influenced the increase of predominantly secular as opposed to religious attitudes or world-views. Science has thus affected, in major ways, the economic, the political and the religious aspects of life, and it is continuing to do so.

Fig. 1.2   Spiral galaxy.

When did all this start? We can't, obviously, pinpoint an exact date, but it is generally agreed that the significant period was some time between the death of Copernicus in 1543 and the work of Isaac Newton in the 1680s. In this period there originated in Europe a series of changes which have come to be called the Scientific Revolution. Professor Herbert Butterfield, in his book *The Origins of Modern Science*, puts it like this:

> The 17th century, indeed, did not merely bring a new factor into history, in the way we often assume—one that must just be added, so to speak, to the other permanent factors. The new factor immediately began to elbow at the other ones, pushing them out of their places—and, indeed, began immediately to seek control of the rest, as the apostles of the new movement had declared their intention of doing from the very start. The result was the emergence of a kind of Western civilisation which when transmitted to Japan operates on tradition there as it operates on tradition here—dissolving it and having eyes for nothing save a future of brave new worlds. It was a civilisation that could cut itself away from the Graeco-Roman heritage in general, away from Christianity itself— only too confident in its power to exist independent of anything of the kind. We now know that what was emerging towards the end of the 17th century was a civilisation exhilaratingly new perhaps, but strange as Nineveh and Babylon. That is why, since the rise of Christianity, there is no landmark in history that is worthy to be compared with this.[1]

This scientific revolution of around 300 years ago may seem very distant in time to us today with our immediate concerns relating, say, to tomorrow's shopping, next month's rent or next year's examinations. Yet 300 years is a very short time indeed compared with some other time-periods we might consider.

First let us look at historical time, that is, at periods from which written records survive. This extends back about 5000 years, beyond the Classical Greek civilisation of about 500 BC, to the civilisations clustered around the Eastern Mediterranean at about 3000 BC. The civilisations of Ancient Egypt, Babylon and Sumeria were all thriving at this time.

Beyond 5000–6000 years ago we enter the era of pre-history, with no written records, and only surviving bones, fossils and tools or other artefacts to tell us what life was like. We know that around 8000 years ago ($\sim$ 6000 BC) bronze and iron were used for tools and implements in the Bronze and Iron Ages. We also know that, about 12,000 years ago ($\sim$ 10,000 BC), some people ceased to live the nomadic life of hunter-gatherers and started to cultivate the land and set up permanent settlements. This was the Agricultural or Neolithic revolution.

If we go further back we reach the cave-dwelling hunter-gatherers who made many striking paintings on the walls of their caves, such as those found at Lascaux in France which date from around 15,000 years ago. It is at this time, or perhaps a little earlier—say 30,000 years ago—that the first statues appear. It would be interesting to know how symbolic objects such as paintings and statues were related to the evolution of language. Unfortunately, verbal utterance leaves no permanent records for us to recover!

The men of this period, the late Stone Age, probably lived in temporary hunting encampments, and we know that they used fairly well-formed stone tools and weapons since many of these implements have been found by archaeologists. Some of these people were of the same species, *Homo sapiens*, as us, but others were of a different but related species called Neanderthal man. A typical Neanderthal skull dates from about 70,000 years ago; it has a different shape from a *Homo sapiens* skull, with an appreciably smaller brain cavity and prominent eyebrow ridges.

Now let us jump back ten times further in time to about three-quarters of a million years ago. We now find a different species of man, Peking man, who used very primitive tools. Many relics—bones and stone tools—of these early Stone Age people have been found in the Olduvai Gorge in Tanzania. The earliest date back to around 2 million years ago.

Now let us jump back yet another ten times further in time to around 20 million years ago, where we find that the dominant species on earth are no longer men but are still mammals such as apes, mammoths and earlier ancestors of the horse.

Now if, for the third time, we jump back in time by another factor of ten, we reach the age of the dinosaurs, around 200 million years (Ma) ago. The dominant species now are enormous reptiles like the *Tyrranosaurus rex* or the *Diplodocus* which might be 30 metres long and weigh many tonnes. We can appreciate their size from Fig. 1.3, which shows some tiny shrew-like mammals, contemporary with the dinosaurs, one of which is drinking from a pool of water in a dinosaur's

Fig. 1.3 Early mammal reconstruction. The *Amblotherium*, little larger than a present-day mouse, is shown drinking from a puddle in a dinosaur's footprint. *From Evolution, Part 2, The Age of Dinosaurs, Filmstrip DW–F59, Diana Wylie Ltd.*

footprint. It is interesting to realise that the earliest mammals looked like this—so small as to be probably beneath the notice of the dinosaurs. And yet all the mammals which now exist, including us, have evolved from creatures like this who lived around 200 Ma ago.

All these different forms of life have existed on earth while the sun has made one revolution about the centre of our spiral galaxy (see Fig. 1.2). And man has only been around for 2 Ma or so, about one per cent of that time.

Let us resume our time-travelling and go back another 100 Ma to the Carboniferous period. Now the earth is largely covered by dense forests which gave rise to most of the coal and oil deposits which now exist.

A further 100 Ma take us to the Devonian period, 400 Ma ago, when life was still confined to the seas. Many different kinds of fish and other sea-creatures existed at this time, some of them quite large.

200 Ma further back and we are in the Cambrian period of 600 Ma ago, when life was still aquatic but much less complex and differentiated, consisting mainly of early shell-fish and other invertebrate creatures.

Before this the fossil record thins out considerably, since earlier forms of life, largely single-celled organisms like algae or bacteria, have no hard parts and so rarely form fossils. However, there are many geological strata beneath the fossil-rich strata and we can date them radioactively. In this way we can push back our terrestrial time-scale a lot further—to around 3800 Ma, the date of the oldest known terrestrial rocks.

Fig. 1.4 sums up much of our time-travelling. Around the outside in a linear scale running clockwise from the origin of the earth (4600 Ma ago), through the origin of life (3400 Ma), the first green algae (1000 Ma), life on land (400 Ma), and the first mammals (200 Ma) during the age of the dinosaurs. On this scale the age of man (2Ma) is represented by a line less than half a millimetre in thickness (top left). In the centre this 2 million years of man's prehistory is shown on an expanded vertical scale. On this scale the agricultural revolution of around 10,000 BC comes only in the top millimetre while recorded history—the last 5000 years or so—occupies less than half a millimetre (see the extended inset on the right). The 17th-century scientific revolution is still too recent to show up clearly even on the scale of the inset for the last 10,000 years.

You might think we have gone back far enough, or as far as we can go. But, with the aid of astronomy, we can go back further yet. By studying the light reaching us from galaxies like the one shown in Fig. 1.2, and from other kinds of evidence such as that provided by radio-astronomy, we can try to trace the history of the universe back beyond the time of the formation of the solar system. This sort of work is more speculative than many of the things we have mentioned above but the current favoured theory amongst cosmologists, the so-called Big Bang theory, gives us a possible date for the origin of the universe of around 20,000 Ma—that is, about four times the age of the solar system.

Now let us pause and ask a question. How do we know all this?

On the whole, all of the information outlined above is accepted, well known, even if you didn't know it all. We accept that it is there and that we can refer to it

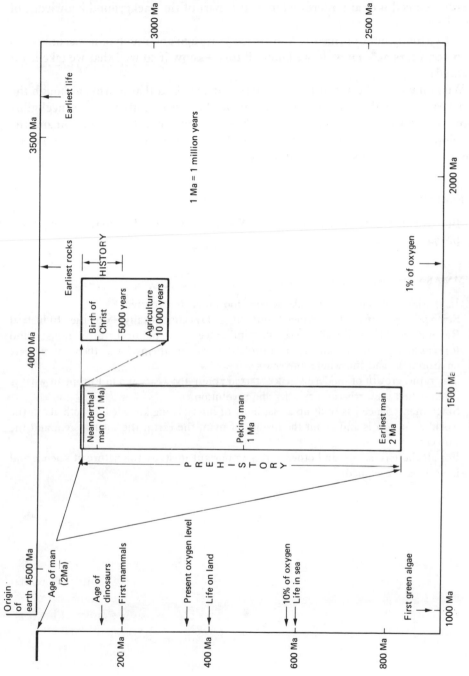

Fig. 1.4 The age of the earth.

if ever we need it or are interested in it. It is part of the background knowledge of our time.

Yet virtually none of this was known 300 years ago and much of it was unknown even 100 years ago. How do we know all this—know it so well that we take it for granted?

We know it on the authority of modern science. And that is why, although the 300 years since the scientific revolution of the 17th century seems negligible compared with the time-spans we have been considering, that revolution and the developments that followed from it are well worth further study. In this book we are going to make a start on that study.

REFERENCE

1   Butterfield, H., *The Origins of Modern Science 1300–1800* (London: G. Bell, 1951) p. 174.

SUMMARY

1   In the late 20th century we take science too much for granted.
2   Science has drastically changed our lives (a) economically—via the Industrial Revolution; (b) politically—via its influence on political revolutions; and (c) religiously—in its effect on people's view of themselves, their immediate environment and the whole universe.
3   The rapid growth of modern science started about 300 years ago in Europe in what is often called the 17th-century scientific revolution.
4   Since then science has built up a vast body of interlocking knowledge, both about the world as it now is and about the history of man, the earth, the solar system and the universe.
5   For all these reasons—and others—it is important to study the nature of science and how it has developed.

PART TWO

The origins of modern science

2.1   Classical Greece and Greek astronomy

In Chapter 1 I tried to indicate *why* it is important to know something about the 17th-century scientific revolution and its development. The next question is *how* are we going to achieve this and, in particular, *how* are we going to deal with such a large topic within the confines of a single book.

You can see from the chapter headings that, very roughly, the first half of the book deals with the origins of the scientific revolution and with some of its consequences up to the end of the 19th century, while the second half covers some of the momentous developments that have taken place in the 20th century in both science and technology together with their influence on society.

The most dramatic developments in the scientific revolution were those in astronomy and dynamics—associated with Copernicus, Galileo and Newton.

The key change probably came with Copernicus and his switch from an earth-centred to a sun-centred cosmology. To us in the 20th century this doesn't seem much of a problem *but* for medieval man it was a major shift in his way of thinking and it took more than 150 years for Copernicus' ideas to be generally accepted.

To understand why it took so long we need to look closely at the astronomical ideas which existed before Copernicus, at the origins of these ideas, and at how they were integrated into a coherent view of the world. This takes us back much further than to the time just before Copernicus—it takes us back nearly another 2000 years, to ancient Greece.

Probably the best known fact today about the ancient Greeks is that they founded the original Olympic Games. These took place every four years, with a few breaks due mainly to wars, from 776 BC to AD 394—a period of well over a thousand years (see Fig. 6.5.1, page 295). By contrast our modern Olympic Games are not yet 100 years old, having first taken place in 1896.

However, there was much more than this to the civilisation of the ancient Greek world which originated in the islands of the Eastern Mediterranean and went through several distinct stages in its long life.

Its early, and some say its greatest, period was that of the independent city states—around 500 BC. The best known of these were Athens and Sparta but other famous names include Corinth, Delphi and Troy. These city states were not just confined to the area of modern Greece. Pythagoras, the famous Greek geometer, lived in Croton, in what is now Italy, while the site of Troy is in modern Turkey. A major characteristic of the Greek city states was their diversity. Although they were united by a common language and culture, they were fiercely independent and had very different political and social structures. Athens, for example, is often described as the birthplace of democracy, while the government of Sparta was more centralised and emphasised military virtues.

In the 4th century BC most of the Greek world and much else besides was conquered by the Macedonians, led first by Philip of Macedon and then by his son, Alexander the Great. The journeys of conquest of Alexander the Great included, as well as ancient Greece, the whole of the ancient Persian empire, parts of Northern India and the major civilisation of Egypt.

One of the effects of Alexander's conquests was that Greek ideas were transmitted over nearly all the civilised world. Alexander also founded the city of Alexandria in Egypt, which was named after him and which was to become in later centuries one of the major centres of Greek culture and learning. When he was young, Alexander had had as his tutor the Greek philosopher Aristotle, and it is sometimes claimed that the great influence which Aristotle's ideas subsequently had was in part due to his having lived at just the time when Alexander's armies were carrying Greek ideas far and wide. Alexander and Aristotle died within a year of one another in 323 and 322 BC respectively.

The later development of Greek ideas took place after Greece, together with the many Greek colonies which existed all over the Mediterranean and Near East, had come under the rule of the Romans in the 2nd century BC. At its greatest extent, around AD 180, the Roman Empire stretched from the Antonine Wall in Scotland in the north to the Sudan in the south, and from Spain in the west to Persia in the east. This empire, or most of it, was to last until around AD 450.

Over this enormous period of time, more than 1000 years, the ancient Greeks created many of the cultural traditions and areas of learning which we take for granted today. And it is striking how many of them date from the early period of the independent city states. The leading Greek thinkers succeeded, to a great extent, in banishing the supernatural from the interpretation of nature. They argued that natural phenomena are governed by unchanging principles which can be understood by man through his powers of reason.

Two famous names from ancient Greece are Homer and Socrates. To Homer, a shadowy figure from around 700 or even 800 BC, is attributed the earliest written literature. Socrates, the famous philosopher who was condemned to death in 399 BC, allegedly for corrupting the young, greatly influenced later Greek philosophers such as Plato and Aristotle. In the century or so before Socrates, the Greeks had virtually created philosophy, theatre and drama, and the writing of history—and had made great innovations in architecture.

Also at this time, around 420 BC, there lived the famous physician, Hippocrates, to whom is attributed a vast body of medical literature and who has given his name to the Hippocratic Oath of the medical profession.

Much of mathematics, too, was created in ancient Greece, both in the theory of numbers and in geometry. Pythagoras, who is remembered for his theorem about the sides of a right-angled triangle, lived around 540 BC, while Euclid, the best known geometer of the ancient world, lived around 300 BC. Euclid's books on geometry, with their formal axioms and rigorous methods of proof, shaped mathematical thinking for more than 2000 years. The word 'geometry' literally means 'earth-measuring' and the Greek Eratosthenes, who lived around 230 BC, applied geometrical ideas in devising one of the earliest estimates of the radius of the earth. This example illustrates the use of mathematics in what we would call scientific work, and this emphasis on mathematics was central to much of Greek science.

For example, the Pythagoreans, the followers of Pythagoras, discovered that simple numerical relationships existed between the lengths of vibrating strings and the harmony of the musical notes they emitted. This led them to try to explain all things in terms of simple mathematical laws and musical harmonies. The Pythagoreans also devised a musical scale which was based on very simple numerical ratios.

The Greeks also tried to apply mathematical methods in devising models of the universe and one of them—Aristarchus of Samos, who lived around 280 BC—even proposed a sun-centred model similar to that of Copernicus. However, it was not this model but a different and much more plausible model of the universe which was most common in the ancient world. This was the ancient Two-Sphere Universe with one sphere, the earth, at the centre and a second vast sphere surrounding it in which all the stars were embedded.

The two-sphere model arises from some very simple observations of the stars which anyone can make. The stars are largely the same night after night. The same groups of stars, or constellations, continually reappear in the same relative positions. For example, the constellation we call the Great Bear reappears every night and rotates around the pole star through 60° every four hours—that is, through 360°, or one complete revolution, every twenty-four hours.

All the constellations revolve in this way and so there arose the simple picture of a fixed sphere of stars rotating once a day about a spherical stationary earth. However, not all the heavenly bodies behave as simply as the stars, the major exceptions being the sun, the moon and the planets. The sun moves through the fixed stars on a path called the ecliptic, taking a year to return to its original position. It is the constellations traversed by the sun in its wanderings—Aries, Gemini, Taurus, and so on—that give rise to the astrological divisions of the calendar year.

The moon and the planets also move about amongst the fixed stars—indeed, the Greek word *planetos* means the wanderer—but they are always found on or very near the ecliptic.

So there arose the model shown in Fig. 2.1.1, which is basically the two-sphere universe with the moon, sun and planets moving on circular orbits of different periods all of which are in the same plane.

The orbits were assumed to be perfect circles and the sphere of the stars was a perfect sphere. These assumptions were based on the Platonic idea that the perfection of the heavens was matched by the perfection of the sphere and the circle. This perfection was often associated with Pythagorean musical harmonies and the orbits of planets were related to musical notes, as can be seen from Fig. 2.1.1. The whole heavenly universe was thought to celebrate its perfection by singing—with the harmony of the spheres.

By the 4th century BC, just before Aristotle, this model was widely accepted. Others models *were* proposed, including one from around 350 BC in which Mercury and Venus go round the sun, which in turn circles the earth. But such models were not common.

The field was held by the modified two-sphere universe which was taken over by Aristotle and integrated into his complex synthesis of ideas covering cosmology, the natural world and philosophy. It is not possible even to outline all of Aristotle's work. His writings were an enormous attempt to synthesise the whole of the existing human knowledge and experience into one coherent system.

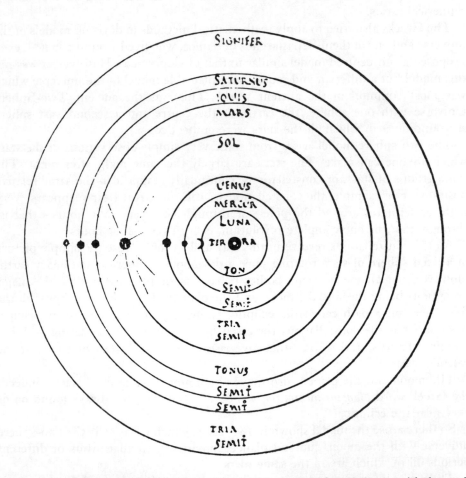

Fig. 2.1.1 The harmony of the spheres in the ancient two-sphere universe with the earth (Terra) at the centre surrounded by concentric, spherical orbits for the moon (Luna), Mercury, Venus, the sun (Sol), Mars, Jupiter (Jovis) and Saturn. Note the musical intervals—tones and semitones—shown between the orbits.

They covered philosophy, logic, cosmology, physics, the natural world, psychology, ethics, politics, and literary criticism. Aristotle also made major contributions to the classification of animals and plants and to the study of social and political questions. However, we do need to describe Aristotle's basic ideas concerning cosmology, matter and motion.

Aristotle divided the universe into two regions—the heavenly region, which was everything from the moon outwards, and the earthly region, which was everything beneath the moon, or the sub-lunary sphere, as it was called.

These two regions were, for Aristotle, totally distinct and totally different. In the heavens everything was perfect, incorruptible and unchanging. The heavenly bodies were perfect spheres, and they were made of a different, more perfect kind of matter, the ether, which was, to quote Aristotle, 'more divine' and 'more precious' than ordinary terrestrial matter. Since it was perfect, this heavenly matter had the property, when it moved, of moving naturally in the most perfect of all figures, the circle. Uniform motion in a circle was at once changing and yet changeless, since it endlessly repeated itself. According to Aristotle, no explanation was needed to explain this uniform circular motion of the heavenly bodies. It was in their nature to move thus. They could do nothing else.

In the sub-lunary sphere, everything was different. Things there were corruptible and subject to change and decay. Aristotle talks of 'the corrupting power of matter' and now he means ordinary matter as it exists on earth. This kind of matter, he thought, was made up of four elements—earth, air, fire and water. All the various substances and bodies to be found in the sub-lunary sphere were thought to be made up of the four elements in varying proportions. Furthermore, Aristotle supposed that each of the four elements had a natural place in the sub-lunary sphere—the earth at the centre, water as a sphere surrounding the earth and concentric with it, air as a sphere surrounding the water, and fire as another concentric sphere above the air and below the moon. The elements, if left alone, would return to these natural places, since it was their nature to do so. In this way he was able to explain qualitatively certain everyday, commonsense observations such as that stones fall to the ground, their natural place; that water settles on the surface of the earth, in lakes or seas, its natural place; that air bubbles rise through water; and fire and flames leap up high into the air. For Aristotle they were only doing what comes naturally.

This cosmology of Aristotle's, together with the associated laws of motion, was enormously appealing. It fitted ordinary commonsense observations. It provided good qualitative or descriptive explanations for a wide variety of phenomena. And it formed part of Aristotle's much larger, coherent system of thought. So Aristotle's system was very attractive to the general philosopher or educated person, and it continued to be a powerful intellectual force for nearly 2000 years after Aristotle's death.

However, for the specialist astronomer it had one ultimately fatal flaw. When you looked at it in detail, it didn't work. It couldn't explain such things as the variations in brightness and apparent distance of the planets. And, perhaps most important, it couldn't explain the so-called retrograde motion of the planets as

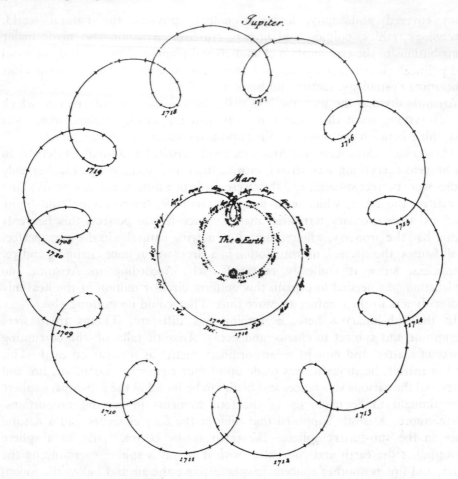

Fig. 2.1.2   The motions of Mars and Jupiter relative to the earth during 1708–10 (Mars) and 1708–20 (Jupiter) showing loops of retrogression.

seen from the earth during which the planets appear to stop and then go backwards in the heavens before resuming their former orbits.

Retrograde motion is illustrated in Fig. 2.1.2, which shows the motions of Jupiter and Mars around the earth. These particular observations of loops of retrogression date from the early 18th century but of course similar loops were also observed by the ancient Greeks.

These retrograde motions were problems for Aristotle's model, and expert astronomers came up with a variety of devices by which they attempted the difficult *dual* task of explaining these observations, or 'saving the appearances' as they called it, and *at the same time* retaining uniform circular motion as the natural motion of the heavenly bodies.

Two suggestions on how to do this were made by the mathematician Apollonius around 200 BC. The first was the eccentric, in which the planets still moved in uniform circular orbits but the earth was eccentrically displaced a short distance away from the centre of the orbits. This could explain how the planets could vary in distance from the earth and hence in brightness. The second suggestion of Apollonius was the epicycle, shown in Fig. 2.1.3(a), in which the earth is at the centre but the planet rotates uniformly around a small circle whose centre itself rotates uniformly around the earth. The epicycle was able to explain, in principle, how retrograde motion could occur, as is shown in Fig. 2.1.3(b).

However, it is one thing to explain in principle how retrogression or variable brightness can occur. It is quite another to devise a system which will explain in detail all the complex wanderings of all the planets, *and* the sun and the moon especially as more and more accurate observations became available.

The most successful attempt to do this was made by Ptolemy, a Greek who lived in Alexandria in the 2nd century AD. His famous book *Mathematical Composition*, or the *Almagest* as it is usually called, contained detailed mathematical models for the motions of the sun, the moon and all the planets using various

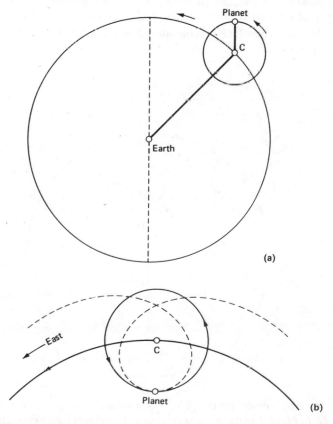

Fig. 2.1.3 (a) The epicyclic model of Apollonius; (b) the loop of retrogression (dotted line) in the actual orbit of the planet is produced by the combined motion of the epicycle and the main circular motion of C.

combinations of epicycles, eccentrics and a further more complex invention of his own called the equant. In outline, Ptolemy's system resembled the two-sphere universe of Fig. 2.1.1, but the details were extremely complex. Fig. 2.1.4 shows just how complex Ptolemy's models for the planets other than Mercury had become. The details of these models are not important for our purposes. The point is that Ptolemy's whole system was very complex. It contains about 80 or so epicycles or their equivalent. So it is a long way from the spherical simplicity of the original two-sphere universe of the 4th century BC. But it had the great advantage that it *worked*. Starting from known observations of planetary positions, it accurately predicted the future positions of all the planets. And it continued to do so very accurately for more than 1000 years, an extraordinary achievement.

Since Ptolemy's system was so complex mathematically, most people couldn't really follow it. Astronomy was probably the first science in which a barrier between professionals and laymen became clear-cut. Nevertheless, to the educated layman who had only an overall grasp of his work, Ptolemy's system had many advantages. Not only did it work, and work supremely well, it also retained the principle of circular motion for the heavenly bodies. And it kept the earth at the centre. It could therefore be considered to have left Aristotle's overall system of thought intact.

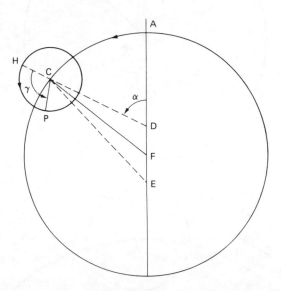

Fig. 2.1.4 Ptolemy's model for the planets other than Mercury. *From Greek Science after Aristotle by G. E. Lloyd, Chatto and Windus, 1973.* The planet P moves on the circumference of an epicycle, centre C. C moves round an eccentric circle, centre F, but the movement is uniform not with respect to F but with respect to D, the equant point drawn so that EF = FD. The angle α increases uniformly. E is the earth.

SUMMARY

1 To understand the situation in astronomy just before Copernicus, we have to go back more than 1000 years to the work of astronomers in classical Greece.

2 Classical Greek civilisation lasted many hundreds of years and had a vast range of cultural and intellectual achievements to its credit.

3 The simple two-sphere universe gave a good first approximation to the motions of the heavens and, when it was incorporated in Aristotle's synthesis, was able to give a plausible qualitative explanation of many phenomena, including those of motion.

4 This simple model was not able to explain or predict the details of the heavenly motions.

5 Ptolemy's system, though complex, was so well constructed that it dominated astronomy for more than 1000 years. It could also be held to be compatible with Aristotle's wider synthesis which was a dominating intellectual force for even longer—around 1800 years.

2.2 The Copernican revolution

In the last chapter we looked at classical Greek ideas in astronomy and in particular at those of Aristotle and Ptolemy. In this chapter we are first going to see how those ideas were transmitted to medieval Europe and how they were received there. Then we will discuss the work of Copernicus and how it was received.

THE TRANSMISSION AND INTEGRATION OF GREEK IDEAS

After the fall of the Roman Empire in the 5th century AD, the classical Mediterranean civilisation collapsed. Over all of Europe a period known as the Dark Ages began. Most of the old learning was lost. For example, by the 7th century AD only a few scraps of Aristotle, on logic, and none of Ptolemy's work had survived. The Dark Ages of Europe lasted for about 600 years, and it was not until the 12th century AD that any of the classical Greek books became known in Europe again.

However, Greek astronomy was preserved, and developed, all this time in the Arab world. The name we use today for Ptolemy's great work, the *Almagest*, is in fact the title of a 9th-century Arab translation and literally it means *The Greatest*.

In the 12th century most of Aristotle, together with the *Almagest*, were translated from Arabic into Latin, at that time the international language of Europe, and the works of Aristotle and Ptolemy became part of the curriculum of the new European universities.

Over the next two centuries Aristotle's ideas were integrated into Christian theology largely through the work of Thomas Aquinas in his massive twelve-volume *Summa Theologica* ($\sim$ 1260). By the 14th century, Aristotle's ideas were becoming part of the common culture of the day, as we can see if we look at the *Divine Comedy* (1314), the masterpiece of the Italian poet Dante which was written in Italian and not Latin.

Fig. 1.1 (page 2) shows Dante's universe, which quite clearly integrates Christian ideas with the ancient two-sphere universe of Aristotle. At the centre is the earth and it is surrounded by the spheres of the moon, the sun and the other planets. Beyond the sphere of the stars lies Paradise and the Home of God. Mount Purgatory is clearly shown, and inside the earth are the circles of Hell, which are so graphically described in Dante's *Inferno*.

One interpretation of the significance of this integration is given by Thomas Kuhn in his book *The Copernican Revolution*:

> For the Christian .... the new universe had symbolic as well as literal meaning, and it was this Christian symbolism that Dante wished most of all to display. Through allegory his *Divine Comedy* made it appear that the medieval universe could have had no other structure than the Aristotelian-Ptolemaic. As

he portrays it, the universe of spheres mirrors both man's hope and his fate. Both physically and spiritually man occupies a crucial intermediate position in this universe filled, as it is, by a hierarchical chain of substances that stretches from the inert clay of the center to the pure spirit of the Empyrean. Man is compounded of a material body and a spiritual soul: all other substances are either matter or spirit. Man's location, too, is intermediate: the Earth's surface is close to its debased and corporeal center but within sight of the celestial periphery which surrounds it symmetrically. Man lives in squalor and uncertainty, and he is very close to Hell. But his central location is strategic, for he is everywhere under the eye of God. Both man's double nature and his intermediate position enforce the choice from which the drama of Christianity is compounded. He may follow his corporeal, earthy nature down to its natural place at the corrupt center, or he may follow his soul upward through the successively more spiritual spheres until he reaches God. As one critic of Dante has put it, in the *Divine Comedy* the 'vastest of all themes, the theme of human sin and salvation, is adjusted to the great plan of the universe'. Once this adjustment had been achieved, any change in the plan of the universe would inevitably affect the drama of Christian life and Christian death. To move the earth was to break the continuous chain of created being.[1]

This was part of the background into which Copernicus came.

ASTRONOMY JUST BEFORE COPERNICUS

The rediscovery of Ptolemy's *Almagest* in the 12th century stimulated interest in observational astronomy in Europe. In the generation before Copernicus, professional astronomers like Peurbach (1423–62) and Regiomontanus (1436–76) spent much time teaching and improving Ptolemy's theory. They also published improved sets of astronomical tables which had at least two practical purposes— to help astrologers with their predictions and to reform the old calendar, which was gradually getting out of step with the seasons.

At this time astronomy was popular amongst laymen, too. One example of this is the marvellous astronomical model shown in Fig. 2.2.1. It was made by an Italian called de Dondi in Padua in about 1350 and it is a superb clockwork model of Ptolemy's system. It has seven faces—one each for the sun, the moon and the five planets known in antiquity, Mars, Jupiter, Saturn, Venus and Mercury. Each face has a separate clockwork mechanism which drives the models of the heavenly bodies along paths which represent their actual paths as seen from the earth. De Dondi's clock makes graphically visible the basic principles of Ptolemy's astronomy, wheels within wheels and a separate mechanism for each of the heavenly bodies.

The general popularity of astronomy is also reflected in the widespread interest in the astrolabe, a simple astronomical instrument which had been much developed by Arabic astronomers. An astrolabe consists of two concentric discs. The back disc is fixed and represents the earth. The front disc rotates and has the positions of many prominent stars marked on it. It represents the sphere of the

Fig. 2.2.1   Reconstruction of De Dondi's clockwork model of Ptolemy's system.

stars rotating about the fixed earth. An astrolabe is the ancient two-sphere universe in microcosm. It can be used to tell you where and at what time you will be able to see particular stars on any night of the year. Or, if you measure the position of a star, you can use the astrolabe to tell the time. Chaucer, the 14th-century English poet, wrote a book explaining how to use an astrolabe.

Fig. 2.2.2   Nicolaus Copernicus (1473–1543).

NICOLAUS COPERNICUS

Copernicus (see Fig. 2.2.2) was born in 1473 and died in 1543. He lived through that great split in the Christian Church which we call the Reformation. This great division in the Church between Protestants and Catholics still exists today, more than 400 years later. The Reformation is most closely associated with Martin Luther and his public defiance of the Church authorities from 1517 onwards.

It was just at this time that Copernicus was working out his astronomical theory in detail. In 1514 he had privately circulated a short book, the *Commentariolus*, in which he set out his main ideas. However, they needed further development before he could publish them more widely. It is thought that he had largely completed his calculations by the 1520s, but he didn't publish his new work then, possibly because he knew that it might bring him into conflict with the Church. He was himself a canon of the Church and may well have been reluctant to arouse further controversy to add to that already raging around Luther. So Copernicus' famous book—*De Revolutionibus Orbium Caelestium* or *On the Revolution of the Heavenly Spheres*—did not appear until 1543, the year of Copernicus' death. And

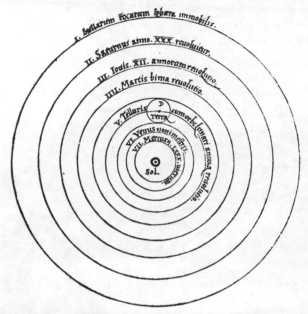

NICOLAI COPERNICI

net, in quo terram cum orbe lunari tanquam epicyclo contineri
diximus. Quinto loco Venus nono menſe reducitur., Sextum
deniҩ locum Mercurius tenet, octuaginta dierum ſpacio circũ
currens. In medio uero omnium reſidet Sol. Quis enim in hoc

pulcherrimo templolampadem hanc in alio uel meliori loco po
neret, quàm unde totum ſimul poſsit illuminare; Siquidem non
inepte quidam lucernam mundi, alȷ mentem, alȷ rectorem uo﹣
cant. Trimegiſtus uiſibilem Deum, Sophoclis Electra intuentē
omnia. Ita profecto tanquam in ſolio re gali Sol reſidens circum
agentem gubernat Aſtrorum familiam. Tellus quoҩ minime
fraudatur lunari miniſterio, ſed ut Ariſtoteles de animalibus
ait, maximā Luna cũ terra cognationē habet. Concipit interea à
Sole terra, & impregnatur annuo partu. Inuenimus igitur ſub
hac

Fig. 2.2.3 Page from the first printed edition (1543) of Copernicus' *De Revolutionibus
Orbium Caelestium*, showing the main features of his system. Note that the book is written
in Latin.

Copernicus is traditionally supposed to have received the first printed copy on his
deathbed.

What did *De Revolutionibus* contain? It was divided into six books and all
Copernicus' key innovations are set out in Book 1, which is almost completely
verbal and contains virtually no mathematics. Fig. 2.2.3 shows a page from Book 1
of the first printed edition of 1543. The diagram outlines the main features of
Copernicus' model of the universe. The sun is now at the centre and all the
planets, including the earth, revolve about it. The earth also revolves about its

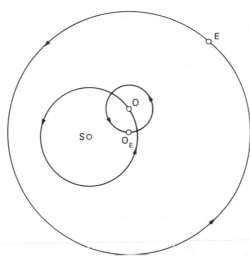

Fig. 2.2.4 Copernicus' model for the orbit of the earth. The sun is at S and the earth E revolves on a circle whose centre, O_E, revolves slowly about a point O, which in turn revolves on a sun-centred circle. *Redrawn from Kuhn, T. S., The Copernican Revolution, Vintage Books/Harvard University Press.*

own axis once a day, thus explaining the apparent motion of the fixed stars. So there is more than one motion for the earth. And the moon revolves around the earth, which means that there is more than one centre of motion. Copernicus' system could, in principle, also explain retrogressive planetary motion without needing to bring in epicycles. For example, as the earth overtakes an outer planet such as Jupiter, that planet will, as seen from the earth, temporarily appear to go backwards against the background of the fixed stars.

So Book 1 of *De Revolutionibus* appears at first sight to put forward a simpler astronomical scheme than that of Ptolemy. But when we open Book 2 and read on, that simplicity all but vanishes. This is because Copernicus' simple scheme can't explain the detail, any more than the ancient two-sphere universe of Aristotle could. So what Copernicus did in Books 2 to 6 of *De Revolutionibus* was to modify his simple scheme so as to try to explain this detail. He produced complex mathematical models for the motion of all the heavenly bodies, including the earth. And he did this almost exactly as Ptolemy had done. He tried to retain uniform circular motions as basic, but modified these by means of a whole series of epicycles, eccentrics and similar devices. Fig. 2.2.4 shows Copernicus' detailed model for the orbit of the earth; his models for the other planets were even more complicated.

Overall, Copernicus' system was almost as complex as Ptolemy's. Roughly speaking, Copernicus used about sixty or so epicycles compared with Ptolemy's eighty. *De Revolutionibus* was a magnificent mathematical performance. It did do some things a little more simply than Ptolemy had done but in many ways it was still very Ptolemaic. And it had several significant disadvantages.

Firstly, at that time there was no direct evidence that the earth was moving.

The evidence from stellar parallax (see page 27), or the Foucault pendulum, did not become available until the 19th century, about 300 years later.

Secondly, Copernicus' system didn't explain all that Aristotle's had done. In particular he had no theory to explain either the motion of the heavens or terrestial motions. So in this area Copernicus' system was significantly inferior to the traditional view.

Finally, and perhaps most important in some of the subsequent controversies, it no longer put man at the centre of all things.

At first, despite Copernicus' fears, the book caused relatively little stir. A second edition didn't appear until 1566, 23 years after the first, and a third not until 1617, 50 years after that. Two factors may have been responsible for its initial lack of impact. Firstly, a preface was inserted by the editor, Osiander, almost certainly without Copernicus' knowledge. This preface implied that Copernicus didn't really mean it. The earth wasn't really moving. It was just that the calculations were easier if we assume that it is. In the words of the original, 'they [these ideas] are not put forward to convince anyone that they are true, but merely to provide a correct basis for calculation'. There is some indication that this was not Copernicus' own view. Secondly, the book itself was inaccessible to the majority of readers. It was written in Latin and the mathematics were too difficult for any but professional astronomers to follow easily. However, this latter group did use it and they found many of Copernicus' mathematical techniques very useful. As T. S. Kuhn puts it:

> Copernicus died in 1543, the year in which *De Revolutionibus* was published, and tradition tells us that he received the first printed copy of his life's work on his deathbed. The book had to fight its battles without further help from its author. But for those battles Copernicus had constructed an almost ideal weapon. He had made the book unreadable to all but the erudite astronomers of his day. Outside of the astronomical world the *De Revolutionibus* created initially very little stir. By the time large-scale lay and clerical opposition developed, most of the best European astronomers, to whom the book was directed, had found one or another of Copernicus' mathematical techniques indispensable. It was then impossible to suppress the work completely, particularly because it was in a printed book and not, like Oresme's work or Buridan's, in a manuscript. Whether intentionally or not, the final victory of the *De Revolutionibus* was achieved by infiltration.[2]

So Copernicus' work was used by the astronomers but its main outlines gradually became known to a wider audience too. Popularisers of Copernicus' system gave public lectures and published books of their own. In one of them, *Perfit Description of the Caelestiall Orbes*, published by Thomas Digges in 1576, there appeared a diagram of Copernicus' system but with the stars extending out into space rather than being restricted to the surface of the celestial sphere.

By the end of the 16th century there were two distinct popular reactions to Copernicus, which were not directly connected with his detailed astronomical work. The pro-Copernicans tended to be hostile to the acceptance of Aristotle's authority and to be in favour of change and innovation. The anti-Copernicans were usually pro-Aristotle and had a strong respect for stability and traditional beliefs. Amongst the latter was John Donne, the English poet and priest. In a letter written in 1611, Donne conceded to the Copernicans that 'these opinions of yours may well be true . . . In any case they are now creeping into every man's mind'. Nevertheless, Donne was deeply worried by this, as is shown by the following quotation from his poem 'The Anatomy of the World', also published in 1611:

> And new Philosophy calls all in doubt,
> The Element of fire is quite put out;
> The Sun is lost, and th'earth, and no man's wit
> Can well direct him where to look for it.
> And freely men confess that this world's spent,
> When in the Planets, and the Firmament
> They seek so many new; then see that this
> Is crumbled out again to his Atomies.
> 'Tis all in pieces, all coherence gone;
> All just supply, and all Relation:
> Prince, Subject, Father, Son, are things forgot,
> For every man alone thinks he hath got
> To be a Phoenix, and that then can be
> None of that kind, of which he is, but he.[3]

Here Donne is clearly linking the overthrow of traditional ideas in astronomy with what he saw as threats to all ordered life.

Now let us leave the popular reception of Copernicanism and turn to some advances in professional astronomy in the late 16th and early 17th centuries and in particular to the work of Tycho Brahe and Johannes Kepler.

TYCHO BRAHE (1546–1601) AND JOHANNES KEPLER (1571–1630)

Tycho Brahe was born in Denmark. He was primarily an observational astronomer who devoted himself to a systematic study of the errors in the astronomical instruments of his day. Tycho, as he is usually known, continually redesigned and rebuilt his instruments until he could improve them no more. He claimed an accuracy of $\pm 1'$ of arc for his fixed star positions and $\pm 4'$ for his planetary positions, all achieved with the naked eye. The accuracy of Tycho's claims have since been confirmed by measurements with modern telescopes. Fig. 2.2.5 shows Tycho seated inside his Great Quadrant and pointing to illustrations of many of his new instruments. These huge instruments were very expensive, so Tycho persuaded the King of Denmark to finance him in a large permanent observatory at Uraniborg in Denmark. When Tycho had set up his instruments and observatory, he used them to compile a vast amount of accurate astronomical data. In particular he made long sequences of observations of the orbits of the planets extending over a period of about 30 years. He also observed a new star or nova in 1572 and, by accurate measurement of a comet's orbit in 1577,

Fig. 2.2.5 Tycho Brahe with his great quadrant and other instruments at his observatory at Uraniborg in Denmark.

he showed that it was much further away than the moon. These two bits of evidence told against the Aristotelians, who had always maintained that the heavens were unchanging.

But in one major respect Tycho was a traditionalist. He didn't believe that the earth moved. This was primarily because, however hard he tried, he couldn't observe any parallax for any of the fixed stars. Parallax is the apparent change in angular position of nearby stars which should occur if they are observed from two different positions in the earth's orbit—that is, at different times of the year. So Tycho argued that the earth couldn't be moving. As he might have said: 'If the earth is moving, stellar parallax should be observable. Here am I, the greatest observational astronomer who has ever lived, and I can't observe it. Therefore the earth can't be moving'.

So he proposed his own planetary system, shown in Fig. 2.2.6, which was a compromise between Ptolemy and Copernicus. The earth was still at the centre and the moon and sun rotated round it. But the other five planets went round the sun in what could be seen as five enormous epicycles.

Yet that was as far as Tycho could go. He could put forward the outline of a system like Aristotle's two-sphere universe or what Copernicus described in Book 1 of *De Revolutionibus*. But as he was no mathematician, he couldn't do what Ptolemy had done for Aristotle or what Copernicus had done in his Books 2 to 6— that is, work out a detailed theory, based on his model, which would fit his observations. So he tried to find a mathematician who would do it for him, and he found him in Johannes Kepler.

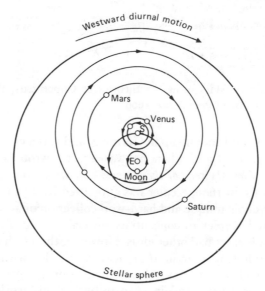

Fig. 2.2.6 Tycho Brahe's system for the planets. The moon and the sun rotate about the earth, which is at the centre of the stellar sphere. All the other planets rotate on enormous epicycles whose common centre is the sun. *Redrawn from Kuhn, T. S., The Copernican Revolution, Vintage Books/Harvard University Press.*

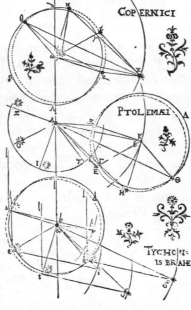

Fig. 2.2.7   The orbit of Mars using the theories of Copernicus, Ptolemy and Tycho Brahe; from Kepler's *Astronomia Nova*, 1609.

Kepler was born and brought-up in Germany and was a convinced Copernican from an early age. In 1596, when he was 25, he wrote a book, *Mysterium Cosmographium* or *The Mystery of the Cosmos*—in which he defends Copernican ideas. Soon after this, in 1600, Kepler was appointed by Tycho to be his assistant and to work on Tycho's unique and hard-won collection of accurate astronomical data. Tycho wanted Kepler to concentrate on elaborating Tycho's own system (see Fig. 2.2.6) but Kepler had other ideas. However, this clash did not last long as Tycho died the following year in 1601, leaving Kepler with free access to all Tycho's precious data. Kepler was in his element. For nearly ten years he worked and reworked Tycho's data, and in 1609 published his work in a book called *Astronomia Nova* or *The New Astronomy*. Fig. 2.2.7 shows a page from *Astronomia Nova* on which Kepler is trying to devise models based on the Copernican, the Ptolemaic and the Tychonic system which would fit Tycho's data for the orbit of

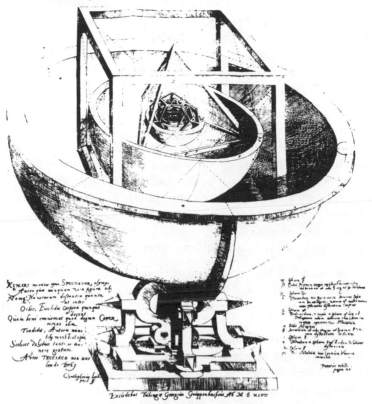

Fig. 2.2.8 Kepler's nesting solids model for the planets.

Mars. This orbit is the least circular of all the planets. It had always been a problem and Kepler thought that if he could get his calculations right for Mars then he could easily deal with the other planets. After years of effort, and hard calculation, Kepler decided that none of these systems could be made to fit the data. Nevertheless, unlike a modern scientist, he wrote down and published all his trials and failures so that we can follow his trains of thought.

This decision of Kepler's, that none of the systems fitted the data, was a very finely balanced one. He could get the data very nearly to fit the models—to an accuracy of $\pm 8'$ of arc, which would have been more than adequate for any previously known data. But Kepler had Tycho's data, accurate to $\pm 4'$. So he rejected orbits based on the systems of Ptolemy, Copernicus and Tycho and went on to devise his own models, based on the ellipse, for the orbits of Mars and the other planets. These calculations also appeared in *Astronomia Nova*, together with what we now call Kepler's first two laws of planetary motion—that planets move in ellipses with the sun at one focus, and that their speeds vary so that the line joining the planet to the sun sweeps out equal areas in equal times.

Kepler rejected circles as the basis for heavenly motions; he didn't have the earth at the centre—indeed, nothing was at the centre of his ellipses although the sun was at one focus; and he used no motions with constant angular velocity. So his ideas were much more of a break with what had gone before than were those of Copernicus, and he seems to be much more 'modern' in his outlook. But there is another side to Kepler. Look at this diagram (Fig. 2.2.8) from the *Mysterium Cosmographium* of 1596. It shows Kepler's famous nesting solids model for the planetary orbits. This is how Kepler explained it:

> The Earth is the measure for all the other spheres. Circumscribe a dodecahedron about it, then the surrounding sphere will be that of Mars; circumscribe a tetrahedron about the sphere of Mars, then the surrounding sphere will be that of Jupiter; circumscribe a cube about the sphere of Jupiter, then the surrounding sphere will be that of Saturn. Now place an icosahedron within the sphere of the Earth, then the sphere which is inscribed is that of Venus; place a octahedron within the sphere of Venus, and the sphere which is inscribed is that of Mercury.
>
> There you have the reason for the number of the planets.[4]

This was not just a youthful aberration of Kepler's, for the same idea appears in his book *Harmonice Mundi* or *The Harmonies of the World* which appeared in 1619, ten years after *Astronomia Nova*. And it was in *Harmonice Mundi* that there appeared what we now call Kepler's Third Law of Planetary Motion—that the square of the period of orbital rotation of a planet is proportional to the cube of the mean distance of the planet from the sun. This law was given no special prominence by Kepler. It was buried away amidst pages of largely unsuccessful attempts to find geometrical or arithmetic harmonies in the world. Kepler devoted much more space to trying to detect what he thought were musical harmonies amongst his data concerning planetary distances and speeds. Like the Pythagoreans, Kepler was always trying to listen to the music of the heavenly spheres.

Kepler's work was written in Latin. It was obscure and difficult to follow, unless you were as good a mathematician as Kepler. And, like Copernicus, Kepler made relatively little immediate impact on his contemporaries.

In these, and in many other respects, he was very different from Galileo Galilei—the man who, more than anyone else, brought the ideas of Copernicus into the arena of public debate.

REFERENCES

1 Kuhn, T. S., *The Copernican Revolution* (New York: Harvard University Press, 1957), pp. 112–13. Reprinted by permission.
2 Ibid., p. 185. Reprinted by permission.
3 Donne, J., *An Anatomy of the World* in *John Donne Poems* (London: J. M. Dent, Everyman's Library, 1976), p. 182.
4 Kepler, J., *Mysterium Cosmographium* (Tubingen: 1596); translated in Koyré, A., *The Astronomical Revolution* (London: Methuen & Co., 1973), p. 146.

SUMMARY

1 After many centuries of ignorance of ancient Greek ideas in Europe, the works of Ptolemy and Aristotle were reintroduced there in the 12th century.
2 In the next 200 years the universe of Aristotle and Ptolemy became firmly integrated into medieval Christian theology.
3 Copernicus' *De Revolutionibus*, which he published somewhat reluctantly in 1543, put forward a new astronomical system in which the sun was at the centre and the earth moved.
4 Copernicus' system had some advantages over Ptolemy's, but the difference is nowhere near as clear-cut as is sometimes claimed. In particular he had no direct evidence that the earth moved.
5 Copernicus' ideas made no immediate impact but gradually, over many years, they became more widely known both among professional astronomers and to the layman.
6 Tycho Brahe's new and very accurate data enabled Kepler to reject both old and new systems based on uniform circular motion and to put forward his own system based on ellipses. Kepler's work was detailed and mathematical, and accessible only to professionals.

PART THREE

The 17th-century scientific revolution

3.1 Galileo and his conflict with the Catholic Church

Galileo (see Fig. 3.1.1) was an almost exact contemporary of Kepler. He was born at Pisa in 1564 in the year that also saw the death of Michelangelo and the birth of Shakespeare.

In 1589, when he was only 25, he was appointed Professor of Mathematics at the University of Padua. While he was there Galileo worked mainly on mathematics and mechanics, showing not only mathematical skill but also considerable ability as a practical craftsman.

GALILEVS GALILEI FLORENTINVS
ANNVM AGENS LXXVIII

Fig. 3.1.1 Galileo Galilei (1564–1642); portrait drawn in 1642.

Fig. 3.1.2   Two early telescopes, probably made by Galileo.

GALILEO'S TELESCOPIC DISCOVERIES

Galileo is probably most famous for his work with the telescope, which began in 1609. A kind of telescope had been invented in Holland, probably in 1608, and Galileo had heard vague rumours about the wonderful things it could do. He immediately tried to make one of his own and soon succeeded in making telescopes which were appreciably better than earlier ones. Fig. 3.1.2 shows some of these telescopes.

One of the earliest things Galileo did with his telescopes was to turn them to the heavens. In a few months of feverish activity he made a whole series of remarkable discoveries. Galileo was very excited about these and he soon published a book called *The Starry Messenger*, describing what he had seen. It was aimed at a wide audience, and you can feel something of Galileo's enthusiasm from the title-page which summarised the contents of the book like this:

THE
STARRY MESSENGER
Revealing great, unusual, and remarkable spectacles, opening these
to the consideration of every man,
and especially of philosophers and
astronomers;

AS OBSERVED BY GALILEO GALILEI
Gentleman of Florence
Professor of Mathematics in the
University of Padua,
WITH THE AID OF A
SPYGLASS
lately invented by him,
In the surface of the Moon, in innumerable
Fixed Stars, in Nebulae, and above all
in FOUR PLANETS
swiftly revolving about Jupiter at
differing distances and periods,
and known to no one before the
Author recently perceived them
and decided that they should
be named
THE MEDICEAN STARS

Venice
1610[1]

The book sold like hot cakes and Galileo capitalised on this success by making and selling telescopes and by trying to get a new and better job at the University of Florence. So he dedicated *The Starry Messenger* to Cosimo di Medici, Grand Duke of Tuscany and ruler of Florence. Here is an extract from the preface:

I likewise name them, calling them the Medicean Stars, in the hope that this name will bring as much honor to them as the names of other heroes have bestowed on other stars. For, to say nothing of your Highness's most serene ancestors, whose everlasting glory is testified by the monuments of all history, your virtue alone, most worthy Sire, can confer upon these stars an immortal name. No one can doubt that you will fulfill those expectations, high though they are, which you have aroused by the auspicious beginning of your reign, and will not only meet but far surpass them. Thus when you have conquered your equals you may still vie with yourself, and you and your greatness will become greater every day.

Accept then, most clement Prince, this gentle glory reserved by the stars for you. May you long enjoy those blessings which are sent to you not so much from the stars as from God, their Maker and their Governor.

Your Highness's most devoted servant,
Padua, March 12, 1610.   GALILEO GALILEI[2]

It worked. Galileo got the job. But it is interesting to compare this fulsome statement with his later determined stand against the authority of the Pope and the Church.

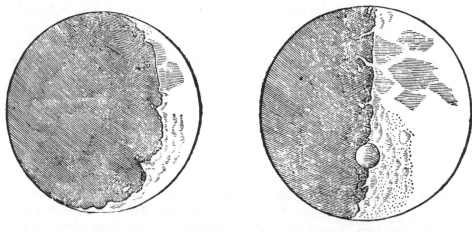

Fig. 3.1.3   Sketches of the moon from *The Starry Messenger*.

What did Galileo see through his telescope? First he saw the surface of the moon as no one had ever seen it before, full of crags and mountains, craters and valleys (see Fig. 3.1.3). And Galileo described what he saw as if the earth and the moon were similar bodies:

> There is a similar sight on earth about sunrise, when we behold the valleys not yet flooded with light though the mountains surrounding them are already ablaze with glowing splendor on the side opposite the sun. And just as the shadows in the hollows on earth diminish in size as the sun rises higher, so these spots on the moon lose their blackness as the illuminated region grows larger and larger.[3]

On another page, Galileo actually compares part of the moon's surface to a region in Bohemia. Modern photographs have shown that Galileo's observations through his telescope were surprisingly accurate, despite all the doubts cast on them at the time.

Galileo also observed hundreds and hundreds of new stars. For example, in the well-known constellation Orion in the southern sky, '... to the three stars in the Belt of Orion and the six in the Sword which were previously known, I have added eighty adjacent stars discovered recently...'[4] Altogether he found '... more than five hundred new stars distributed among the old ones within limits of one or two degrees of arc'.[5] This happened wherever he pointed his telescope. Everywhere he found hundreds of new stars which no-one had ever seen before. And nowhere did this happen more dramatically than when he looked at the Milky Way. With his telescope Galileo was able to resolve this familiar bright streak in the night sky into an enormous number of individual stars.

And that wasn't all. Galileo also looked carefully at the planet Jupiter and discovered that it had four satellites of its own. He made drawings of these on successive nights as they shifted position around Jupiter. Galileo thought these moons of Jupiter were so important that he set out his detailed nightly observations in *The Starry Messenger*. These observations start on 7th January 1610 and go on until 2nd March, only ten days before the book was published.

All this Galileo had achieved in a few months of telescopic observations. And there was more to come. In the next few years Galileo used his telescope to make a detailed study of sunspots in which he described where they were and how they came and went. He also discovered that the planet Venus varied dramatically in apparent size at different times and that it showed phases which were similar to the phases of the moon. Galileo thought that this last discovery was very significant because it was inexplicable on Ptolemy's earth-centred theory but easily explained using the sun-centred model of Copernicus.

Galileo had always been pro-Copernican and even more strongly anti-Aristotelian and he thought that his telescopic evidence had virtually clinched the case for Copernicus and against Aristotle and Ptolemy. After all, he had shown that the heavens weren't incorruptible and unchanging—the rough surface of the moon, sunspots which came and went, all his new stars. He had shown that a third centre of motion existed—Jupiter with its own planets and satellites—in addition to the sun and the earth. And he had discovered a phenomenon, the phases of Venus, which Ptolemy's system could not explain but which was perfectly understandable on that of Copernicus.

Galileo was very excited about all this and took steps to make it as widely known as possible—with books, letters, public lectures and discussion groups. In this he was totally unlike Copernicus and Kepler.

But in his excitement Galileo overlooked one vital fact. He had no convincing *proof* that the earth moved. With all the discoveries he had made with his telescope in so short a time, he may have thought that it wouldn't be long before he found such proof. But the proof never came in his lifetime and this led him into all sorts of trouble with the Church.

A fair assessment, by about 1615, just after his telescopic discoveries, would be that Galileo had made the Aristotelian and Ptolemaic view of astronomy virtually untenable, but that he had not convincingly established the Copernican model in its place.

GALILEO AND THE CATHOLIC CHURCH

The conflict between Galileo and the Catholic Church is often described as a straightforward battle between science and religion—what historians of science call the Warfare thesis. However, this conflict, like the conflict between Copernicus and Ptolemy, is not as clear-cut as is often claimed. As early as 1611, the Jesuit astronomer Clavius, working in Rome, had confirmed some of Galileo's early telescopic discoveries. And Galileo himself always maintained that he was a faithful Catholic and merely wished to save the Church from falling into the defence of error.

However, even as early as 1611, there were those in the Church who didn't see it like that and who were even more concerned at the way Galileo was carrying the debate about Copernicanism into the public arena. Moves were made to call

Galileo to account and, after much discussion in both public and private, Galileo was summoned to Rome in 1616 to defend himself and his opinions. He did so eloquently and at great length. His main point was to claim effective autonomy for what he called 'experimental philosophy'. He asked the Church fathers to consider that:

> . . . it is not in the power of the professors of demonstrative sciences to change their opinions at will and apply themselves first to one side and then to the other. There is a great difference between commanding a mathematician or a philosopher and influencing a lawyer or a merchant, for demonstrated conclusions about things in nature or in the heavens cannot be changed with the same facility as opinions about what is or is not lawful in a contract, bargain, or bill of exchange.[5]

He warned the Church how difficult it would be to stop discussion of Copernicus' ideas:

> If in order to banish the opinion in question [Copernicanism] from the world it were sufficient to stop the mouth of a single man . . . then that would be very easily done. But things stand otherwise. To carry out such a decision it would be necessary not only to prohibit the book of Copernicus and the writings of other authors who follow the same opinion but to ban the whole science of astronomy. Furthermore, it would be necessary to forbid men to look at the heavens . . .[6]

Galileo also stressed how damaging it would be to other parts of the Church's teaching if it now condemned something which was later shown to be true. On this point Galileo quoted the influential 4th-century Christian theologian St Augustine in his defence:

> And why should the Bible be believed concerning the resurrection of the dead, the hope of eternal life, and the Kingdom of Heaven, when it is considered to be erroneously written as to points which admit of direct demonstration or unquestionable reasoning?[7]

Throughout Galileo maintained that he was a staunch Catholic:

> As to rendering the Bible false, that is not and never will be the intention of Catholic astronomers such as I am; rather, our opinion is that the Scriptures accord perfectly with demonstrated physical truth. But let those theologians who are not astronomers guard against rendering the Scriptures false by trying to interpret against it propositions which may be true and might be proved so.[8]

The arguments were long but in the end the decision went against him. The key weakness in Galileo's case remained. However effectively he had demolished Aristotle and Ptolemy, he had not established the case for Copernicanism, and in particular for the movement of the earth, *beyond all reasonable doubt*. So Galileo was forbidden to defend Copernicus publicly, and he turned to other matters for the next few years.

DIALOGO

DI

GALILEO GALILEI LINCEO
MATEMATICO SOPRAORDINARIO

DELLO STVDIO DI PISA.

E Filofofo, e Matematico primario del

SERENISSIMO

GR.DVCA DI TOSCANA.

Doue ne i congreſſi di quattro giornate ſi diſcorre
ſopra i due

MASSIMI SISTEMI DEL MONDO
TOLEMAICO, E COPERNICANO;

*Proponendo indeterminatamente le ragioni Filoſofiche, e Naturali
tanto per l'vna, quanto per l'altra parte.*

CON PRI   VILEGI.

IN FIORENZA, Per Gio:Batiſta Landini MDCXXXII.

CON LICENZA DE' SVPERIORI.

Fig. 3.1.4 Title-page of Galileo's *Dialogue Concerning the two Chief World Systems*,
Florence 1632.

Then, in 1623, the Pope died and his successor was Cardinal Barberini, one of
the Church leaders who had been sympathetic to Galileo in 1616. So Galileo
sought, and was granted, permission to reopen the discussion. The result was a
large book, *The Dialogue Concerning the Two Chief World Systems*. This famous
book was not a technical astronomical treatise written in Latin like Copernicus'
De Revolutionibus. It was written in Italian and aimed at the widest possible
audience. Fig. 3.1.4 shows the title-page. The title itself shows that Galileo was
primarily concerned with the systems of Ptolemy and Copernicus. Unlike
Kepler, he ignored the plausible compromise system of Tycho Brahe, and he
made no mention of Kepler's *Astronomia Nova* of 1609.

Galileo's *Dialogue* went successfully through the papal censorship and was
published in Florence early in 1632. It immediately sold very well. But the Pope's
approval didn't last. The book appeared in February 1632 and was banned in
August, probably because of the way it was written. The Pope thought he had
agreed to a book which put the arguments on both sides of the question. And so
Galileo had written the book as a dialogue between three people—an Aristotelian,
a Copernican, possibly meant to be Galileo himself, and an independent who was
open to persuasion by the other two. All three are shown in the frontispiece of the

Fig. 3.1.5 Frontispiece of Galileo's *Dialogue* showing the dialogue in progress.

book (see Fig. 3.1.5). But Galileo had loaded the argument. Throughout the book he makes the Copernican very persuasive but is less than fair to the Aristotelian. He often makes his arguments seem ridiculous. Even his name, Simplicio, could be taken in two ways. It could be a reference to Simplicius, the great 6th-century authority on Aristotle. Yet it could also be seen as partly derogatory, with the implication that he is a bit of a simpleton. And once or twice Galileo even put some of the Pope's own words into Simplicio's mouth.

So Galileo, now an old man of 68, was called to Rome again to face the Inquisition. Again he had the better of the argument but again he couldn't clinch his case with *proof* of the earth's motion. So he was forced to recant, this time in public. His book was banned and he was sentenced to house arrest for the rest of his life.

Yet this defeat was in many ways a victory for Galileo. Firstly, by its very public action against Galileo, the Church drew much more attention to Galileo's

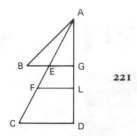

221

plane AC at F; then AF will be the mean proportional between CA and AE. And since the time through AC is to the time through AE as line FA is to AE, and the time through AE is to the time through AB as AE is to AB, it follows that the time through AC is to the time through AB as AF is to AB. Thus it remains to be proved that the ratio of AF to AB is compounded from the ratio of CA to AB and from the ratio of GA to AL, which [latter] is the ratio of the square roots of heights DA and AG taken inversely. But this is also evident: if CA is taken with respect to FA and AB, the ratio of FA to AC is the same as the ratio of LA to AD, or GA to AL, which is the square root of the ratio of the heights GA and AD; and the ratio of CA to AB is the ratio of the [corresponding] lengths; therefore the proposition holds.

PROPOSITION VI. THEOREM VI

If, from the highest or lowest point of a vertical circle, any inclined planes whatever are drawn to its circumference, the times of descent through these will be equal.

Let a circle be erect to the horizontal GH, and from its lowest point (that is, from its contact with the horizontal) let the diameter FA be erected. From the highest point, A, draw any inclined planes AB and AC out to the circumference; I say that the times of descent through these are equal.

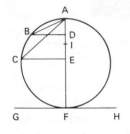

Draw BD and CE perpendicular to the diameter, and let AI be the mean proportional between the heights of the planes EA and AD. Since the rectangles FA–AE and FA–AD are equal to the squares on AC and AB; and since also as rectangle FA–AE is to rectangle FA–AD, so EA is to AD; then as square CA is to square AB, so line EA is to AD. But as line EA is to DA, so square IA is to square AD; hence the squares on lines CA and AB are to each other as the squares on lines IA and AD. Therefore as line CA is to AB, so IA is to AD. Now, it was demonstrated in the preceding [proposition] that the ratio of the time of descent through AC to the time of descent through AB is compounded from the ratios of CA to AB and of DA to AI, which [latter] is the same as the ratio of BA to AC; therefore the ratio of the time of descent through AC to the time of descent through AB is compounded from the ratios of CA to AB and of BA to AC. Therefore the ratio of their times is the ratio of equality ; hence the proposition holds.

Fig. 3.1.6 A page from *The Two New Sciences* in which Galileo is using Euclidean geometry to prove an important property of motion down an inclined plane. *From Galileo Galilei, Two New Sciences, translated by S. Drake, © University of Wisconsin Press, 1974.*

book than it otherwise might have received. It was not long before it was translated and published in other parts of Europe, even in parts of Europe that were Catholic such as France.

Perhaps more important, Galileo devoted his remaining years to writing another book, *The Two New Sciences* or, to give its full title, *Discourses and Mathematical Demonstrations Concerning Two New Sciences Pertaining to Mechanics and Local Motions.*

The manuscript was smuggled out of Italy and published in Holland in 1638.

Once again it was written in Italian as a dialogue between the same three people. But this book was not just written for a popular audience. Within the book is another book, written in Latin—a mathematical treatise in which Galileo states his axioms, sets out his propositions and proves them deductively, in the style of Euclid and Archimedes (see Fig. 3.1.6).

In *The Two New Sciences* Galileo narrowed his view. Instead of dealing with the grand sweep of the whole universe, he set down his mature views on the much more mundane topics on which he had worked 40 years earlier in Padua. In particular he dealt with what he called Local Motions, or the motion of bodies near the earth's surface.

It is in this book that we find the work which still forms a central part of our 20th-century ideas on motion. On Day 3 of the dialogue Galileo introduced for the first time the basic categories of motion which are found in any modern elementary textbook on dynamics—uniform motion at constant speed, or equable motion as he called it, and uniformly accelerated motion. Galileo described these motions in minute mathematical detail.

When Galileo came to discuss uniformly accelerated or free-fall motion, he had a brilliant idea. He considered first the motion of a body rolling down an inclined plane rather than that of a freely falling body. In this way he slowed down the motion, and so was able to measure what was happening and then describe it mathematically. Then he took this motion down an inclined plane to its two limits. On the one hand it shades into free-fall motion when the plane is vertical. On the other hand, when the plane is horizontal it shades into continuous motion at constant speed, or inertial motion as we now call it.

The Two New Sciences also contained Galileo's famous work on the paths of projectiles, whether they be stones or balls thrown in the air or, with more practical potential, cannon-balls or shells fired from guns. Again Galileo had a brilliant idea. He separated the complex trajectory of a projectile into two simpler components—a horizontal, uniform or equable motion and a vertical, free-fall motion. And he showed that the result was a parabolic path. This is how Galileo describes his idea:

I mentally conceive of some moveable projected on a horizontal plane, all impediments being put aside. Now it is evident from what has been said elsewhere at greater length that equable motion on this plane would be perpetual if the plane were of infinite extent; but if we assume it to be ended and situated on high, the moveable (which I conceive of as being endowed with heaviness), driven to the end of this plane and going on further, adds to its previous equable and indelible motion that downward tendency which it has from its own heaviness. Thus there emerges a certain motion, compounded from equable horizontal and from naturally accelerated downward motion, which I call 'projection'. We shall demonstrate some of its properties.[9]

This Galileo proceeded to do, starting by showing that the path of any projectile is a parabola, a familiar result to us but not to his contemporaries.

Throughout the book Galileo proves nearly all his propositions using the classical geometrical methods of Euclid and the ancient Greeks. Although the book *is* very mathematical, it is misleading to think of it as solely mathematical. The whole book is based on a complex and subtle interweaving of careful experiment and mathematical reasoning. This is apparent in the way Galileo establishes his initial concepts and axioms, in the way he estimates and allows for errors of measurement and in the way he uses agreement with experience and observation as a final test for the physical validity of his mathematical deductions.

What Galileo did in *The Two New Sciences* was to describe his *experiments* on motion with great *mathematical* precision. He thus made a start on the construction of a genuinely new science of motion, more quantitative and more firmly based than anything which had come before. Galileo knew it was only a start and that others would need to build on it. For example, his work described *how* bodies moved but didn't try to explain *why* they moved as they did.

In *The Two New Sciences* Galileo says that through his work 'there will be opened a gateway and a road to a large and excellent science of which these labours of ours shall be the elements, a science into which minds more piercing than mine shall penetrate to recesses still deeper'.[10] In time, Galileo's work did lead, in the hands of Newton, to a consistent dynamics which answered the questions *how* and *why* for motions both on earth and in the heavens. Newton thus provided a new synthesis to replace the one which Copernicus had destroyed when he removed the earth from the centre of the Aristotelian cosmos. And, by a strange coincidence, the year in which Galileo died at the age of 78 was the year in which Newton was born.

REFERENCES

1   Galileo, *The Starry Messenger* (Venice: 1610) in Drake, S., *Discoveries and Opinions of Galileo* (New York: Doubleday, 1957), p. 21.
2   Ibid., pp. 25–6.
3   Ibid., p. 32.
4   Ibid., p. 47.
5   Galileo, *Letter to the Grand Duchess Christina: Concerning the Use of Biblical Quotations in Matters of Science* (Strasbourg: 1636) in Drake, S., op. cit., p. 194; this letter was written in 1615 but not published until more than 20 years later.
6   Ibid., p. 195.
7   Ibid., p. 208.

8   From Galileo's notes made before going to Rome in 1616; quoted in Drake, S., op. cit., p. 168.

9   Galileo, *The Two New Sciences*, translated by Drake, S. (Madison: University of Wisconsin Press, 1974), p. 217.

10 Ibid., p. 147.

SUMMARY

1   Galileo's telescopic discoveries revealed a whole mass of descriptive evidence which helped to undermine the traditional Aristotelian world-view. This evidence was easy to understand and publicly available to all who could read.

2   By about 1615 the new astronomical observations of the late 16th and early 17th centuries had made the ideas of Copernicus very plausible both for professional astronomers and the public.

3   Galileo was somewhat unfair in always arguing *for* the Copernican system and *against* that of Aristotle and Ptolemy. By 1632 he should have been trying to distinguish between the systems of Copernicus and Tycho Brahe—a much more difficult task.

4   The famous dispute between Galileo and the Church was much more complex than is often stated. There were many in the Church who supported Galileo, at least initially. And Galileo, an articulate and forceful controversialist, always maintained that he was a faithful Catholic.

5   Galileo's case had a central contradiction which he couldn't escape. On the one hand he claimed that scientific authority should prevail over the authority of the Pope when something could be demonstrated beyond reasonable doubt. On the other hand, he was never able to provide such a demonstration on the really crucial issue 'Does the earth move?'.

6   The banning of Galileo's work not only led to it becoming more widely known; it also led to his writing a book on motion which set out for the first time many of the essential concepts and methods which were later used by Newton and are still used today.

3.2 Developments in the experimental tradition: new instruments and medical knowledge

Up till now we have focused our attention on astronomy. This has taken us into mathematics, into the study of motion, and has made us look at the development of some specific scientific instruments such as Tycho Brahe's great quadrant or Galileo's telescopes. In this chapter we will take a broader look at some other changes which took place in the late 16th and 17th centuries. First we will discuss the emergence of new instruments and see how the astronomical instruments we have mentioned were only a part of a much wider development in what was called 'the practical arts'. Then we will describe some of the changes in medical knowledge associated primarily with Vesalius and William Harvey. Finally we will assess the importance of these new instruments for the methods of science.

First let us consider the development of the practical arts which had been taking place in Europe over the whole period of the Renaissance in the 15th and 16th centuries. Two major landmarks were the introduction of the magnetic compass in the 13th century and the development of printing from around 1450 onwards. It is worth noting that neither of these major developments originated in Europe since they were well known in China at considerably earlier dates (see Part 5: Science and technology in traditional China).

The following passage was written in the middle of the 16th century, just after the death of Copernicus:

> The world sailed round, the largest of Earth's continents discovered, the compass invented, the printing-press sowing knowledge, gun-powder revolutionising the art of war, ancient manuscripts rescued and the restoration of scholarship, all witness to the triumph of our New Age.[1]

It sums up the mixture of practical and intellectual achievements which characterised the period and which is such a complete contrast to the separation between practical and intellectual activities made by the ancient Greeks whom the men of the Renaissance so revered.

Fig. 3.2.1 is a picture, *The Ambassadors*, painted by Holbein in 1533 and now in the National Gallery in London. It shows the French Ambassador to the court of Henry VIII and a bishop who was ambassador to the Holy See in Rome. On the shelf behind them there are a number of scientific instruments, amongst them a celestial and a terrestrial globe and various astronomical instruments, including a portable quadrant. The appearance of scientific instruments in pictures like these emphasises the high status given to practical objects in the 16th century. It contrasts dramatically with the almost complete dominance of religious themes in the paintings of the medieval period.

Fig. 3.2.1   *The Ambassadors* painted by Holbein in 1533 (National Gallery, London); note the display of globes and scientific instruments.

NEW INSTRUMENTS IN THE 17TH CENTURY

In the 17th century the development of the practical arts accelerated. Many new instruments were invented, some of which caused considerable changes in scientific work. Eventually their influence spread into many trades and occupations. The 17th century is sometimes characterised as the period of six great new instruments—the telescope, the microscope, the barometer, the vacuum pump, the pendulum clock and the thermometer. The first two of these, the telescope and the microscope, are often called the greatest artefacts in history. They opened up whole new fields of observation, of both the very large and the very small. Let us look at these six new instruments in turn.

The telescope

The telescope we have already mentioned. It was first made into an effective instrument by Galileo, a man who possessed both the practical skills to make it and the intellectual skill to know what to do with it. Yet even Galileo's telescopes were far from perfect. The images they gave were distorted by imperfectly ground lenses or by intrinsic limitations such as spherical and chromatic aberration. Nor did Galileo, despite his claims, have any theory of optics to explain how his telescope worked. This theory was principally provided by two men—Kepler, and a Frenchman, René Descartes. In 1611 Kepler worked out a theory for spherical aberration and a method of correcting it using hyperbolic instead of spherical lens surfaces. Then, in 1637, Descartes in his *La Dioptrique* (*Optics*) gave a general theory for lens combinations and for the elimination of spherical aberration. However, unlike Galileo, neither Kepler nor Descartes were practical men, able to grind their own lenses. So it was not until a little later, from about 1650 onwards, that improved practical telescopes became available. When this happened there followed, in the period from about 1650 to 1680, a wave of astronomical discoveries about the planets, their satellites and surface markings, including the recognition of the rings of Saturn by Huyghens in 1656. Other important developments were an increase in the accuracy of observation of planetary orbits and those of other wandering bodies like comets, and a beginning of the systematic study of the enormous number of new stars whose existence the telescope had revealed. The telescope, in effect, brought about an extension of a familiar world—the world of planetary and stellar astronomy which had been studied since antiquity.

The microscope

With the microscope, things were very different. Microscopes were developed early in the century. Fig. 3.2.2 shows a compound microscope designed by Descartes. The large mirror is collecting light to illuminate the object.

Like the early telescopes, these early microscopes gave relatively poor images and it was not until the 1660s that improved instruments became available. The man whose name was most closely associated with the microscope at that time was Robert Hooke, who was the Curator of the newly-formed Royal Society of London.

At each meeting of the Royal Society Hooke had to present at least one microscopic observation. On 22nd April 1663 he 'brought in two microscopical observations, one of leeches in vinegar; the other of a bluish mould upon a mouldy piece of leather'. On 6th May he reported 'a microscopical observation of a female gnat'; on 20th May three observations, on the head of an ant, on the point of a needle, and on 'a strange fly like a gnat'. The following week Hooke was 'charged to look upon sage with a microscope, and to observe whether there lurked any little spiders in the cavities of the leaves, that might make them noxious'.[2]

The society's patron, Charles II, was invited to a meeting to look through this marvellous new instrument and, following this visit, the society sponsored the publication of Hooke's *Micrographia* in 1664. Fig. 3.2.3 shows one of the many magnificent illustrations from that book—a louse together with a human hair.

The extraordinary new world revealed by the microscope led to a great upsurge

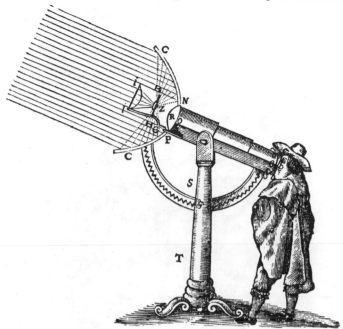

Fig. 3.2.2 Compound microscope due to Descartes—the large parabolic mirror is collecting light to illuminate the object at Z.

in enthusiasm for the new instrument, and for a while new observations came thick and fast. For example, the Italian, Malpighi, used microscopes of comparatively low power as a natural extension of his work on comparative anatomy. Malpighi made detailed studies of the development of chick embryos, and he demonstrated for the first time the existence of capillary blood vessels. But after a few years the enthusiasm for the microscope faded as rapidly as it had grown in the 1660s. In 1691 Hooke wrote:

> But tho' there has been some life left in the grinders of glasses, yet the warmth of those, that should have used them, has grown cool; and little of new discoveries hath been made by them . . . Microscopes are now reduced almost to a single votary, which is Mr. Leeuwenhoek; besides whom, I hear of none that make any other use of that instrument, but for diversion and pastime, and by that reason it is become a portable instrument, and easy to be carried in one's pocket.[3]

Unlike the telescope, which had extended and brought nearer a familiar world, that of planetary and stellar astronomy, the microscope revealed a completely new world. By extending observations far beyond what was possible with the unaided eye, the microscope made it difficult for people even to start to think coherently about this new world. Nobody had any experience of what was to be found there, except, of course, for what could be seen at the kind of very low magnification used by Malpighi. It wasn't until a century or more later, from the 1820s onwards, that people began to explore the microscopic world with any confidence.

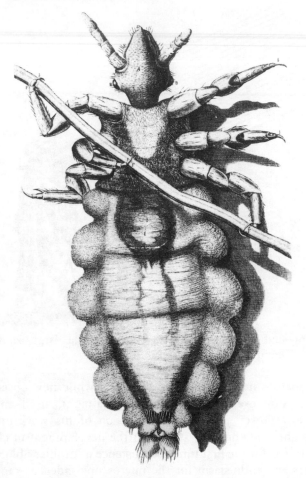

Fig. 3.2.3   Drawing of louse, with human hair, from Hooke's *Micrographia* of 1664.

The main exception was the Dutchman mentioned by Hooke, Antony van Leeuwenhoek. Leeuwenhoek is one of the most remarkable figures in the history of science. He was a man with no scientific training who published no books or scientific papers.[4] What he did have was uncanny eyesight and extraordinary skill in making microscopes. These were mostly simple microscopes consisting of a single tiny lens. Fig. 3.2.4 shows one of them. With microscopes like this, Leeuwenhoek made a series of observations in the 1670s which eventually were to transform medicine and biology. In his own words, he discovered in a pail of lake water 'little animals or animalcules'. Here is how he described them:

> And the motion of most of these animalcules in the water was so swift, and so various, upwards, downwards, and round about, that 'twas wonderful to see; and I judge that some of these little creatures were above a thousand times smaller than the smallest ones I have ever yet seen, upon the rind of cheese, in wheaten flour, mould and the like.'[5]

Fig. 3.2.4   Simple microscope made by Antony van Leeuwenhoek about 1670. The single tiny lens is shown in the centre of the flat board. The rest of the instrument is a device for mounting the specimen and adjusting its position with respect to the lens. The specimen is viewed by placing the eye close to the lens on the opposite side of the board to that shown.

Later he described 'little animals' in the human body—in fact in a specimen of his own saliva:

> I now saw very plainly that these were little eels or worms, lying all huddled up together and wriggling; just as if you saw, with the naked eye, a whole tubful of very little eels and water, with the little eels a-squirming among one another; and the whole water seemed to be alive with these multifarious animalcules. This was for me, among all the marvels that I have discovered in nature, the most marvellous of all.[6]

But Leeuwenhoek's work on what were in fact bacteria and protozoa at first led nowhere. There were a number of reasons for this. For a start, Leeuwenhoek was not very good at drawing what he saw. Fig. 3.2.5 shows some of his drawings of bacteria. These vague sketches were not anything like as convincing as the detailed illustrations from Hooke's *Micrographia* (see Fig. 3.2.3). And nobody else could make microscopes like Leeuwenhoek. Other people even found it difficult to repeat his observations with Leeuwenhoek's own instruments. So he remained an isolated figure and became the butt of the satirists of the time. And yet his microscopes really were extraordinarily good. Many of them have survived, and modern work on them has confirmed their high magnification, up to about 300 times, and high resolution, down to about one-millionth of a metre. Microscopes of this quality were not made by anybody else until the 19th century.

The barometer
The next instrument we will consider is the barometer. This developed from some work on water suction pumps described by Galileo in *Two New Sciences*. It had long been known that such pumps could raise water through a height of no more than about 32 or 33 feet. There was, however, no satisfactory explanation for this limitation. In the 1630s, experiments were made in Italy with what were, in effect, water barometers. However, these were extremely cumbersome devices. They involved setting up long pipes, like drainpipes, on the outside of houses with

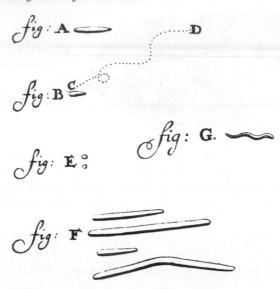

Fig. 3.2.5   Drawings of bacteria made by Leeuwenhoek. Compare these drawings with the illustration from Hooke's *Micrographia* shown in Fig. 3.2.3.

great globes on the top. It was difficult with this kind of apparatus to make reliably repeatable observations. They were always leaking or even collapsing. The main advance in the understanding of air pressure came with the work of another Italian, Torricelli, in the 1640s. Torricelli worked with mercury rather than water and this meant, of course, that his apparatus was much smaller and therefore more manageable. Fig. 3.2.6 shows one of his experiments which demonstrated that the height of the mercury column is always the same even though the tubes vary in size and cross-section.

It is clear from Torricelli's writings that he interpreted his experiments in terms of the balancing of the pressure due to a column of air extending from the earth's surface to the top of the atmosphere against the hydrostatic pressure of the column of mercury. Torricelli was also sufficiently skilful to be able to detect the small changes in the height of his mercury column from day to day. Further confirmation of the central principle came in 1648 when the French mathematician Pascal did his famous experiment on the Puy de Dôme mountain with two identical mercury barometers. One he carried up the mountain and observed its mercury column decreasing in length the higher he went. No change was observed in the length of the mercury column in the other barometer, which remained at the foot of the mountain throughout the experiment.

Torricelli's experiments raised another interesting question. What was in the space above the mercury? Traditional ideas, deriving largely from Aristotle, had denied the possibility of an empty space or vacuum. So people were reluctant to say that there was nothing there. The question continued to intrigue them, however, and it became possible to study these possibly 'empty spaces' after the development of another new instrument, the air pump.

Fig. 3.2.6 One of Torricelli's experiments with barometers of different size and cross-section.

The air pump

One of the earliest air pumps was developed by Von Guericke in Germany around 1650. Von Guericke devised a dramatic public demonstration of his air pump and of the magnitude of the pressure exerted by the atmosphere. Fig. 3.2.7 shows this demonstration at Magdeburg in Germany in which two teams of eight horses are shown failing to separate two hollow hemispheres when they were emptied of air. The so-called Magdeburg hemispheres were easily separated by hand when they were full of air.

Von Guericke's work attracted the interest of the Hon. Robert Boyle in England who improved the design of the air pump and put apparatus into the space which was being pumped. Boyle carried out a whole series of experiments in the partial vacuum such as the extinction of a flame, the death of a bird or mouse and the loss of transmission of sound. Such experiments with the air pump, and other new pieces of scientific apparatus, were later often demonstrated in public.

Fig. 3.2.7 Public demonstration of air pressure with the Magdeburg hemispheres. Sixteen horses are shown failing to separate the hemispheres when they were emptied of air using von Guericke's air pump.[7] The objects in the sky are not flying saucers but diagrams showing the hemispheres in detail.

The pendulum clock
The fifth new instrument is the pendulum clock which was invented by the Dutchman, Christian Huyghens, in the 1660s. Before this no accurate method existed for measuring short time intervals. For example, Galileo used a water clock to measure the time taken for bodies to roll down his inclined plane.

The thermometer
It was Galileo too who devised what he called a thermoscope which consisted of an air-filled glass bulb with a fine tube attached dipping into a reservoir of liquid. As the air in the bulb was heated or cooled so the level of liquid in the tube fell or rose. But since the reservoir was open to the atmosphere, the instrument responded to changes in atmospheric pressure as well as temperature. Thermoscopes were used as clinical thermometers in the 1640s and soon evolved into the kind of sealed liquid-in-glass thermometer with which we are familiar today.

*   *   *

Other instruments were also invented in the 17th century including a large number of mathematical instruments such as slide rules and other calculating devices of various kinds.

Considered separately, these new instruments of the 17th century are interesting but perhaps not crucial in their impact. Taking them all together, however, they do represent a dramatic change in what was possible. They greatly extended the range of possible observations. They made it easier for these observations to become more quantitative and not just descriptive. And they made it much more feasible for people to make experiments in controlled environments.

MEDICAL KNOWLEDGE AND THE EXPERIMENTAL TRADITION

We have seen how astronomy in the 15th and 16th centuries was dominated by the work of two Greeks—Aristotle from the early classical period and Ptolemy from the 2nd century AD. Interestingly enough, the same is true of the medical knowledge of the period. The dominating figures in medicine were Hippocrates from around 420 BC (see Chapters 2.1 and 6.5) and Galen who lived from AD 129 to 200. Both men emphasised observation as the prerequisite for medical knowledge, with Galen laying particular emphasis on dissection as the basis for anatomical studies. Galen possessed very great dissecting skill but unfortunately he was seldom able to dissect a human body.

Much of the work that went into his masterpiece, *On Anatomical Procedures*, was done on the Barbary ape. Consequently he sometimes described structures which are not present in man. However, his writings were so comprehensive, coherent and well-organised that they, like the works of Ptolemy in astronomy, became the authoritative medical texts for almost 1500 years. Fig. 3.2.8 shows the title-page of a collected edition of Galen's works published anew in 1562.

While Galen's anatomy was reasonably accurate, his physiology was very different from our modern ideas. For example, like many earlier medical writers, Galen thought that blood was made in the liver and veins from nourishment received from the stomach and that the blood was then distributed to other parts of the body which were made up of matter contained in the blood.

Galen's work began to be superseded with the rise of a school of anatomists in 16th-century Italy. These men *were* able to dissect human bodies and the most famous of them was Vesalius (1514–64) who became professor of surgery and medicine at the University of Padua. In 1543, the same year as Copernicus' *De Revolutionibus* appeared, Vesalius published his book *De Humani Corporis Fabrica* (On the Fabric of the Human Body), the frontispiece of which shows him publicly demonstrating the dissection of a human corpse. Fig. 3.2.9 shows Vesalius demonstrating the anatomy of the arm. A major feature of *De Fabrica* was the quality of its many illustrations, which were drawn by a pupil of Titian. Two illustrations are shown in Figs. 3.2.10 and 3.2.11. The dramatic poses and the background landscapes are very different from the illustrations in later medical texts.

CL▸
GALENI
PERGAMENI
OMNIA, QVAE EXTANT,
IN LATINVM SERMO‐
NEM CONVERSA.

QVIBVS POST SVMMAM ANTEA ADHIBITAM
diligentiam, multum nunc quoꝗ splendoris accessit, quòd lo‐
ca quàmplurima ex emendatorum exemplarium
collatione & illustrata fuerint & castigata.

HIS ACCEDVNT NVNC PRIMVM CON. GESNERI
Præfatio & Prolegomena tripartita, De uita GALENI,
eiusꝗ libris & interpretibus.

EX III. OFFICIN. FROBENIANAE EDITIONE.

M. D.   LXII.

Fig. 3.2.8   Title-page of Galen's collected works published in 1562.

Fig. 3.2.9   Vesalius demonstrating the anatomy of the arm.

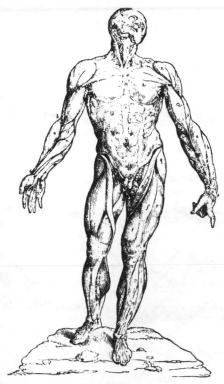

Fig. 3.2.10   First muscle tabula from Vesalius' *De Fabrica*.

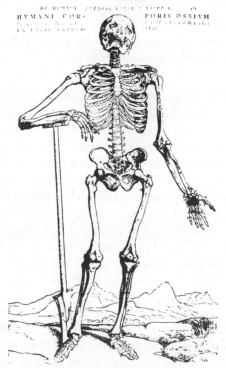

Fig. 3.2.11   First skeleton from Vesalius' *De Fabrica*.

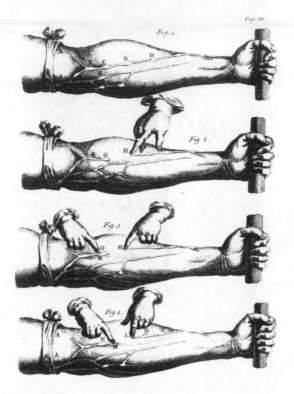

Fig. 3.2.12   Diagrams from William Harvey's *De Motu Cordis*, showing that the valves in veins only allow blood to flow towards the heart.

It was in the school of anatomy established by Vesalius at Padua that William Harvey (1578–1657) received the training in anatomy which underlay his famous work on the circulation of the blood. Harvey was at the University of Padua from 1600 to 1602, at the same time as Galileo. Fig. 3.2.12 shows some diagrams from Harvey's book *Exercitatio Anatomica de Motu Cordis et Sanguinis in Animalibus* (An Anatomical Essay on the Movement of the Heart and Blood in Animals). These diagrams illustrate experiments in which Harvey showed that the valves in the veins only allowed blood to flow towards the heart and not away from it, as Galen had argued.

Harvey established his case for the circulation of the blood by combining a detailed knowledge of anatomy, and in particular of the valves of the heart and veins, with simple experiments on living subjects (such as the one illustrated in Fig. 3.2.12) and the use, probably for the first time in biology, of a semi-quantitative argument. He calculated, very roughly, the amount of blood pumped by each heartbeat into the aorta, the major artery, and showed that in half-an-hour or so more blood was pumped by the heart than is contained in the whole body. This meant that the heart pumped about 400 pints of blood in 24 hours—a vast quantity which clearly could not have been made from the amount of food eaten in the same period. All his evidence pointed to the circulation of the blood but at first these unorthodox ideas were not well received and Harvey lost many patients after *De Motu Cordis* was published.

We thus see developing in medicine in the 17th century the combination of the same three features we saw emerging in the physical sciences—careful observation, deliberate experimentation, and the use of quantitative arguments, although of course the last two are rather more difficult to apply in this field of study. These three features have remained essential components of the methods used in nearly all the sciences right down to the present day.

REFERENCES

1 Fernel, J., *De Abditis Rerum Causis*, 1548; quoted in Boas, M., *The Scientific Renaissance, 1450–1630* (London: Fontana, 1970), p. 13.
2 These descriptions by Hooke are to be found in the Philosophical Transactions of the Royal Society; they are also given in Nicolson, M., *Science and Imagination* (Cornell: Cornell University Press, 1956), p. 162.
3 Quoted in Nicolson, M., op. cit., pp. 168–9.
4 Fortunately Leeuwenhoek was a prolific letter-writer, see Dobell, C., *Antony van Leeuwenhoek and his 'Little Animals'* (London: John Bale and Danielsson, 1932).
5 Nicolson, M., op. cit., p. 165.
6 Ibid., p. 166.
7 See, however, Cardwell, D. S. L., *From Watt to Clausius* (London: Heinemann, 1971), p. 13 for a recent calculation of the forces involved which suggests that von Guericke's horses were rather feeble specimens!

SUMMARY

1 In the 17th century the 'practical arts' of the Renaissance developed rapidly and gave rise to many new instruments—the telescope, the microscope, the barometer, the air pump, the pendulum clock and the thermometer.
2 These instruments extended the range of possible observations, made them more quantitative, and made it easier to conduct experiments in controlled environments.
3 These trends in the physical sciences were also reflected in the biological sciences—for example, in the careful anatomical observations of Vesalius and the work of William Harvey on the circulation of the blood.

3.3 Descartes, the mechanical philosophy and the rise of mathematics

In this chapter we are going to look at the work and influence of the Frenchman, René Descartes (1596–1650); Fig. 3.3.1 shows a portrait of Descartes. We will first consider Descartes' work in mathematics and then look at some other ideas of his which were inspired by his mathematical work.

RENATUS DESCARTES, NOBILIS GALLUS, PERRONI DOMINUS, SUMMUS MATHEMATICUS & PHILOSOPHUS

F. Hals pinxit   L. Suiderhoef sculpsit   P. Clay excudit

Fig. 3.3.1 Portrait of Descartes.

DESCARTES' MATHEMATICS

We mentioned at the end of Chapter 3.1 that Galileo's work on local motion was later developed by Newton into a consistent dynamics for motions both on earth and in the heavens. This development was made much easier by new work in mathematics which linked the ancient study of geometry with the much more recent emergence of algebra.

As we have seen, the mathematics Galileo used was classical geometry. Look again at Fig. 3.1.6 (page 40) which shows a page from Galileo's *Two New Sciences*. On this page Galileo is proving some propositions which he needs for his

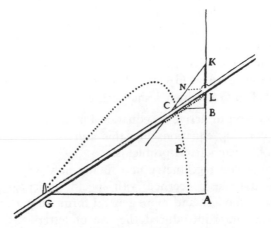

Aprés cela prenant vn point a diſcretion dans la courbe,
comme C, ſur lequel ie ſuppoſe que l'inſtrument qui ſerc
a la deſcrire eſt appliqué, ie tire de ce point C·la ligne
C B parallele a G A, & pourceque C B & B A ſont deux
quantités indeterminées & inconnuës , ie les nomme
l'vne y & l'autre x. mais affin de trouuer le rapport de
l'vne à l'autre ; ie conſidere auſſy les quantités connuës
qui determinent la deſcription de cete ligne courbe,
comme G A que ie nomme a, K L que ie nomme b, &
N L parallele a G A que ie nomme c. puis ie dis, comme
N L eſt à L K, ou c à b, ainſi C B, ou y, eſt à B K, qui eſt
par conſequent $\frac{b}{c}$ y : & B L eſt $\frac{b}{c}$ y --b, & A L eſt x +
$\frac{b}{c}$ y -- b. de plus comme C B eſt à L B, ou y à $\frac{b}{c}$ y--b, ainſi
a, ou G A, eſt á L A, ou x + $\frac{b}{c}$ y -- b. de façon que mul-
S ſ
tipliant

Fig. 3.3.2   Page from Descartes' *La Géométrie* (1637) in which he describes the use of
Cartesian coordinates.

work on motion. The style and the structure of the argument would have been totally familiar to Euclid or any other mathematician of ancient Greece.

Now look at Fig. 3.3.2 which shows a page from Descartes' *La Géométrie*, published in 1637. Here the style of the mathematics is totally different, as we can see if we translate part of the text:

> Then I take on the curve an arbitrary point, as C, at which we will suppose the instrument applies to describe the curve. Then I draw through C the line CB parallel to GA. Since CB and BA are unknown and indeterminate quantities, I shall call one of them y and the other x. To the relation between these quantities I must consider also the known quantities which determine the description of the curve, as GA, which I shall call a: KL which I shall call b: and NL parallel to GA, which I shall call c.'[1]

Descartes then goes on to derive an equation for the curve in terms of y, x, a, b and c. It is by deriving equations like this and then manipulating them that Descartes solves his geometrical problems.

Descartes here is doing geometry in a totally new way. In the 16th century algebra had ceased to be concerned with specific practical problems and had begun to take on its modern and more general form. In particular, a Frenchman called Viete (1540–1603) introduced the use of letters to represent unknown variable quantities. Viete used capital letters—vowels for unknowns and consonants for given quantities or constants. In *La Géométrie* Descartes is doing the same thing except that he uses small letters—those at the end of the alphabet for unknowns and those at the beginning for constants. The pages of *La Géométrie* are full of algebraic equations for curves, written in a notation which is very similar to that used in modern textbooks on algebra. Descartes thus united algebra and geometry and was the founder of coordinate geometry; we still call the axes and coordinates he used *Cartesian* axes and *Cartesian* coordinates in recognition of his momentous mathematical innovation.

DESCARTES AND THE MECHANICAL PHILOSOPHY

Outside France, Descartes is now mainly remembered amongst scientists for his mathematics and in particular for his Cartesian coordinates. However, in the 17th century he was an enormously influential figure due to his work in philosophy and in all the sciences. Yet even then mathematics was his dominant interest and his inspiration.

Early in his life, around 1620, Descartes had written but not published a short book called *Rules for the Direction of the Mind*. The key rule was Rule 4—'There is need for a Method for finding out the Truth'. The method advocated by Descartes was deductive mathematics which he suggested could be generalised into a universal mathematics of very wide application.

Nearly 20 years later, in 1637, Descartes set out a more considered version of these same ideas in his famous book *Discours de la Méthode pour bien conduire la Raison et Chercher la Vérité dans les Sciences* or *Discourse on the Method of Reasoning Well and Seeking Truth in the Sciences*.

The spirit of Descartes' *Discourse on Method* is captured in this passage:

These long chains of reasonings which geometers are accustomed to using to teach their most difficult demonstrations, had given me cause to imagine that everything which can be encompassed by man's knowledge is linked in the same way, and that, provided only that one abstains from accepting any for true which is not true, and that one always keeps the right order for one thing to be deduced from that which precedes it, there can be nothing so distant that one does not reach it eventually, or so hidden that one cannot discover it.[2]

That was Descartes' claim and, in order to illustrate his method, he published three *Essays in this Method* as appendices to the book. The first was called *La Dioptrique* which dealt with the eye and light; the microscope shown in Fig. 3.2.2 is from *La Dioptrique*. The second appendix was *Les Météores*, a discussion of various meteorological phenomena which is hardly ever read today. The third appendix was *La Géométrie* which, as we have seen, contained Descartes' fundamental contribution to mathematics, the fusion of algebra and geometry in coordinate geometry.

These appendices show up the major problems of Descartes' mathematically deductive method. How do you establish the starting points or premises of your argument? And how do you carry through the deductions so that each step is as clear and as certain as a mathematical proof?

As one might expect, Descartes' method works fairly well in *La Géométrie* but in his work on optics Descartes has to introduce observation and experiment to supplement his mathematics and to enable him to reach some definite conclusions. And in the appendix on meteorology the observations and experiments he would need are well beyond his, or anyone else's, resources at that time, or even today.

Descartes also tried to apply his method to cosmological problems and in 1634 he wrote a book on this subject called *Le Monde* or *The World*. However, Descartes decided not to publish *Le Monde* immediately as it contained some discussion of Copernicus' ideas and Galileo had just been condemned by the Inquisition.

In fact *Le Monde* wasn't published until 1664, 14 years after Descartes' death, but many of its ideas and much else besides was included in his *Principia Philosophica* or *The Principles of Philosophy* in 1644.

The *Principia Philosophica* is in four main parts. Throughout the book Descartes tries to argue deductively starting from what he calls *First Philosophy* or the principles of sense perception; then he moves on to what he calls *Principles of Material Things*, that is of ordinary solid bodies; then to *The Visible World* including light; and finally to *The Earth* in which he includes the solar system and the rest of the universe, or what we would call cosmology. Throughout Descartes asks the question How? rather than the question Why? He finds his answers in terms of his ideas on matter and motion. For Descartes matter consists of myriads of tiny particles or corpuscles which fill his universe. These particles are in constant motion, and all the phenomena we observe, from the smallest to the

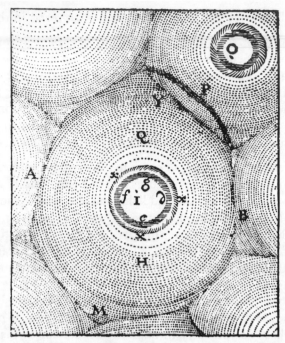

Fig. 3.3.3 A diagram from Descartes' *Principia Philosophica* showing corpuscles swirling in the vortices which, he claimed, were carrying the planets and other heavenly bodies through space.

largest scale, are due to the collisions between them. Descartes thought he had deduced the properties of the whole material universe from the motion and collisions of particles.

Fig. 3.3.3 is one of the diagrams from *Principia Philosophica* which shows Descartes' corpuscles whirling in vortices. At the centre of each vortex is a sun and, according to Descartes, it is the motion of these vortices which carries the planets and other heavenly bodies through space.

Descartes claimed to have deduced the properties of the whole universe using similar mechanical principles to those involved in the design of machines:

> My assigning definite shapes, sizes and motions to insensible particles of bodies, just as if I had seen them, and this in spite of admitting that they are insensible, may make some people ask how I can tell what they are like ... I have been greatly helped by considering machines. The only difference I can see between machines and natural objects is that the workings of machines are mostly carried out by apparatus large enough to be readily perceptible by the senses ... whereas natural processes almost always depend on parts so small that they utterly elude our sense ... So, just as men with experience of machinery, when they know what a machine is for, and can see part of it, can readily form a conjecture about the way its unseen parts are fashioned; in the same way, starting from sensible effects and sensible parts of bodies, I have tried to investigate the insensible causes and particles underlying them.'[3]

Descartes' *mechanical philosophy*, as it came to be called, was enormously influential. The historian Thomas Kuhn sums up its influence like this:

> It is today, childishly easy to discover errors and inadequacies in Descartes' discussion of vortex cosmology and in the astronomy, optics, chemistry, physiology, geology, and dynamics that he derived from it. His vision was inspired, and its scope was tremendous, but the amount of critical thinking devoted to any one of its parts was negligibly small. . . . But in the development of seventeenth-century science the parts of Descartes' system were far less important than the whole. Descartes' brilliant successors . . . could and did change his laws of collision, his description of vortices, and his laws for the propagation of light. But they did not compromise his conception of the universe as a corpuscular machine governed by a few specified corpuscular laws. For half a century that conception guided the search for a self-consistent Copernican universe.[4]

Yet Descartes' *Principia Philosophica* had one extraordinary feature, particularly as it was written by a man whose whole work was inspired by mathematics. There was virtually *no* mathematics in it. It *was* deductive but it was nearly all *verbal*.

It was this feature of Descartes' writings that profoundly dissatisfied Isaac Newton (1642–1727; see Fig. 3.4.1, page 65). When young, Newton was impressed with Descartes' work, but he came to think that its lack of any detailed mathematical working out of its principles was a crippling weakness. So, when Newton wrote his own major work on motion in 1687, the title he chose was *Philosophiae Naturalis Principia Mathematica* or *The Mathematical Principles of Natural Philosophy* (see Fig. 3.4.4, page 72). Newton's *Principia*, as the book is usually called, did work out in great mathematical detail a full theory for the motions of bodies both on the earth and in the heavens. But in one major respect Newton *was* deeply indebted to Descartes. It was developments stemming from Descartes' work on coordinate geometry that provided Newton with some of the essential mathematical tools for his own momentous work.

REFERENCES

1 Smith, D. E. and Latham, M. L. (trans.) *The Geometry of René Descartes* (New York: Dover Publications, 1954), p. 52.

2 Descartes, *Discourse on Method*, Part 2; quoted in Smith, A. G. R., *Science and Society in the Sixteenth and Seventeenth Centuries* (London: Thames and Hudson, 1972), p. 75.

3 Descartes, *Principia Philosophica* in Anscombe, E. and Geach, P. T., (trans.) *Descartes: Philosophical Writings* (London: Thomas Nelson, 1970), p. 236. By permission of Van Nostrand Reinhold (UK).

4 Kuhn, T. S., *The Copernican Revolution* (New York, Harvard University Press, 1957), p. 242. Reprinted by permission.

SUMMARY

1   New mathematical techniques were needed before the study of motion could be developed beyond the stage at which Galileo had left it.

2   The most important of these techniques was the fusion of algebra and geometry in the coordinate geometry of Descartes.

3   Descartes took the deductive style of mathematical proof as his model for all forms of enquiry.

4   Descartes used his deductive method in developing a *mechanical philosophy* in which everything in the universe was explained in terms of the motions and collisions of particles.

5   Newton was dissatisfied with the lack of mathematical rigour in Descartes' major work *The Principles of Philosophy* and devoted much effort to his own *Mathematical Principles of Natural Philosophy* or the *Principia*, as it is usually known.

3.4 Newton and his *Principia*

Isaac Newton (see Fig. 3.4.1) is the man who, more than any other single figure, symbolises the scientific revolution of the 17th century.

In this chapter we will consider some of the better-known achievements and episodes in his life, emphasising in particular his masterpiece, the *Principia*. Then in Chapter 4.1 we will describe some of the lesser-known aspects of this many-sided man.

After the time of Galileo, the main centres of scientific activity were to be found in Northern Europe rather than in Italy. Galileo's trial may have contributed to this—remember that his *Two New Sciences* was published at Leyden in Holland rather than in Italy. Descartes, for example, was born in France and spent much of his adult life in Protestant Holland, as of course did Huyghens, who was born there and educated at the University of Leyden. Even during Galileo's lifetime, Kepler had worked mainly in Prague where he had used Tycho Brahe's data to discover three laws of planetary motion (see Chapter 2.2) which turned the complex planetary orbits of Copernicus into precise and simple mathematical

Fig. 3.4.1 Portrait of Sir Isaac Newton (1642–1727).

formulae. Kepler's laws were to be of central importance in Newton's work, and we shall pay particular attention to Kepler's first law which stated that all the planets move in elliptical orbits with the sun at one focus. Newton was to replace Kepler's precise mathematical descriptions of the planetary orbits with dynamic and causal explanations of the planetary motions which were just as mathematically precise as Kepler's descriptions but which were applicable to any moving body, planet or projectile, in the heavens or on earth.

NEWTON'S EARLY LIFE AND WORK

Isaac Newton was born at Woolsthorpe in the English county of Lincolnshire on Christmas Day, 1642, twelve years after Kepler's death and in the same year that Galileo died. Fig. 3.4.2 shows an 18th-century sketch of Newton's mother's house.

Little is known of Newton's early life except that Newton was always a solitary child. He was brought up mainly by his grandmother, for his father had died before Newton was born and his mother soon remarried.

Newton entered Cambridge University in 1661, the year after the English monarchy was restored in the person of Charles II. Charles II's father, Charles I, had been beheaded in 1649 and England had been a republic for the following eleven years under Oliver Cromwell.

Fig. 3.4.2   An early 18th-century sketch of Newton's mother's house at Woolsthorpe in Lincolnshire, England.

At Cambridge, Newton studied mainly mathematics and he graduated in 1664. The next two years, 1665 and 1666, were years of a great plague in England. Nearly a quarter of the population of London died in 1665 alone, about 100,000 out of a population of nearly 450,000. Because of the plague the University closed, and Newton left Cambridge and went home to Woolsthorpe for 18 months.

During this brief period, at the age of 23 or 24, Newton made many of his most famous discoveries. Here is how he described these discoveries in a letter which was written many years later, near the end of his life:

I found the Method (of fluxions) by degrees in the years 1665 and 1666. In the beginning of the year 1665 I found the method of approximating Series and the Rule for reducing any dignity of any Binomial into such a series. The same year in May I found the method of tangents of Gregory and Slusius, and in November had the direct method of fluxions, and the next year in January had the Theory of colours, and in May following I had entrance into ye inverse method of fluxions. And the same year I began to think of gravity extending to ye orb of the Moon, and having found out how to estimate the force with w^{ch} (a) globe revolving within a sphere presses the surface of the sphere, from Kepler's Rule of the periodical times of the Planets being in a sesquialterate proportion of their distances from the centers of their Orbs I deduced that the forces wch keep the Planets in their Orbs must (be) reciprocally as the squares of their distances from the centers about wch they revolve: and thereby compared the force requisite to keep the Moon in her Orb with the force of gravity at the surface of the earth, and found them answer pretty nearly. All this was in the two plague years of 1665 and 1666, for in those days I was in the prime of my age for invention, and minded Mathematicks and Philosophy more than at any time since.[1]

In other words, Newton claimed to have discovered the Binomial Theorem, invented both the differential and the integral calculus—which he called the direct and inverse method of fluxions—started his work on colours and, most important, to have conceived his inverse square law of universal gravitation and worked out its application to the motion of the moon and of bodies on the earth, all in 18 months.

Recent historical research indicates that Newton's memory may have been at fault and that not all of this work dates from 1665–6. In particular no direct evidence has been found for the calculation involving the moon's orbit and the force of gravity at the earth's surface. Even so, it was a remarkable 18 months.

Yet at the time Newton published nothing about all these discoveries. Almost anybody else would have published them at once. Galileo might well have shouted them from the rooftops of the Vatican. But not Newton. When the plague was over in 1667 he quietly returned to Cambridge and continued with his work. But Newton nevertheless made a considerable impression, and in 1669 Isaac Barrow, the Professor of Mathematics who had taught Newton, resigned and Newton at once succeeded him as Lucasian Professor of Mathematics at the age of 26.

NEWTON ON LIGHT

Surprisingly, Newton's first lectures and publications were not on mathematics or mechanics or gravitation, but on optics. Newton, like Galileo before him, was just as skilled in the 'practical arts' as he was in mathematics. In 1672 he designed and built a new kind of telescope which used a large concave mirror to collect the light rather than the large lenses common at that time (see Fig. 3.4.6). Newton made and polished his own mirrors in his own workshop. Then he sent one of his telescopes to the Royal Society in London, who tried it out at one of their weekly meetings and promptly elected Newton a Fellow of the Society.

Soon after this Newton wrote to Oldenburg, the Secretary of the Royal Society, that he had made:

> a Philosophical discovery which induced me to the making of the said Telescope, and which I doubt not but will prove much more grateful then the communication of that instrument, being in my Judgment the oddest if not the most considerable detection which hath hitherto beene made in the operations of Nature.[2]

There was little that was modest about Newton!

This discovery concerned the colours which always arise when white light is refracted by lenses or prisms, and Newton wrote an account of it for the Royal Society. This is how he described his key experiments:

> I procured me a Triangular glass-Prisme, to try therewith the celebrated *Phaenomena of Colours*. And in order thereto having darkened my chamber, and made a small hole in my window-shuts, to let in a convenient quantity of the Suns light, I placed my Prisme at his entrance, that it might be thereby refracted to the opposite wall. It was at first a very pleasing divertisement, to view the vivid and intense colours produced thereby; but after a while applying myself to consider them more circumspectly, I became surprised to see them in an *oblong* form; which, according to the received laws of Refraction, I expected should have been *circular*.

> And I saw . . . that the light, tending to (one) end of the Image, did suffer a Refraction considerably greater then the light tending to the other. And so the true cause of the length of that Image was detected to be no other, then that *Light* consists of *Rays differently refrangible*, which, without any respect to a difference in their incidence, were, according to their degrees of refrangibility, transmitted towards divers parts of the wall.[3]

Newton thus explains the elongation of the spectrum which occurs because the prism separates the white light into colours and some of them are bent or refracted more than others.

Newton continued his experiments, thus:

> Then I placed another Prisme . . . so that the light . . . might pass through that also, and be again refracted before it arrived at the wall. This done, I took the

first Prisme in my hand and turned it to and fro slowly about its Axis, so much as to make the several parts of the Image . . . successively pass through . . . that I might observe to what places on the wall the second Prisme would refract them.

When any one sort of Rays hath been well parted from those of other kinds, it hath afterwards obstinately retained its colour, notwithstanding my utmost endeavours to change it.[4]

Newton thought that this was a *critical experiment* which proved that refraction separates the colours and that, once they are separated, they cannot be changed further, thus contradicting the traditional view which was that light became coloured because it was modified by the prism or lens. Newton goes on:

I have refracted it with Prismes, and reflected with it Bodies which in Daylight were of other colours; I have intercepted it with the coloured film of Air interceding two compressed plates of glass; transmitted it through coloured Mediums, and through Mediums irradiated with other sorts of Rays, and diversely terminated it; and yet could never produce any new colour out of it.

But the most surprising, and wonderful composition was that of *Whiteness*. There is no one sort of Rays which alone can exhibit this. Tis ever compounded, and to its composition are requisite all the aforesaid primary Colours, mixed in a due proportion. I have often with Admiration beheld, that all the Colours of the Prisme being made to converge, and thereby to be again mixed, reproduced light, intirely and perfectly white.

Hence therefore it comes to pass, that *Whiteness* is the usual colour of *Light*; for, Light is a confused aggregate of Rays indued with all sorts of Colors, as they are promiscuously darted from the various parts of luminous bodies.[5]

Fig. 3.4.3 shows an experiment in which Newton splits a beam of white light into its constituent colours, recombines the colours to give white light and then splits the light a second time into a coloured spectrum.

Newton thought that his arguments and experimental demonstrations on the

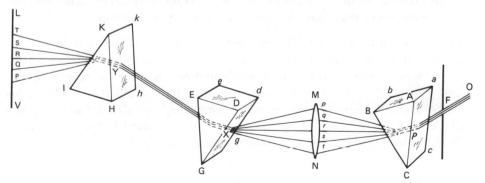

Fig. 3.4.3   Newton's three-prism experiment in which a beam of white light OF is split by a prism ABC into a coloured spectrum which in turn is recombined by lens MN and prism DEG to form white light again, XY. Finally this reconstituted beam of white light is split again into its component colours by a third prism HIK.

nature of light were convincing and conclusive. But his contemporaries didn't agree, and he was plunged into controversy. The most forceful of his opponents was Robert Hooke, who had put forward his own theory of light in his *Micrographia* of 1664 (see Chapter 3.2 and Fig. 3.2.3, page 48). Hooke was older than Newton, he was already Curator of the Royal Society, and he regarded himself as the authority on light. For four years Newton spent himself in trying to answer Hooke and other critics. Then, wearying of the time it took, he retired to Cambridge and got on with his work alone. As he wrote at the time:

> I was so persecuted with discussion arising from the publication of my theory of light that I blamed my own impudence for parting with so substantial a blessing as my quiet to run after a shadow.[6]

For Newton was a solitary figure, as he had been all his life, and he possessed enormous powers of concentration. When asked how he made his discoveries he is said to have replied, 'By always thinking unto them'.

This description, by a servant of his, gives some idea of the sort of man he was:

> I never knew him to take any recreation or pastime either in riding out to take the air, walking, bowling, or any other exercise whatever, thinking all hours lost that was not spent in his studies, to which he kept so close that he seldom left his chamber except at term time, when he read in the schools as being Lucasianus Professor . . . He very rarely went to dine in the hall, except on some public days, and then if he has not been minded, would go very carelessly, with shoes down at heels, stockings untied, surplice on, and his head scarcely combed. At some seldom times when he designed to dine in the hall, he would turn to the left hand and go out into the street, when making a stop when he found his mistake, would hastily turn back, and then sometimes instead of going into the hall, would return to his chamber again.[7]

Mostly Newton would have meals sent to his rooms and forget them. His secretary would ask whether he had eaten. 'Have I?' Newton would reply.

At this time in his life Newton was also working on chemical experiments, and how he went about these has also been described by his servant:

> His Brick Furnaces (as to the manner born) he made and alter'd himself without troubling a Bricklayer . . . He very rarely went to bed, till 2 or 3 of ye clock, sometimes not till 5 or 6, lying about 4 or 5 hours, especially at Spring and Fall of ye Leaf, at wch times he us'd to imploy about 6 weeks in his Elaboratory, the Fire scarcely going out either night or day, he siting up one Night as I did another, till he had finished his Chymical Experiments, in ye Performance of wch he was ye most accurate, strict, exact . . .[8]

THE WRITING OF THE *PRINCIPIA*

All this time, as far as we know, Newton wasn't working on mechanics or gravitation, but in 1679 Hooke wrote to him on these matters and in the course of the correspondence it became clear that Hooke had arrived at some of the same ideas as Newton. But Hooke was unable to work out his ideas mathematically. For example, on 6th January 1680, Hooke writes:

... my supposition is that the Attraction (to the centre) always is in duplicate proportion to the Distance from the Center Reciprocall (i.e. as $1/d^2$), and consequently that the Velocity will be in subduplicate proportion to the Attraction and consequently as Kepler supposes Reciprocall to the Distance ...[9]

Eleven days later, on 17th January 1680, Hooke wrote asking Newton to calculate the curve which a body would describe if subject to an inverse square law attractive force:

I doubt not that by your excellent method you will easily find out what this curve must be, and its proprietys, and suggest a physical reason of this proportion.[10]

But Newton took offence at this suggestion and the correspondence ceased.

Others in London, such as Christopher Wren, the architect of St Paul's Cathedral, and Edmund Halley, the discoverer of Halley's comet, were also trying to solve Hooke's problem, but again without success. Then in 1684 Halley, who was younger than Hooke and more tactful, went to Cambridge to see Newton. This is a contemporary account of what happened:

Without mentioning either his own speculations, or those of Hooke and Wren, he at once indicated the object of his visit by asking Newton what would be the curve described by the planets on the supposition that gravity diminished as the square of the distance. Newton immediately answered, *an Ellipse*. Struck with joy and amazement, Halley asked him how he knew it? Why, replied he, I have calculated it; and being asked for the calculation, he could not find it, but promised to send it to him.[1]

Newton at once set to work to redo his calculation but as he did so he realised that it could hardly stand by itself. For it to be understood, he would have to set out carefully the dynamical principles on which it was based. A few months later he sent some of this work to Halley—and Halley immediately rushed to Cambridge again to persuade Newton to write up all his calculations on dynamics and gravitation and let them be published by the Royal Society. This was how the *Principia* came to be written. Over the next 18 months, constantly encouraged by Halley, Newton completed his great work and it appeared in 1687. Fig. 3.4.4 shows the title-page. Even then the *Principia* nearly didn't appear. Hooke claimed that he had given Newton the idea of the inverse square law and demanded that Newton acknowledge this. Newton was furious. He wrote to Halley thus:

(Hooke) has done nothing, and yet written in such a way, as if he knew and had sufficiently hinted all but what remained to be determined by the drudgery of calculations and observations, excusing himself from that labour by reason of his other business, whereas he should rather have excused himself by reason of

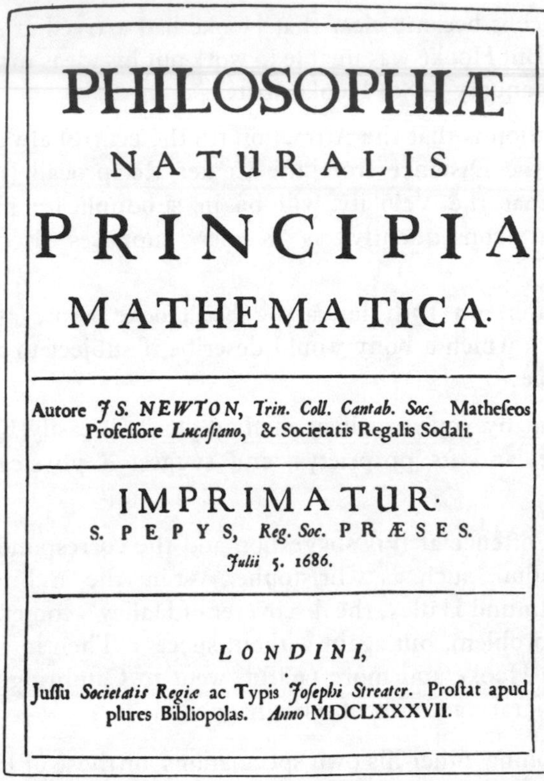

PHILOSOPHIÆ

NATURALIS

PRINCIPIA

MATHEMATICA.

Autore *JS. NEWTON*, *Trin. Coll. Cantab. Soc.* Matheseos
Professore *Lucasiano*, & Societatis Regalis Sodali.

IMPRIMATUR·
S. PEPYS, *Reg. Soc.* PRÆSES.
Julii 5. 1686.

LONDINI,

Jussu *Societatis Regiæ* ac Typis *Josephi Streater.* Prostat apud
plures Bibliopolas. *Anno* MDCLXXXVII.

Fig. 3.4.4   Title-page of the first edition of Newton's *Principia*, 1687.

his inability. For 'tis plain, by his words, that he knew not how to go about it.
Now is not this very fine? Mathematicians, that find out, settle, and do all the
business, must content themselves with being nothing but dry calculators and
drudges; and another, that does nothing but pretend and grasp at all things,
must carry away all the invention, as well as those that were to follow him, as of
those that went before.[12]

Newton threatened to suppress a major part of the book, even though it was
now complete and ready for printing. Halley talked him out of this and the
Principia appeared but without acknowledgement to Hooke. What did it contain?
It consisted of three Books, all written in Latin, highly mathematical and full of
interesting and difficult calculations.

Book 1 is called *The Motion of Bodies* and starts with Newton's definitions of
quantity of matter and quantity of motion or, in our modern nomenclature, of
mass and momentum:

DEFINITION I
*The quantity of matter is the measure of the same, arising from its density and bulk
conjointly*

Thus air of double density, in double space, is quadruple in quantity; in
triple space, sextuple in quantity . . .

DEFINITION II

The quantity of motion is the measure of the same, arising from the velocity and quantity of matter conjointly

The motion of the whole is the sum of the motions of all the parts; and therefore in a body double in quantity, with equal velocity, the motion is double; with twice the velocity, it is quadruple. . . .[13]

Following his definitions, Newton then sets out his familiar Laws of Motion:

AXIOMS, OR LAWS OF MOTION

Law I

Every body continues in its state of rest, or of uniform motion in a right line, unless it is compelled to change that state by forces impressed upon it.

Law II

The change of motion is proportional to the motive force impressed; and is made in the direction of the right line in which that force is impressed.

Law III

To every action there is always opposed an equal reaction: or, the mutual actions of two bodies upon each other are always equal, and directed to contrary parts.[14]

Newton then uses these definitions and laws to calculate the motions of point-masses or particles when various forces act on them. Fig. 3.4.5 shows one of the important results he obtained. On this page Newton shows that, if a point-mass moves in an ellipse, then it must be acted on by an inverse square law force directed towards one focus of the ellipse. In other words Newton is here solving the problem Halley had set him and in effect deriving Kepler's first law of planetary motion. The diagram on this page of the *Principia* also appears on the reverse side of an English £1 note (see Fig. 3.4.6), together with Newton's reflecting telescope and a portrait of Newton with a copy of the *Principia* open in front of him. The £1 note also has a drawing of the sun which unfortunately has been placed at the centre of the ellipse C and not at the focus S—an error that would have infuriated Newton who later became Master of the Royal Mint.

Book 1 of the *Principia* ends with an important series of theorems which show that a spherical body, in respect of any gravitational force which it exerts on any external body, behaves *exactly* as if all its mass were concentrated at its centre. Therefore the motions of the planets, the sun and the moon can all be calculated just as if these bodies were in fact point-masses.

In Book 2 of the *Principia* Newton discusses the motions of bodies in fluids in considerable mathematical detail. The Book ends with an attempt to refute Descartes' theory of planetary motion in which the planets are swept round by the motion of swirling vortices of fluid (see Chapter 3.3 and Fig. 3.3.3, page 62).

Book 3 of the *Principia* is called *The System of the World*. As Newton himself puts it on the first page:

In the preceding books I have laid down the principles of philosophy; principles not philosophical but mathematical: such, namely, as we may build

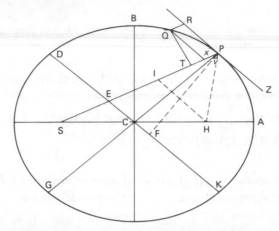

Fig. 3.4.5 Newton's proof that a body moving in an ellipse must be acted on by an inverse square law force directed towards one focus of the ellipse; with this proof Newton is, in effect, deriving Kepler's first law of planetary motion. *Redrawn from Cajori, F. (ed), Sir Isaac Newton's Mathematical Principles of Natural Philosophy, vol. 1, University of California Press, 1962.*

Fig. 3.4.6 Reverse of English £1 note showing a diagram from the *Principia* (see Fig. 3.4.5), Newton's reflecting telescope and Newton with a copy of the *Principia* open in front of him.

our reasonings upon in philosophical enquiries . . . It remains that, from the same principles, I now demonstrate the frame of the System of the World.[15]

Once again Newton is far from modest, but Book 3 lives up to his claim. In it Newton uses the results derived in Book 1 to deduce almost every detail of the motions of the planets round the sun, of the moon round the earth, of the moons of Jupiter (discovered by Galileo) around Jupiter, and of comets. All the time he compares his predictions with the best available observations, and in many cases adds new observations of his own. The agreement between Newton's theoretical predictions and the observations was remarkably close over all this very wide range of phenomena. Fig. 3.4.7 shows one such comparison for the motion of Halley's comet of 1683. The discrepancy between theory and observation is

1683 Equatorial time	Sun's place	Comet's longitude computed	Latitude north computed	Comet's longitude observed	Latitude north observed	Difference longitude	Difference latitude
d h m	° ′ ″	° ′ ″	° ′ ″	° ′ ″	° ′ ″	′ ″	′ ″
July 13.12.55	♌ 1.02.30	♋ 13.05.42	29.28.13	♋ 13. 6.42	29.28.20	+1.00	+0.07
15.11.15	2.53.12	11.37.48	29.34. 0	11.39.43	29.34.50	+1.55	+0.50
17.10.20	4.45.45	10. 7. 6	29.33.30	10. 8.40	29.34. 0	+1.34	+0.30
23.13.40	10.38.21	5.10.27	28.51.42	5.11.30	28.50.28	+1.03	−1.14
25.14. 5	12.35.28	3.27.53	24.24.47	3.27. 0	28.23.40	−0.53	−1. 7
31. 9.42	18.09.22	♊ 27.55. 3	26.22.52	♊ 27.54.24	26.22.25	−0.39	−0.27
31.14.55	18.21.53	27.41. 7	26.16.57	27.41. 8	26.14.50	+0. 1	−2. 7
Aug. 2.14.56	20.17.16	25.29.32	25.16.19	25.28.46	25.17.28	−0.46	+1. 9
4.10.49	22.02.50	23.18.20	24.10.49	23.16.55	24.12.19	−1.25	+1.30
6.10. 9	23.56.45	20.42.23	22.47. 5	20.40.32	22.49. 5	−1.51	+2. 0
9.10.26	26.50.52	16. 7.57	20. 6.37	16. 5.55	20. 6.10	−2. 2	−0.27
15.14. 1	♍ 2.47.13	3.30.48	11.37.33	3.26.18	11.32. 1	−4.30	−5.32
16.15.10	3.48. 2	0.43. 7	9.34.16	0.41.55	9.34.18	−1.12	−0. 3
18.15.44	5.45.33	♉ 24.52.53	5.11.15	♉ 24.49. 5	5. 9.11	−3.48	−2. 4
			South		South		
22.14.44	9.35.49	11. 7.14	5.16.58	11.07.12	5.16.58	−0. 2	−0. 3
23.15.52	10.36.48	7. 2.18	8.17. 9	7. 1.17	8.16.41	−1. 1	−0.28
26.16. 2	13.31.10	♈ 24.45.31	16.38. 0	♈ 24.44.00	16.38.20	−1.31	+0.20

Fig. 3.4.7 Table from *Principia* which compares the calculated and observed positions of Halley's comet of 1683. *Redrawn from Cajori, F. (ed), Sir Isaac Newton's Mathematical Principles of Natural Philosophy, vol. 2, University of California Press, 1962.*

seldom more than 1 or 2 minutes of arc and is often less. In Book 3 Newton also gave a quantitative and precise theory of the tides as being due to the combined gravitational attraction of the sun and moon on the waters of the earth.

NEWTON'S ACHIEVEMENT

What Newton had done in the *Principia*, for the first time since Aristotle, was to provide a complete theory for the structure and motion of the whole known universe. Moreover, he unified the phenomena of earth and heavens in a way that Aristotle had never done. Fig. 4.1.2 shows a diagram from the *Principia* which sums up the way in which Newton unified the study of terrestrial and planetary motion. In the *Principia* Newton had underpinned Copernicus' system, explained Kepler's laws of planetary motion, incorporated and extended Galileo's work on local motion and projectiles, and totally superseded Descartes' *Principia Philosophica*. As the historian, Thomas Kuhn, puts it:

> These mathematical derivations were without precedent in the history of science. They transcend all the other achievements that stem from the new perspective introduced by Copernicanism . . . From Newton's inverse-square law and the mathematical techniques that related it to motion, both the shape and the speed of celestial and terrestrial trajectories could be computed for the first time with immense precision. The resemblance of cannon ball, earth, moon and planet was now seen, not in a vision but in numbers and measurement. With this achievement seventeenth century science reached its climax.[16]

But despite this, Newton's work was not immediately accepted. For a start, its mathematics was extraordinarily difficult. Very few people, even today, can read the *Principia* easily. And Newton had not made things easier by setting out all his proofs in the classical geometrical style of Euclid, Apollonius and Galileo and not in terms of his own method of fluxions (the differential and integral calculus) which he had used to arrive at many of his results.

Others objected to Newton's law of gravitation on philosophical grounds. In their view, it was not a proper mechanical law at all. It was just the sort of magical 'action at a distance' that many people in the 17th century had been trying to eliminate from natural philosophy.

But gradually the extraordinary scope and power of Newton's theories were recognised and Newton came to be held in enormous respect, reverence and even awe.

After the *Principia* Newton did very little further science except to publish his book on *Opticks* in 1704. Hooke had died in 1703, and perhaps Newton had delayed publication to avoid further controversy.

For the rest of his life, Newton basked in the glory and recognition he was granted and, when he died in 1727 at the age of 85, he was buried in Westminster Abbey. The poet Alexander Pope wrote this celebrated 'Epitaph intended for Sir Isaac Newton':

Nature and Nature's laws lay hid in night:
God said, Let Newton be! and all was light.

But Newton did not always see himself like that. Late in life he wrote this:

To explain all nature is too difficult a task for any one man or even for any one age. 'Tis much better to do a little with certainty, and leave the rest for others that come after you, than to explain all things.[17]

And, less precisely, in a famous passage:

I do not know what I may appear to the world; but to myself I seem to have been only like a boy playing on the sea-shore, and diverting myself in now and then finding a smoother pebble or a prettier shell than ordinary, while the great ocean of truth lay all undiscovered before me.[18]

REFERENCES

1   Newton, *Catalogue of Portsmouth Collection* (Cambridge, 1888), Section I, Division xi, number 41; quoted in Herivel, J., *The Background to Newton's Principia*, (Oxford: Clarendon Press, 1965), pp. 66–7; the most probable date for this letter is 1718, more than 50 years after the events being described.

2   Newton to Oldenburg, January 18 1672, in Turnbull, H. W. (ed.), *Correspondence of Isaac Newton, I*, (Cambridge: Cambridge University Press, 1959), pp. 82–3.

3   Newton in a communication to the Royal Society, 1672; quoted in Bronowski, J., *The Ascent of Man*, (London: BBC Publications, 1973), p. 226.

4   Ibid., pp. 226–7.

5   Ibid., p. 227.

6   Letter from Newton to Leibniz, quoted in ibid., p. 226.

7   Quoted in Gillispie, C. C., *The Edge of Objectivity*, (Princeton: Princeton University Press, 1960), p. 138.

8   Quoted in Hall, A. R., *From Galileo to Newton 1630–1720*, (London: Fontana, 1970), p. 328.

9   Hooke to Newton, Turnbull, H. W. (ed.), *Correspondence of Isaac Newton II*, (Cambridge: Cambridge University Press, 1960), p. 309.

10   Ibid., p. 313.

11   Quoted in Gillispie, C. C., op. cit., p. 137.

12   Turnbull, H. W. (ed.), op. cit., Vol. II, p. 438.

13   Cajori, F., (ed.), *Sir Isaac Newton's Mathematical Principles of Natural Philosophy: Vol. I The Motion of Bodies* (Berkeley: University of California Press, 1962), p. 1.

14   Ibid., p. 13.

15   Ibid., *Vol II The System of the World*, p. 397.

16   Kuhn, T. S., *The Copernican Revolution*, (New York, Harvard University Press, 1957), p. 256. Reprinted by permission.

17   Quoted in Bronowski, J., op. cit., p. 236.

18   Ibid., p. 236.

SUMMARY

1   When he was young Newton made significant discoveries in mathematics, dynamics, optics and gravitation—but he left these discoveries unpublished.

2   Newton's earliest published work, on the nature of light and of colours, led to so much controversy that he kept on working privately but published little.

3   In the 1680s Newton was persuaded to publish his work on motion and gravitation in his *Philosophiae Naturalis Principia Mathematica* or *Mathematical Principles of Natural Philosophy*.

4   For the first time since Aristotle, the *Principia* provided a complete theory for the motions of bodies both in the heavens and on earth. It underpinned Copernicus' system, explained Kepler's laws of planetary motion, incorporated and extended Galileo's work on local motion and projectiles, and totally superseded Descartes' *Principia Philosophica*.

5   Although very few fully understood its mathematics, the *Principia* brought Newton great public recognition and acclaim.

3.5 Communication between scientists: the origins of the scientific community

In the 20th century there has been an enormous growth in the visible superstructure of the scientific community—scientific societies, specialist journals, research institutions, international conferences. The organised scientific community is now so large and so well established that we take it for granted. It is difficult for us to imagine a time when it did not exist. Yet in the 17th century there was virtually no organised science at all. Fig. 3.5.1 shows the number of scientific journals which have been founded over the last three hundred years. Now there are tens of thousands of such journals: in 1700 there were fewer than ten, and before 1650 there were none.

Of course this does not mean that scientists did not communicate with one another before 1650.[1] We have already mentioned many books published before

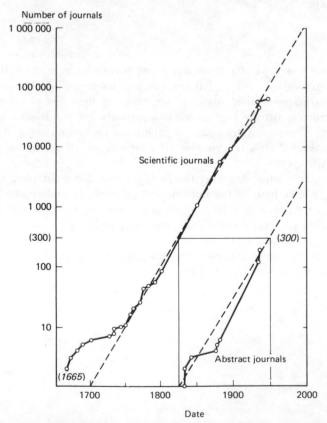

Fig. 3.5.1 Total number of scientific journals and journals of abstracts founded, 1650–1950. *Redrawn from de Solla Price, D. J., Little Science, Big Science, Columbia University Press, 1961.*

that date—for example, Copernicus' *De Revolutionibus*, and Vesalius' *De Fabrica* in the 16th century, and the works of Kepler, Descartes, Galileo and others in the early 17th century. This period also saw the beginnings of two characteristic institutions of the modern scientific community—the scientific society or academy and the scientific journal.

Galileo was a member of probably the earliest scientific society, the *Accademia dei Lincei* or Academy of the Lynxes, which was founded in 1603 but collapsed in 1630 when its patron died. Galileo was so proud to be a member of the *Accademia dei Lincei* that he gave his membership precedence, on the title-pages of his *Two New Sciences* and his *Dialogue Concerning the Two Great World Systems*, over his post as Chief Philosopher and Mathematician to the Grand Duke of Tuscany. Another academy, the *Accademia del Cimento* or Academy of Experiments flourished for only a short time, 1657 to 1667, under the patronage of the Duke of Florence. However, this academy did publish an account of its proceedings, the *Saggi di Naturali Esperienze* or *Essays on Natural Experiments*. Although the *Saggi* only appeared as a single volume in 1666, it was the forerunner of those proceedings of scientific societies which became the first scientific journals.

Two contemporaries of Galileo, Marin Mersenne (1588–1648) in France, and Francis Bacon (1561–1626) in England, were also influential in the development of cooperation in science, although neither made major contributions to scientific knowledge.

MARIN MERSENNE

Mersenne was educated at the same Jesuit academy at La Flèche that Descartes was to attend a few years later. Mersenne then studied theology and, after he became a Minim friar in 1611, he devoted his life to the development of scientific knowledge as a major contribution to the glory of God. In 1625 he published a book of over 1000 pages—*La Vérité des Sciences, contre les Sceptiques ou Pyrrhoniens* or *The Truth of the Sciences, against the Sceptics or Pyrrhonians*. In this book, Mersenne attacked those sceptics who doubted the possibility of rational knowledge. He argued that man could attain a body of knowledge which could not seriously be doubted, and tried to convince his readers by devoting more than 800 pages to a summary of the vast body of scientific and mathematical knowledge which existed in his day.

From about 1623 onwards, Mersenne began to develop contacts with philosophers and scientists from all over Europe. Some came in person to visit him in Paris but Mersenne's main channel of communication was the vast correspondence he carried on for more than 20 years. He became the unofficial secretary of the new republic of science, and his correspondents included Galileo, Descartes, the chemist van Helmont, the philosopher Hobbes and the mathematicians Fermat and Pascal. Later in the century letters like those exchanged by Mersenne, containing news of scientific and mathematical discoveries, were to form a major part of the first scientific journals.

Mersenne also 'marked a notable step in the organisation of experimental science in the seventeenth century' by his 'insistence on the careful specification of experimental procedures, repetition of experiments, publication of the numerical results of actual measurements as distinct from those calculated from theory, and recognition of approximations'.[2]

FRANCIS BACON

Francis Bacon was an almost exact contemporary of William Shakespeare (1564–1616) and was a patient of William Harvey's on more than one occasion. He was a prominent figure at the court of Queen Elizabeth I and later, under James II, he became Lord Chancellor of England, a much more powerful position then than now. Bacon did no significant scientific work himself but, relatively late in life, he wrote two books about science which were to become extremely influential. The first, published in 1620, was called *Instauratio Magna* or *The Great Instauration*; Fig. 3.5.2 shows the title-page. In this book Bacon set out his ideas on how knowledge should be advanced. For Bacon the first step was the collection of as many empirical observations as possible about all kinds of phenomena, objects,

Fig. 3.5.2   Title-page of *Instauratio Magna* by Francis Bacon, 1620. The book describes Bacon's views on scientific method and the organisation of science.

arts and crafts. He then gave a series of rules for proceeding inductively from these observations to the discovery of what he calls the 'forms of simple natures'. And Bacon was in no doubt why knowledge was to be sought—'The seal and legitimate goal of the sciences is the endowment of human life with new inventions and riches.'[3]

Bacon thought that, once he had set out his method, anyone who followed the rules could do science—he claims to have 'levelled men's wits'.[4] This meant that 'with proper organisation and financial support, it should be possible to complete the edifice of science in a few years and to gather all the practical fruit that it promised for the good of men'.[5] And in his last book *The New Atlantis*, published in 1627, a year after he died, Bacon described a Utopia in which science and scientists were organised by the government and in which both had a very high status. In this book, the organised society of scientists was called 'Solomon's House' and Bacon's account of it begins like this:

> The End of our Foundation is the knowledge of Causes, and secret motions of things; and the enlarging of the bounds of Human Empire, to the effecting of all things possible.[6]

Bacon was not highly regarded by his contemporaries. For example, Harvey said that, although Bacon was 'esteemed for his wit and style', Harvey 'would not allow him to be a great philosopher'. 'He writes philosophy like a Lord Chancellor.'[7] For many years Bacon's ideas were neglected but, in time, his views on empirical observation, on the usefulness of science and on the necessity of organising scientific activity came to be quoted again and again as authoritative statements on how and why science should be fostered.

We see in Bacon's writings, just as we do in those of Descartes (see Chapter 3.3), a concern for establishing a method of obtaining true and certain knowledge. It is interesting, however, that their common concern for a unique philosophy of science should have led them to such different conceptions of scientific method.

Descartes' advocacy of the deductive mathematical method for seeking knowledge is clearly at the opposite pole from Francis Bacon's method of induction from systematic observations and experiments. Yet in practice Descartes' work was not as completely lacking in experiments as Bacon's scheme was in mathematics.

Galileo, too, often claimed to have produced 'necessary demonstrations' when in fact his combination of mathematics, experiment, observation and verbal arguments left room for legitimate disagreement, even amongst other experts.

These examples show how important it is to try to establish what scientists actually do rather than simply to accept what they say they do or even what philosophers of science say they do. The methods actually used by scientists in the 17th century were more complex and varied than many of the simple models advocated at that time or since.

THE ESTABLISHMENT OF SCIENTIFIC SOCIETIES AND JOURNALS

One thing that scientists unquestionably did do as the century progressed was to develop new ways of communicating with one another both formally and informally.

The 1660s saw the foundation of two national scientific societies which are still flourishing today—the Royal Society in London and the Académie des Sciences in Paris. Both these societies had some of their roots in informal meetings of scientists over the preceding 20 years.

Many of the men who founded the Royal Society had been meeting informally but regularly in various groups in both London and Oxford since the 1640s. These groups discussed scientific and philosophical matters and conducted experiments, but they avoided 'matters of theology and state affairs'[8]—probably a wise omission given the violent disagreements on these matters in England during the Civil War of 1642–9 and Cromwell's republican commonwealth in the 1650s. One of these groups was called the 'Invisible College' by Robert Boyle. After the king, Charles II, was restored to the throne in 1660, moves were made to obtain royal approval for a new scientific academy. In July 1662, Charles II granted a royal charter to the Royal Society of London for the Improvement of Natural Knowledge. Today more than 300 years later, it is still the foremost scientific society in Britain. The founder fellows included Robert Boyle, Christopher Wren and Edmund Halley, together with literary men like Samuel Pepys and John Evelyn. Robert Hooke was appointed Curator of Experiments and in 1663 he wrote that:

> The business and design of the Royal Society is:
> To improve the knowledge of naturall things, and all useful Arts, Manufactures, Mechanick practises, Engynes and Inventions by Experiments—(not meddling with Divinity, Metaphysics, Moralls, Politicks, Grammar, Rhetorick, or Logick).
> To attempt the recovering of such allowable arts and inventions as are lost.
> To examine all systems, theories, principles, hypotheses, elements, histories, and experiments of things naturall, mathematicall, invented, recorded, or practised, by any considerable author ancient or modern. In order to the compiling of a complete system of solid philosophy for explicating all phenomena produced by nature or art, and recording a rationall account of the causes of things.
> All to advance the glory of God, the honour of the King, the Royall founder of the Society, the benefit of his Kingdom, and the generall good of mankind.[9]

Fig. 3.5.3 is the frontispiece to Thomas Sprat's *History of the Royal Society*. It shows the patron of the society, Charles II, being crowned with a laurel wreath. Prominently placed on the right is Francis Bacon, whose 'Solomon's House' from the *New Atlantis* closely resembled the early Royal Society.

In its early years the Royal Society owed a great deal to its secretary, Henry Oldenburg, one of the very few paid officials of the society. Oldenburg was

Fig. 3.5.3 Frontispiece of Thomas Sprat's *History of the Royal Society* (1667). The society's patron Charles II is being crowned with a laurel wreath; Francis Bacon is on the right.

secretary from the society's foundation in 1662 until he died in 1677. Like Mersenne, Oldenburg maintained a vast correspondence with foreign scientists which enabled the society to be in contact with what was going on in science all over Europe. And it was Oldenburg who, in 1665, started the *Philosophical Transactions of the Royal Society*, a monthly journal which is still being published. In addition to papers from members and reports of the society's meetings, the journal also printed letters from abroad summarising the investigations of many European scientists. Fig. 3.5.4 shows a contents page from one issue of *Philosophical Transactions* in 1685. Notice the wide range of subjects covered— including what we today would call physics, medicine, geology and anatomy— and the book reviews and letters from abroad. Notice, in particular, item 8—*A Discourse proving from Experiments, that the larger the wheels of a Coach, etc., are, the more easily they are drawn over stones, etc., lying in the way*. This item illustrates some of the down to earth, practical concerns of the early Royal Society.

The origins of the Académie Royale des Sciences of Paris, founded in 1666, also lay in informal meetings of scientists, at some of which Oldenburg had been present. In 1663 these scientists asked for state help in setting up an academy, and

[835]   *Numb.* 167.
Beginning the fiveteenth Volume.

PHILOSOPHICAL
T R A N S A C T I O N S.

Jan. 28th. 168⅘.

The CONTENTS.

1.   *Some Experiments about* Freezing, *and the Difference betwixt common* fresh-water Ice, *and that of* Sea-water ; *also a probable Conjecture about the Original of the* Nitre *of* Ægypt *: by* Dr. Lifter.   2. *A Letter from* Dr. Turbervile *of* Salisbury, *containing some considerable Observations in the Practise of Physic.*   3. *Observations on the* Cicindela Volans *; by* R. Waller Esq. *F. of the* R. S.   4. *A Letter from* Mr. J. Davis *concerning the* Wurtemburg Siphon.   5. *The Description of a Siphon, made to perform all the Effects express to be done by the* Wurtemburg Siphon : *by* Dr. Papin ; *F. of the* R. S. 6. *A Catalogue of Experiments drawn up by* Sr. W. Petty, *præs. of the* Phil. Soc. *of* Dublin ; *and by him presented to that* Society.   7. *A Letter from* Mr. J. Beaumont *jun, concerning a new way of* Cleaving Rocks.   8. *A Discourse proving from Experiments, that the larger the wheels of a* Coach, &c. *are, the more easily they are drawn over Stones, &c. lying in the way.*   9. *An account of a large præternaturall Glandulose Substance, found between the* Pericardium, *and* Heart *of an* Ox.   10. *Accounts of three Books ;* 1. De Origine Fontium *per* R. Plott LL. D. *Musæi* Ashmoleani Oxoniæ Præpositum, & R. S. S.   2. *Medicina septentrionalis Collatitia ,* Opera Theoph. Boneti.   3 Pechlini *Dialogus de* Potu Theæ.

Fig. 3.5.4   Contents page from *Philosophical Transactions of the Royal Society*, 1684–5, Vol. 15, p. 835.

in 1666 Louis XIV (see Fig. 4.3.1, page 118) agreed to support the new institution rather more generously than Charles II had done in England. Sixteen paid academicians were appointed; some foreign scientists, for example Huyghens, were included; and, somewhat later, laboratories were provided too. The Académie des Sciences was given formal responsibility for maintaining standards of scientific work and for the improvement of French industry.

However, the Académie did not initially publish its own journal, probably because the independent *Journal des Sçavans* had been founded in 1665. This journal mainly contained book reviews but also reported on scientific meetings held both in France and abroad. It is still appearing today under the title *Journal des Savants*. Later, the Académie des Sciences did publish its own regular series of *Mémoires*.

In the 18th century the Royal Society and the Académie des Sciences became models for the establishment of many other national academies of science—for example, in Berlin in 1700, in St Petersburg (now Leningrad) in Russia in 1725, in Stockholm in 1741 and in Copenhagen and Philadelphia in 1743. Many of these academies soon produced their own journals and by 1750, as can be seen from Fig. 3.5.1, the number of scientific journals began the exponential growth that has continued ever since. By 1800 about a hundred journals were being published and the earliest specialist journals were beginning to appear. Examples include the *Chemisches Journal* in Germany in 1778 and *The Botanical Magazine* in 1787. The Linnaean Society, devoted to natural history, became England's first specialist scientific society in 1788.

By the end of the 18th century the informal networks of communication established by Mersenne and Oldenburg in the 17th century had been transformed into institutions which were well on the way towards becoming the highly organised scientific community of today.

REFERENCES

1   The word 'scientist' was first used by William Whewell in his *History of the Inductive Scientists* (1837); so, strictly speaking, it is inaccurate to refer to those who lived in the 17th and 18th centuries as scientists. However, since no other generic term is wholly satisfactory, I have nevertheless sometimes used the word in this chapter and elsewhere.
2   Crombie, A. C., *Marin Mersenne* in Gillispie, C. C. (ed.), *Dictionary of Scientific Biography*, (New York: Charles Scribner, 1974), Vol. IX, p. 318.
3   Bacon, F., *The New Atlantis* in Spedding, J., Ellis, R. L. and Heath, D. D. (eds.), *The Works of Francis Bacon*, (London: 1857–9).
4   Bacon, F., *Novum Organum* in Spedding, J., Ellis, R. L. and Heath, D. D. (eds.), ibid., Vol. IV, p. 109.
5   Hesse, M. B., *Francis Bacon* in Gillispie, C. C. (ed.), op. cit., Vol. I, p. 375.
6   Bacon, F., *The New Atlantis* quoted in Hesse, M. B., op. cit., p. 376.
7   Attributed to Harvey by John Aubrey; quoted in Keynes, G., *The Life of William Harvey*, (Oxford: Clarendon Press, 1978), p. 160.
8   Wallis, J., quoted in Weld, C. R., *History of the Royal Society*, Vol. I, (London: 1848), p. 35.
9   Hooke, R., quoted in Weld, C. R., ibid., Vol. I, pp. 146–7.

SUMMARY

1   The organised scientific community of today has its roots in the informal correspondence networks maintained by men like Mersenne and Oldenburg in the 17th century.

2   The importance of empirical observation in the development of science was emphasised by Francis Bacon who also wrote enthusiastically of the potential usefulness of scientific knowledge and of how science should be organised and sponsored by governments.

3   The methods actually used by scientists in the 17th century were more complex and varied than the pure induction from observation of Bacon or the mathematical deductions advocated by Descartes.

4   The late 17th century saw the establishment of the first national scientific societies, the Royal Society in London and the Académie des Sciences in Paris—together with the earliest surviving scientific journals, the *Philosophical Transactions* and the *Journal des Sçavans*.

5   By about 1750 the scientific community—journals and scientific societies—had begun the exponential growth which has continued ever since.

PART FOUR

The development of the scientific revolution in the 18th and 19th centuries

A WIDER CONSEQUENCES IN THE 18TH CENTURY

4.1 Newton and the Enlightenment

In this chapter we will look again at Isaac Newton—the central figure in the scientific revolution of the 17th century. We will first mention some little known aspects of Newton's life and work. Then we will outline some of the ways in which Newton influenced the whole climate of thought in the 18th century.

NEWTON'S UNPUBLISHED WORK

It is only one of many extraordinary facts about Newton that he spent probably little more than a quarter of his active life on the work for which he is best known—on dynamics in the *Principia* and on light in the *Opticks*. We now know that he devoted just as much time to his public activities, to alchemy, and to the analysis of the scriptures and other historical documents.

Newton's public activities after he moved from Cambridge to London in 1696 are well-known. He became Master of the Royal Mint in 1699, where he supervised a major reform of the coinage. He was elected President of the Royal Society in 1703, a post he held for the rest of his life. He was knighted in 1705, and when he died in 1727 at the age of 85, he was buried with great public ceremony in Westminster Abbey, the only English scientist ever to be honoured in this way. Fig. 3.4.1 (page 65) shows a portrait of Newton in a formal wig which dates from this late public period of his life; it was painted in 1702 and hangs in the National Portrait Gallery in London.

Privately, however, Newton worked on alchemy and on his historical and scriptural studies much earlier in his life, while he was still at Cambridge, although he published little or nothing about these activities to which he devoted so much time. Fig. 4.1.1 shows an informal portrait of Newton, this time without a formal wig, which dates from around 1689. It is the earliest portrait of Newton that we have and gives us some idea of what Newton looked like when he was at the height of his powers.

Newton's unpublished writings on alchemy were really extensive—more than a million words including at least 100,000 words which were records of alchemical

Fig. 4.1.1   Portrait of Newton around 1689.

experiments done by Newton himself. These writings date mainly from around 1675 up to the early 1690s—that is, during the same years in which he made most of his pioneering discoveries in dynamics and wrote the *Principia*. One of Newton's alchemical manuscripts was the *Index Chemicus*—a kind of encyclopedia of alchemical terms compiled from all the existing works on alchemy; it contained about 5000 separate references to more than 150 treatises on alchemy. Writing of the *Index Chemicus*, R. S. Westfall says that: 'I have spent a week dissecting the Index to the extent that I have. It is hard for me to imagine that anyone could have composed it in less than a thousand times that week—although I hasten to add that I cannot find room in Newton's career for any period approaching a thousand weeks.'[1] A thousand weeks is twenty years!

What was alchemy? It was an ancient art which became widely known and practised in 15th-century and 16th-century Euorpe. Alchemists dealt with what is in some sense the equivalent of modern chemistry. They prepared substances by a combination of methods—solution, distillation, sublimation—in the hope that they would be able to tap a source of miraculous power. Probably the best-known of their aims was the search for the philosopher's stone which would transmute base metals into gold. Their language was obscure and deliberately so as they were secretive about their discoveries and only wanted to pass them on to a select few.

The name vitriol, for example, is still familiar to us in the early name for sulphuric acid—oil of vitriol—from which we derive our word vitriolic! Vitriol was an alchemical term used to describe shiny, crystalline substances. It was derived from the initials of the Latin sentence *Visita Interiora Terrae Rectificando Invenies Occultum Lapidem* or Visit the Interior of the Earth and by Right Action you will find the Occult Stone.

It was easy to satirise alchemists as cheats and frauds, as for example the English playwright Ben Jonson did in a popular play *The Alchemist* which was first performed in 1610 (the year of Galileo's *Starry Messenger*), but many people took alchemy seriously. For example, Tycho Brahe (see Chapter 2.2) writes that he has 'been occupied by the subject as much as by celestial studies from my twenty-third year, trying to gain knowledge'.

So there is precedent for the voluminous notes and summaries made by Newton of the works of alchemists and for the large number of experiments, chemical or alchemical, that Newton's notebooks reveal that he performed from 1670–96, however strange such activities may seem to us.

As well as spending much time on alchemy, Newton was also a considerable biblical scholar and commentator, with a particular interest in the fulfilment of prophecy. He believed that biblical prophecies had been fulfilled in or before his time, and he spent a considerable amount of effort in attempting to interpret the meanings of ancient scriptural documents and writings.

Newton also showed a similar interest in what he thought might be anticipations of some of his own discoveries in the writings of the ancient Greeks. For example, in a draft of a new edition of the *Principia*, Newton asserts that Pythagoras discovered an inverse-square law relation in the harmony of vibrating strings (see Chapter 2.1) and that this led Pythagoras, by analogy with the harmony of the celestial spheres, to anticipate Newton's own inverse square law of gravitation. Newton claims that this knowledge was expressed so cryptically in the Pythagorean writings that it was misunderstood by later generations. Newton also here refers approvingly to those 'ancient Philosophers' who:

> . . . . . called God Harmony and signified his actuating power harmonically by the God Pan's playing upon a pipe and attributing musick to the spheres, made the distance and motion of the heavenly bodies to be harmonical and represented the Planets by the seven strings of Apollo's Harp.[2]

Given all this, together with his alchemy, it is perhaps not surprising that Lord Keynes, the famous economist, saw Newton not as 'the first and greatest of the modern age of scientists' but rather as 'the last of the magicians'.[3]

This is something of an overstatement, although we should remember that, for all the modernity that we read into him, Newton was a man of the 17th century who was passionately interested in any sources of evidence he could find which would help to make sense and coherence of the universe. That this led him to take seriously some approaches which nowadays we would not entertain probably only underlines the gulf which separates the 17th from the 20th century.

NEWTON'S INFLUENCE IN THE 18TH CENTURY

It was only by his published works, and in particular by the *Principia* and the *Opticks*, that Newton influenced the 18th century. His unpublished letters, manuscripts and notebooks were only discovered much later, as a byproduct of the intense interest aroused by the successes of his published work.

Newton's influence on the 18th century is complex. Naturally enough his work affected other scientists and the development of science but it also had a considerable popular effect.

The greatest impact was made by the *Principia*. Fig. 4.1.2 shows a diagram from the *Principia* which sums up the way Newton unified the study of terrestrial and planetary motion. The same laws of motion govern both the path of the kind of projectile studied by Galileo and the motion of a satellite circling the earth. It was the vastness of Newton's conception together with the mathematical precision of his predictions which so impressed his contemporaries. It is worth remembering that the calculations made today for the orbits of artificial satellites and spacecraft are, in all essentials, exactly the same as those made by Newton in the 1680s.

The systematic mathematical work on mechanics, initiated by Newton in the *Principia*, was developed and refined in the 18th century primarily by the French

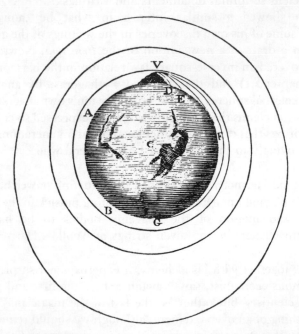

Fig. 4.1.2   Diagram from the *Principia* illustrating how Newton unified the study of terrestrial and planetary motion. A object projected horizontally from the mountain top V with a low velocity behaves as a simple Galilean projectile VD. As the initial velocity is increased, the object carries further round the earth, VE, VFG, until eventually it becomes a satellite of the earth.

mathematicians D'Alembert (1717–83), Lagrange (1736–1813) and Laplace (1749–1827). In the words of Lagrange, in his book *Mécanique Analytique* or *Analytical Mechanics*, published 100 years after the *Principia* in 1788, their aim was:

> .... to reduce the theory of mechanics, and the art of solving the associated problems, to general formulae, whose simple development provides all the equations necessary for the solution of each problem .... to unite, and present from one point of view, the different principles which have, so far, been found to assist in the solutions of problems in mechanics, by showing their links and mutual dependence and making a judgement of their validity and scope possible.[4]

The work of these mathematicians seemed to suggest that the whole universe reflected in some way the logical structure of a mathematical theory. This tendency towards grand system-building reached its culmination in the fully deterministic mechanics of Laplace. In his *Mécanique Céleste* or *Celestial Mechanics* of 1825, Laplace wrote:

> We must envisage the present state of the universe as the effect of its previous state, and as the cause of that which will follow. An intelligence that could know, at a given instant, all the forces governing the natural world, and the respective positions of the entities which compose it, if in addition it was great enough to analyze all this information, would be able to embrace in a single formula the movements of the largest bodies in the universe and those of the lightest atom: nothing would be uncertain for it, and the future, like the past, would be directly present to its observation.[5]

This is surely the ultimate statement of the 'mechanical philosophy'—the universe as an enormous deterministic machine. It is little wonder that this kind of science aroused hostility and caused the poets to shudder.

But this reaction came later. Initially it was not only the expert mathematicians who were impressed by Newton's *Principia*. In the 18th century many writers produced popular accounts of Newton's work, and some of the titles of these books included *Astronomy Explained upon Sir Isaac Newton's Principles, and Made Easy to Those who have not Studied Mathematics* and *The Newtonian System of the World the Best Model of Government, an Allegorical Poem* and even *Il Newtonianismo per le dame*![6]

VOLTAIRE ON NEWTON

The most famous of these popularisers of Newton was Voltaire (see Fig. 4.1.3), a Frenchman who lived from 1694 to 1778, that is for most of the 18th century. Voltaire was a writer who was exiled from Paris in 1726 for his outspoken criticisms of the French government. During this time he came to London and while he was there he attended Newton's funeral in Westminster Abbey. Voltaire developed an enthusiasm for all things English and in 1733 he published a book

Fig. 4.1.3   Statue of Voltaire by Houdon.

called *Letters concerning the English Nation*. A French edition was published in Paris in 1734 under the title *Lettres Philosophiques* or *Philosophical Letters* and it was immediately condemned to be burnt as 'likely to inspire a license of thought most dangerous to religion and civil order'. The book contained chapters on the various religious denominations which coexisted in England, on the English system of government, and on the arts and sciences in England. There are no less than four chapters devoted to Newton and his ideas.

What attracted Voltaire was the whole pluralistic pattern of English life at that time, with its religious and political tolerance and with its relative freedom of expression in the arts and the sciences. In all these areas England made a striking contrast to the rigidities of the absolute monarchy in France, and Voltaire saw Newton's work as an example of man's powers if he were set free. So when he returned to France, Voltaire tried to popularise Newton, publishing his *Elements of the Philosophy of Newton* in 1738. Voltaire was thus contributing to the popular interest in the latest scientific ideas which was typical of the 18th century. Fig. 4.1.4 shows one example of this popular interest. It is a public demonstration of a device called an orrery—a mechanical model of the Newtonian solar system. And Fig. 4.1.5 shows an 18th-century philosopher's table—designed for

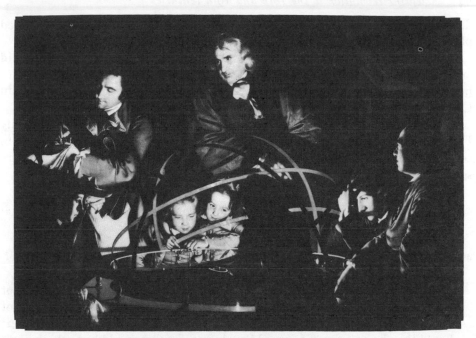

Fig. 4.1.4 A philosopher giving a lecture on the Orrery, a mechanical model of the Newtonian solar system, painted by Wright of Derby in about 1765.

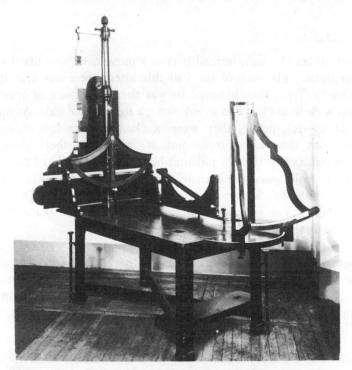

Fig. 4.1.5 18th-century Philosopher's Table for demonstrating Newton's Laws of Motion.

demonstrating Newton's Laws of Motion by means of experiments with inclined planes, pendulums and the like.

Voltaire may well have had a philosopher's table like this. He certainly did experiments in physics to demonstrate Newton's laws. And he had a laboratory built for the purpose in the chateau of one of his wealthy patrons, Madame du Chatelet, with whom he lived for a number of years.

Voltaire certainly regarded himself as a member of a group of writers, mainly French, who have since become known as the *philosophes* and who tried to combine an enthusiasm for the new science with a respect for the arts. Voltaire says of himself in 1735 that:

> I will move from an experiment in physics to an opera or a comedy, and never let my taste be blunted by studies . . .[7]

and three years later, in 1738, Voltaire writes:

> I confess that I don't see why the study of physics should crush the flowers of poetry. Is truth such a poor thing that it is unable to tolerate beauty? The art of thinking well, of eloquent speech, of a lively sentiment and self-expression; should they be enemies of science? Surely not, for that would be to think like a barbarian.[8]

THE ENLIGHTENMENT

Voltaire, and others like him, formed part of a movement now usually known as the Enlightenment. The men of the Enlightenment were not principally great original thinkers. What they believed in was the importance of spreading new ideas far and wide in as clear and well-written a form as possible. As one of them, Concorcet (1743–94), put it, they were 'a class of men less concerned with discovering truth than with propagating it' who 'find their glory rather in destroying popular error than in pushing back the frontiers of knowledge'.[9]

Many of the *philosophes*, the men of the Enlightenment, were hostile to the organised Catholic Church as it existed in France at that time. But very few of them were atheists. For example, Voltaire, in his *Elements of the Philosophy of Newton*, wrote this:

> The whole philosophy of Newton leads of necessity to the knowledge of a Supreme Being, who created everything, arranged all things of his own free will. . . . If the planets rotate through empty space in one direction rather than another, their creator's hand, acting with complete freedom, must have guided their course in that direction.

Above all, most of the *philosophes* took Newton's *Principia* as an inspiration and a model for rational thought in many other fields of enquiry.

However, Newton's other great scientific work, the *Opticks*, also had a

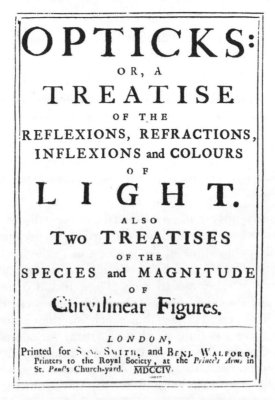

OPTICKS:
OR, A
TREATISE
OF THE
REFLEXIONS, REFRACTIONS,
INFLEXIONS and COLOURS
OF
L I G H T.
ALSO
Two **TREATISES**
OF THE
SPECIES and MAGNITUDE
OF
Curvilinear Figures.

L O N D O N,
Printed for S. Smith, and Benj. Walford.
Printers to the Royal Society, at the *Prince's Arms* in
St. *Paul's* Church-yard. MDCCIV.

Fig. 4.1.6   Title-page of Newton's *Opticks*, First edition, 1704.

considerable influence both on 18th-century science and on popular thought. Fig. 4.1.6 shows the title-page of the first edition which was published in 1704.

The *Opticks* is very different from the *Principia*. It is written in English, not Latin, so it was much easier for it to be widely read. And its emphasis was experimental rather than mathematical. The book begins like this:

> My Design in this Book is not to explain the Properties of Light by Hypotheses, but to propose and prove them by Reason and Experiments: In order to which I shall premise the following Definition and Axioms.[10]

Newton sets out his first five axioms thus:

AXIOMS.
AX. I.

The Angles of Reflexion and Refraction, lie in one and the same Plane with the Angle of Incidence.

AX. II.

The Angle of Reflexion is equal to the Angle of Incidence.

AX. III.

If the refracted Ray be returned directly back to the Point of Incidence, it shall be refracted into the Line before described by the incident Ray.

AX. IV.

Refraction out of the rarer Medium into the denser, is made towards the Perpendicular; that is, so that the Angle of Refraction be less than the Angle of Incidence.

AX. V.

The Sine of Incidence is either accurately or very nearly in a given Ratio to the Sine of Refraction.[11]

Notice that these axioms are all established experimentally and the propositions in the rest of the book are also established primarily by experiments and by verbal reasoning with only minimal use of the simplest of mathematics.

The success of the *Opticks*, and Newton's general prestige, gave an enormous stimulus to all kinds of experimental work. These included studies of electrical machines, of the phenomena of heat, of developments in chemistry and also in what we now call biology and geology but which was then called natural history. In all these areas there was little scope at that time for complex mathematical theories in the style of the *Principia*. But there was ample opportunity for careful observation and experiment, using quantitative techniques where feasible but never being dominated by them. This meant that developments in these areas in the 18th century were not confined to the experts. It was perfectly possible for large numbers of relative amateurs to do interesting work and to make useful contributions. So, in this sense, experimental Newtonianism was a more popular kind of science than Newton's abstruse mathematical works could ever be.

This emphasis on experiment and practice also had a wider popular effect in a major project of the Enlightenment which was most closely associated with two Frenchmen, the writer Denis Diderot (1713–84) and the mathematician Jean D'Alembert (1717–83), whom we have already mentioned for his work on the development of Newton's mechanics.

Diderot and D'Alembert were joint editors of a massive work which was called simply *The Encyclopedia*. It was one of the first books of its kind, and after many struggles with the censors it was published in 17 volumes between 1751 and 1765. Nearly all the main writers of the Enlightenment contributed to the *Encyclopedia* and it included articles on many social, political, philosophical and religious topics—Natural Rights, Reason, Government, History, the Bible, Atheists, even Encyclopedia! But it also contained detailed and up-to-date accounts of all the sciences, arts and trades of the time.

The *Encyclopedia* has been described as 'a vehicle for propaganda as much as for disseminating knowledge', which 'set out to familiarise the layman with the new discoveries in science, and with the kind of rational thinking of which they were the fruit. By this means, it strove to inculcate a critical approach to the problems of human relationships and . . . to spread the belief that human behaviour, like the material universe, was amenable to scientific investigation, and that society and government should be studied scientifically in the interests of human happiness.'[12]

The title-page of the *Encyclopedia* is shown in Fig. 4.1.7 and its various purposes have been summarised by C. C. Gillispie in these words:

> Diderot's *Encyclopedia*, so much the most famous venture of the Enlightenment that the words 'Encyclopedist' and 'philosophe' became almost interchangeable, was itself a natural history of industry. Its sub-title was 'Analytical Dictionary of the Sciences, Arts and Trades'. In Diderot's definition a good dictionary should have 'the character of changing the general way of thinking . . . .' Diderot's masterstroke was to make the technology carry the ideology. It is the latter which has naturally enough monopolized the attention of intellectual historians, for it is the ideology of progress and liberalism which . . . burst passionately into life in the French Revolution and matured into the verities of modern democratic government. But . . . at the time, it was not the liberalism of the *Encyclopedia* which was popular. It was the technology, taking seriously the way people made things and got their livings, dignifying common pursuits by the attention of science. For the *Encyclopedia*, and with it a multitude of industrial studies by the foremost scientists of the Academy, were attempts, Baconian in inspiration but informed by sophisticated method, to lift the arts and trades out of the slough of ignorant tradition—and by rational description and classification to find them their rightful place in the great unity of human knowledge.[13]

The Encyclopedia was beautifully and lavishly illustrated with large numbers of plates, which have made it a valuable original source of information about the state of European technology just at the moment when the Industrial Revolution was about to start.

Fig. 4.1.8 shows a plate dealing with agriculture; Jethro Tull's improved plough is shown in the inset Fig. 2. Fig. 4.1.9 shows a charcoal-fired blast furnace for making iron; the bellows were operated by the waterwheel. Fig. 4.1.10 shows some of the 18 separate processes involved in manufacturing pins; a few years later, Adam Smith in *The Wealth of Nations* used the operation of a pin factory to illustrate the advantages of the division of labour. Finally, Fig. 4.1.11 shows part of the sequence of operations in making a glass. The figures shown here are just a small fraction of the vast amount of detailed information about practical arts and crafts contained in the *Encyclopedia*.

However, there was another aspect to the enthusiasm for all things practical of the Encyclopedists. This was a developing hostility to the aims of the great system-builders like Newton and Laplace and to the very mathematics they used in constructing their systems.

Listen to Diderot, writing in 1754:

> We are at the present time living in a great revolution in the sciences. To judge by the inclination that our writers have towards morals, fiction, natural history

ENCYCLOPEDIE,

OU

DICTIONNAIRE RAISONNÉ

DES SCIENCES,

DES ARTS ET DES MÉTIERS,

PAR UNE SOCIETÉ DE GENS DE LETTRES.

Mis en ordre & publié par M. *DIDEROT*, de l'Académie Royale des Sciences & des Belles-Lettres de Pruſſe ; & quant à la PARTIE MATHÉMATIQUE, par M. *D'ALEMBERT*, de l'Académie Françoiſe, de l'Académie Royale des Sciences de Paris, de celle de Pruſſe, de la Société Royale de Londres, de l'Académie Royale des Belles-Lettres de Suede, & de l'Inſtitut de Bologne.

Tantùm ſeries junĉturaque pollet,
Tantùm de medio ſumptis accedit honoris ! HORAT.

TOME CINQUIEME.

A PARIS,

Chez
{
BRIASSON, *rue Saint Jacques, à la Science.*
DAVID l'aîné, *rue & vis-à-vis la Grille des Mathurins.*
LE BRETON, Imprimeur ordinaire du Roy, *rue de la Harpe.*
DURAND, *rue du Foin, vis-à-vis la petite porte des Mathurins.*

M. DCC. LV.

AVEC APPROBATION ET PRIVILEGE DU ROY.

Fig. 4.1.7 Title-page of the fifth volume of the *Encyclopedia* of D'Alembert and Diderot, 1755.

Agriculture, Labourage.

Fig. 4.1.8 Plate from *Encyclopedia* showing agricultural techniques; Jethro Tull's plough is shown in inset Fig. 2.

and experimental physics, I feel almost certain that before 100 years are up, one will not count three great geometers in Europe. This science will very soon come to a stop where the Bernoullies, the Eulers . . . and D'Alemberts have left it. They will have erected the columns of Hercules. We shall not go beyond that point. . . . When we come to compare the infinite multitude of phenomena of nature with the limits of our understanding and the weakness of our sense-organs, can we ever expect from our work . . . anything but a few broken pieces, separated from the grand chain which unites all things? . . . Even if experimental philosophy should be at work during centuries of centuries, yet the material which it amasses, having become incomparable through sheer size, would still be far from exact enumeration.[14]

Fig. 4.1.9 Plate from *Encyclopedia* showing charcoal-fired blast-furnace for making iron; the bellows are driven by the waterwheel on the left.

Fig. 4.1.10 Plate from *Encyclopedia* showing some of the 18 separate operations involved in manufacturing pins.

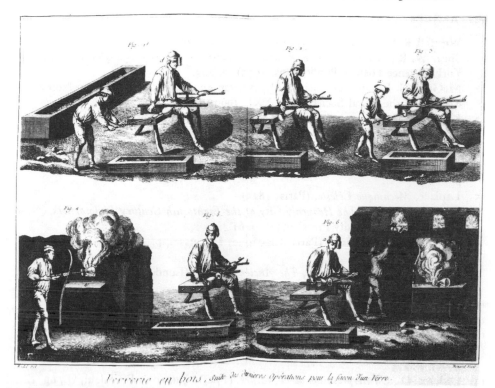

Vérrerie en bois, Suite des Dernieres Opérations pour la façon d'un Verre.

Fig. 4.1.11   Plate from *Encyclopedia* showing part of the sequence of operations involved in making a glass.

Here Diderot is claiming that mathematics will soon grind to a halt—a claim that doesn't look very impressive given the hindsight of another 200 years of rapid growth in mathematics. But he is also recognising the impossibility of dealing mathematically with the vast range and enormous complexity of natural phenomena.

Later Diderot came to reject mathematics on more fundamental grounds. He came to think that mathematics falsifies not only because it idealises but because it deprives phenomena of any connection with feeling or sentiment. Mathematics is inhumane and, even worse, arrogantly inhumane.

With this response Diderot is enunciating, possibly for the first time, the spirit of romantic revulsion against what he saw as the excessive claims of science. This spirit has recurred regularly since then—in the French Revolution, in the Romantic movement of the 19th century, and, in our own time, in the revolt of many people against science and all things scientific.

REFERENCES

1 Westfall, R. S., *The Role of Alchemy in Newton's Career* in Righini Borelli, M. L. and Shea, W. R., *Reason, Experiment and Mysticism in the Scientific Revolution*, (New York: Science History Publications, 1975), p. 203.

2 McGuire, J. E., and Rattansi, P. M., *Newton and 'The Pipes of Pan'* in Notes and Records of the Royal Society, Vol. XXI, 1966, p. 118.

3 Keynes, J. M., *Newton, the Man* in *Essays in Biography*, (London: Macmillan, 1972), pp. 363–4.

4 Lagrange, *Mécanique Analytique*, (Paris: 1788) p.v; quoted in Buchdahl, G., *The Image of Newton and Locke in the Age of Reason*, (London: Sheed and Ward, 1961), p. 6.

5 Laplace, *Mécanique Céleste*, (Paris: 1825).

6 See Becker, C. L., *The Heavenly City of the Eighteenth Century Philosophers*, (New Haven: Yale University Press, 1932), p. 61.

7 Voltaire, *Lettres Choisies*, (Paris: Classiques Larousse), p. 36.

8 Ibid., p. 41.

9 Quoted in Behrens, C. B. A., *The Ancien Régime*, (London: Thames and Hudson, 1967), p. 123.

10 Newton, *Opticks*, (New York: Dover Publications, 1952), p. 1.

11 Ibid., p. 5.

12 Behrens, C. B. A., op. cit., pp. 123–4.

13 Gillispie, C. C., *The Edge of Objectivity*, (Princeton: Princeton University Press, 1960), p. 174.

14 Diderot, D., *L'Interpretation de la Nature*, (1754); quoted in Buchdahl, G., op. cit., pp. 80–1.

SUMMARY

1 Newton spent approximately as much time on alchemy and on biblical criticism and related historical studies as he did on what we regard as his major works, the *Principia* and the *Opticks*.

2 Newton published little or nothing about these other activities and so they had virtually no influence on his contemporaries and successors.

3 The dramatic success of Newton's *Principia* had an enormous influence on the climate of thought in the 18th century; it stimulated further mathematical work of the same type and, in the movement known as the Enlightenment, it encouraged the application of scientific principles to all aspects of life.

4 Newton's *Opticks* also had a dual effect; in the sciences it inspired many empirical studies and, in society at large, it led to greater respect for technology and practical crafts.

5 The *Encyclopedia* of Diderot and D'Alembert led to the widespread dissemination of the reforming ideas of the *philosophes* and of the practical techniques of the day.

4.2 The origins of the social sciences

The 18th century saw the beginnings of systematic work in what we now call the social sciences. In this chapter we will describe the origins of four major social sciences—psychology, sociology, economics and demography or the study of populations. And, since the social sciences are directly concerned with people and society in a way that the natural sciences are not, we will also briefly discuss the social and historical background of the period and the political revolutions which were inspired, in part, by the ideas of the early social scientists.

THE SOCIAL AND HISTORICAL BACKGROUND

We who live in the 20th century find it almost impossible to imagine the enormity of the changes in knowledge and belief, in attitudes to God, man, the world and the universe, which took place in the 18th century. We may perhaps get some feeling for the scale of these changes if we recollect the battle between Galileo and the Church. For what was happening was the break-up of a whole theological and philosophical system, of a way of understanding reality and of giving meaning to life, death and all aspects of human experience. For centuries Europeans had been accustomed to accepting without question a religious explanation of the natural world and of society. Christian teachings had reigned supreme and people were content to live with mystery. And, although Christianity encouraged many humane and compassionate developments, for example in the care of the sick and the poor in monasteries and convents, the Church had also proved itself capable of cruelty when confronted by unorthodox beliefs, as demonstrated by the horrors of the Inquisition. Moreover, it allowed superstition to develop into the atrocities of the great witch hunts of the 16th and early 17th centuries. For example, in the Languedoc region of France alone, 400 so-called witches were burned in the one year of 1577. Such were the penalties for challenging orthodox belief.

Also, at that time, life was very harsh. Death rates were high, with hazards to life and health a feature of everyday experience. Diseases were rampant—the dreaded childbed fever, the killer diseases of infancy, sporadic outbreaks of plague and other epidemic diseases like cholera. The imminent possibility of death and suffering lurked around the corner for everyone all the time, made worse for most of the population by poverty and recurring famine.

Fig. 4.2.1 shows how death could suddenly hit local villages and communities in the 17th century. Notice how death rates could suddenly jump to about six times their normal levels, presumably due to epidemic diseases. And Fig. 4.2.2 illustrates how a particular family could be affected. In 1646 the family of Solomon Bird, an English weaver, was wiped out in an epidemic. The mother died in July, followed in August by the father and three sons aged 21, 18 and 14; two other children, one aged 19 and the other only three weeks old, had died in previous years. Such disasters must have been even more common when

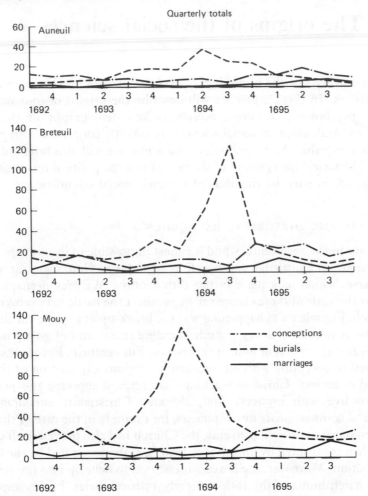

Fig. 4.2.1   Changes in the number of conceptions, burials and marriages in three French parishes in the late 17th century. *Redrawn from Wrigley, E. A., Population and History, London: Weidenfeld and Nicolson, 1969.*

epidemics affected towns. For example, deaths in London from the plague rose to 35,417 in 1625, compared with 11, 17 and 16 in the preceding three years.[1]

With a religion which taught that God would send sinners to everlasting hell, most people could only pick their way precariously through a world they couldn't understand, accepting as divine judgement whatever life might bring. With such deep-seated feelings of powerlessness, the possibility of trying to understand the world or to influence the course of events was generally unthinkable. Even to try to do so would have seemed blasphemous.

It is only by trying to imagine this frame of mind and way of life that we can begin to appreciate the importance of the changes which occurred in the 17th and 18th centuries when people began to challenge the Church's doctrine and to question, investigate and analyse not only the physical world but also human

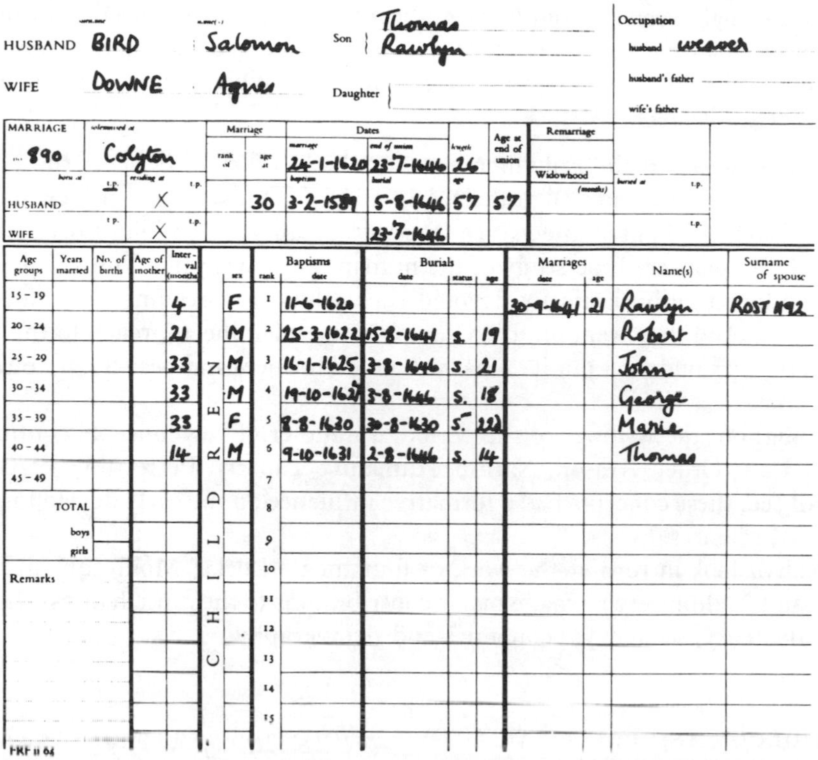

HUSBAND **BIRD** : *Salomon*   Son | *Thomas Rawlyn*   Occupation husband *weaver*

WIFE **DOWNE** : *Agnes*   Daughter | husband's father _____   wife's father _____

MARRIAGE	*solemnised at*	Marriage		Dates			Age at end of union	Remarriage	
...**890**	*Colyton*	rank of	age at	marriage 24-1-1620	end of union 23-7-1646	length 26		Widowhood (months)	buried at t.p.
HUSBAND t.p. *residing at* t.p.	X		30	baptism 3-2-1589	burial 5-8-1646	57	57		t.p.
WIFE t.p.	X t.p.				23-7-1646				t.p.

Age groups	Years married	No. of births	Age of mother	Inter-val (months)	sex	rank	Baptisms date	Burials date	status	age	Marriages date	age	Name(s)	Surname of spouse
15 – 19			4		F	1	11-6-1620				30-9-1641	21	*Rawlyn*	*ROST #92*
20 – 24			21		M	2	25-7-1622	15-8-1641	s.	19			*Robert*	
25 – 29			33	Z M		3	16-1-1625	3-8-1646	s.	21			*John*	
30 – 34			33	M		4	19-10-1627	3-8-1646	s.	18			*George*	
35 – 39			33	F		5	8-8-1630	30-8-1630	s.	22			*Marie*	
40 – 44			14	R M		6	9-10-1631	2-8-1646	s.	14			*Thomas*	
45 – 49					7									
TOTAL					8									
boys					9									
girls					10									
Remarks					11									
					12									
					13									
					14									
					15									

FRF II 64

Fig. 4.2.2   Solomon Bird's family— an example of family reconstitution from records in parish registers. *From Wrigley, E. A., Population and History, Weidenfeld and Nicolson, 1969.*

nature and human society. What had seemed holy and sacrosanct became available for study and what had seemed unchanging was seen as open to change.

NEWTON, THE ENLIGHTENMENT AND THE ORIGINS OF THE SOCIAL SCIENCES

The universal laws of motion and gravitation set out by Newton in his *Principia* had an enormous impact on 18th-century thought (see Chapter 4.1). The discovery of such powerful and versatile 'Laws of Nature' was of immense significance. Instead of being mysterious, nature was now seen as an orderly and predictable system which could be known and understood. And such knowledge could provide the basis for prediction and for control.

The down-to-earth methods Newton used to gain much of his scientific knowledge, particularly in his *Opticks*, were of great importance too. As one writer puts it:

Certainly this new philosophy ravished the 18th Century into admiration; and not the least most astonishing thing about it was the commonplace methods employed to discover such marvellous truth. That Newton discovered the nature of light seemed even less significant to his contemporaries than that he did so by playing with a prism. It was as if nature had for the first time been brought close to men, close enough to be tangible and visible in all its wonderful details.[2]

The *philosophes* of the Enlightenment, or the 'century of light' as the French now call it, freely acknowledge their debt to two Englishmen—to Newton who had shed so much light on the nature of physical reality and to John Locke who, as we shall see, pioneered the study of the human mind. Men like Voltaire, Diderot and Condorcet enthusiastically devoted themselves to spreading new ideas and knowledge. And they went on to try to apply the scientific approach to the study of society. As Condorcet put it 'What we can do for bees and beavers, we ought to do for man'.

Throughout the writings of the period, a number of key concepts continually recur—Law, Order, Reason, Nature, Humanity, Liberty, Perfectibility. And, as we shall see, these concepts had a formative influence on the early development of the social sciences.

We shall look in turn at the work of four men—Locke, Montesquieu, Adam Smith and Malthus—who made major contributions to the growth of psychology, political theory, sociology, economics and demography.

PSYCHOLOGY AND POLITICAL THEORY—JOHN LOCKE (1632–1704).

John Locke was one of the most influential early writers to attempt to apply the new scientific approach to the study of man and society. In fact he was so early that all but the last four years of his life actually fall in the 17th rather than the 18th century. His book *An Essay Concerning Human Understanding* (1690) is one of the first modern contributions to psychology. In this book Locke challenged prevailing beliefs about human nature and the gloomy Christian doctrine which suggested that man was inherently wicked, and he used data from different societies to illustrate his views. He suggested that a person started life with a mind which was a blank sheet or *tabula rasa* which was then shaped by experience, being influenced for good or for ill by personal experiences and by society.

And Locke's psychology did not remain an abstract, academic subject. He translated his psychological theories into politics—putting forward recommendations about how society should be governed. His political theory is particularly important since it expressed the principles which inspired Liberalism in England and Republicanism in America. According to Locke, the function of the state is limited to the protection of the rights of individuals against infringement by other people. Each individual has a Right, according to Natural Law, to the safety of his person and of his property, and to liberty of action—provided that he in turn does not violate the rights of other people. The state's responsibility is to ensure that these rights are respected, and it should do no more than this.

Throughout Locke's work there is a happy belief in the essential goodness and harmony which would occur if the state left individuals the freedom to pursue their own interests. This cheerful picture is startlingly different from the more gloomy view which prevailed in earlier centuries.

Locke also has a very different view of the state from that which was to develop in the 19th and 20th centuries when many thinkers argued that the state should be much more involved in the organisation of society. These later ideas found practical expression in the way in which contemporary socialist societies are organised.

In brief, Locke was important for sowing the seeds of psychology, and for his associated beliefs concerning the nature of society and politics. These beliefs had direct political consequences in their effects on the English and French revolutions and on the American Declaration of Independence. In particular Locke was the philosopher of the English bloodless revolution of 1688 which gave more power to Parliament and less to the King and ended the civil and religious battles which had divided England for so long. This revolution set the scene for the development of liberalism—with, for example, its emphases on tolerance, especially of different religious beliefs; on a system of government based on separation of powers, so that no one person or group should have absolute power; and on a commitment to respect and to protect individual liberty and personal property. It was the climate of tolerance created by this revolution that so impressed Voltaire when he fled to London in 1726. His *Lettres Philosophiques*, as well as the chapters on Newton already mentioned (see Chapter 4.1), also contained a chapter 'On Mr. Locke'.

SOCIOLOGY—MONTESQUIEU (1689-1755)

The 19th-century French writer, Auguste Comte, is usually acclaimed as the founder of sociology partly because it was he who invented the word. But Comte had an important forerunner in the 18th century, in the person of the Frenchman, the Baron de Montesquieu. Like Voltaire, Montesquieu lived in London for a while and was even elected a Fellow of the Royal Society.

Montesquieu's most famous book is *L'Esprit des Lois* or *The Spirit of the Laws* published in 1748. In this book Montesquieu's aim was to make history intelligible—a formidable task as his studies showed that history consisted of an almost limitless diversity of laws, customs, morals, ideas and institutions. Throughout his vast work we can see two ways in which he tried to impose some order on this apparently chaotic diversity: one was by pointing to ways in which social events have historical causes; the other was by classifying the enormous range of laws, manners, customs and ideas into a small number of social types.

In the preface to *L'Esprit des Lois* Montesquieu claims, with an audacity not totally uncharacteristic of certain sociologists of the present day, that:

> We have first of all considered mankind and the result of my thoughts has been, that amidst such an infinite diversity of laws and manners, they were not solely conducted by the caprice of fancy.[3]

In his survey of mankind, Montesquieu distinguishes 'three species of government: republican, monarchical and despotic.'[4] He describes and analyses their different laws, manners and customs and even writes about the effects of climate. He obviously found English life peculiar and associated English government with the climate of the British Isles:

> In a nation so distempered by the climate as to have a disrelish of everything, nay, even of life, it is plain that the government most suitable for the inhabitants is that in which they cannot lay their uneasiness to any single person's charge, and in which, being under the direction of the laws rather than of the prince, it is impossible for them to change the government without subverting the laws themselves.[5]

More fundamentally, Montesquieu is one of the first to discuss religion as a social fact, without regard to its truth or falsity. The essence of Montesquieu's political philosophy, like that of Locke, is liberalism and the balance of powers. As Raymond Aron describes it, for Montesquieu:

> The goal of the political order is to insure the moderation of power by the balance of powers, by the equilibrium of people, nobility and king in the French or the English monarchy, or the equilibrium of the people and the privileged, plebs and patriciate, in the Roman Republic . . . and to seek, by the balance of powers, the guarantee of moderation and liberty.[6]

Here again are the resounding themes of the age, in particular liberty and harmony. It is interesting to reflect here, as we shall again, that ideas have consequences, and that sometimes they influence events in ways which are very different from the intentions of the original thinkers. Montesquieu is a good example of such unintended consequences. He neither foresaw nor would he have desired the French Revolution but his ideas may well have helped to prepare the way for it. And if he had survived until then, he would probably have had to choose, like other liberals of his kind, between the guillotine or emigration.

In conclusion, there is no doubt that Montesquieu was one of the first sociological thinkers, for he sought to explain human behaviour in terms of social institutions and social experiences—which is precisely what sociology is all about.

ECONOMICS—ADAM SMITH (1723–90)

Adam Smith was, like many creative British people, a Scotsman. He had a varied career, studying mathematics and literature and becoming Professor of Logic and then of Moral Philosophy at Glasgow University. He later travelled to Europe to work as a private tutor to the Duke of Buccleuch. Apparently he became very bored, for he wrote to his friend, the philosopher David Hume, 'I have begun to write a book to pass away the time'. This book was his famous *An Inquiry into the Nature and Causes of the Wealth of Nations* published in 1776. While on his travels, Smith had been impressed by the economic writings of the Frenchman, François Quesnay and a group of men called the Physiocrats. Their motto, the famous 'Laissez faire', signified their opposition to the kind of government control of trade and industry which had been advocated by the mercantilists and practised up till then by many governments.

In the *Wealth of Nations*, Smith developed and elaborated the work of the Physiocrats and contributed a number of major ideas of his own.

First, he introduced a historical perspective into economic discussion. Like Montesquieu, he studied history to see what patterns emerged and what could be learnt from them. The *Wealth of Nations* is an immense study which includes many historical accounts of earlier societies like ancient Rome, of the origins of freedom and of the origins of money and of price movements. Some of Smith's themes recur in the work of Karl Marx (see Chapter 4.10), particularly the idea that the main forces which bring about social change are economic.

Adam Smith's second major contribution to the development of economics was his theory of wealth and of the division of labour. He explained how the division of labour, by which he means different people specialising in particular skills and tasks, helps productivity and therefore increases wealth. His most famous example describes how the manufacture of pins can be made cheaper by breaking down the process into different parts and allocating each part to a different person (see Fig. 4.1.10, page 100). This system allows each worker to develop great skill and speed in his part of the process thus saving valuable time—which would otherwise be spent in one person moving from one job to another—and making more efficient use of expensive machinery. There are, of course, great problems associated with this system. Smith himself recognised that it denies a person the satisfaction of producing a craftsman's article and that it may lead to people being regarded as part of a system of production rather than as creative individuals in their own right. But Smith's justification for the division of labour, despite these dangers, was that it increased the overall wealth of a nation, which in turn could lead to an improvement in the standard of living of the workers. And if we think back to the widespread poverty which existed before the Industrial Revolution, we may feel that this argument has some force.

Adam Smith's third major contribution to economic thought was his idea of an 'invisible hand' by means of which the separate actions of many individuals can, without central direction, be coordinated through a price system. This is how Smith puts it:

> As every individual, therefore, endeavours as much as he can both to employ his capital in the support of domestic industry, and so to direct that industry that its produce may be of the greatest value; every individual necessarily labours to render the annual revenue of the society as great as he can. He generally indeed, neither intends to promote the public interest, nor knows how much he is promoting it . . . he intends only his own gain, and he is in this, as in many other cases, led by an invisible hand to promote an end which was no part of his intention. Nor is it always the worse for society that it was no part of it. By pursuing his own interest he frequently promotes that of the society more effectually than when he really intends to promote it.[7]

The economist Milton Friedman describes Adam Smith's idea of an 'invisible hand' in this way:

> This is a highly sophisticated and subtle insight. The market, with each individual going his own way, with no central authority setting social priorities, avoiding duplication, and co-ordinating activities, looks like chaos to untutored eyes. But through Smith's eyes we see that it is a finely ordered and effectively tuned system, one which arises out of men's individually motivated actions, yet is not deliberately created by men. It is a system which enables the dispersed knowledge and skill of millions of people to be co-ordinated for a common purpose. Men in Malaya who produce rubber, in Mexico who produce graphite, in the state of Washington who produce timber, and countless others, co-operate in the production of an ordinary rubber-tipped lead pencil . . . though there is no world government to which they all submit, no common language in which they could converse, and no knowledge of or interest in the purpose for which they co-operate.[8]

Smith's twin ideas of the division of labour and of the 'invisible hand' together provided a powerful account of how individuals in a complex society could simultaneously serve their own interests, and those of other people and of society as a whole. Here we have an economic case against the government-controlled mercantilism of the earlier era and for the philosophy of liberalism as it was developing in Britain and America. For, according to this theory, no statesman or government could achieve what the free market would generate; human behaviour and society are too complex to be efficiently controlled by a central agency. So, according to Smith, the only function of the state was to provide order, justice and certain public goods such as the building of roads and bridges. There are thus clear parallels between Smith's economic philosophy and the political philosophies of Locke and Montesquieu.

A final example of one of Smith's main themes is his deep belief in the existence of a benevolent Providence. Here we see reflections of the same religious faith which had inspired the work of Copernicus, Kepler, Galileo and Newton—the faith that nature moves in a regular and harmonious fashion. But now the belief is harnessed to the ordering of human nature and human society rather than just the inanimate order of the physical world. Thus, although people had reacted against some of the Church's teachings, the new natural and social sciences were not necessarily incompatible with religion.

Smith's work was admired and appreciated in his own lifetime. His friend, the philosopher David Hume, welcomed it warmly and many influential people in government such as Pitt, the English Prime Minister, openly acknowledged his influence on their political policies. There is a story that Adam Smith was late one evening for a dinner party at which many leading politicians were present. When he arrived and tried to apologise, everyone rose to their feet and Pitt himself said: 'We will stand until you are seated, for we are all your scholars'.

Smith's influence on the development of economics is still widely acknowledged and, as we shall see (Chapter 9), some of his ideas are relevant in fields far removed from economics.

DEMOGRAPHY OR THE STUDY OF POPULATIONS—THOMAS MALTHUS (1766–1834)

Demography has its roots in the late 18th and early 19th centuries in the work of the Reverend Thomas Malthus. Malthus was the first man to apply the principles of scientific investigation to trends in population growth and decline. He developed a theory which caused a great stir and anxiety in his own time and has remained of interest and concern ever since. The 'population explosion' of the 20th century can be seen as a contemporary application of his theory (see Fig. 6.1.8, page 249).

Malthus lived at a time when the Industrial Revolution was gathering momentum. Many people, especially in England and France, were very impressed by the opportunities for improvements in the standard of living, even for the labouring masses or the peasants who had previously endured dreadful living conditions in both town and country. It was thus a period of great optimism and no-one seemed to be aware of the problems of balance in the relationships between population, resources and technology. But Malthus tried to work out what might happen if the existing equilibrium between population size and natural resources were to be upset. Up to this period there had been natural checks on population growth such as illnesses and famines leading to the high death rates we mentioned at the beginning of this chapter (see Figs. 4.2.1 and 4.2.2). But Fig. 4.2.3 shows that, in the years preceding Malthus, the population of England had already started on the rapid rise which has continued to the present day.

But it wasn't population growth, in itself, which concerned Malthus. He was primarily worried by the possibility that populations would tend to grow at a rate which could not be matched by increases in the production of food. In other words, people would literally swallow up the food supplies.

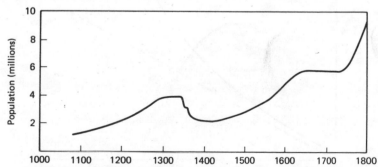

Fig. 4.2.3 Population in England from AD 1000–1800. Note the fall in the 14th century due to the Black Death and the rise in the middle of the 18th century at the start of the Industrial Revolution. *Redrawn from Wrigley, E. A., Population and History, London, Weidenfeld and Nicolson, 1969.*

In 1798 Malthus published his *Essay on the Principle of Population*, which contained all his main ideas and evidence from different countries to support them. As a careful scientist Malthus continued to gather evidence for his theory by observations and travel, and later editions of his book contained details of population changes in most parts of the world as well as in ancient Greece and Rome.

Malthus also tried to give some mathematical precision to his theory by contrasting two types of progression or rates of growth. He argued that populations tend to increase by geometric progression so that over 30 years, roughly a generation, a population of 1000 might double, then double again over the next 30 years to 4000 and reach 8000 after 90 years. Thus, a given population could increase by a factor of eight in three generations.

Conversely, Malthus argued that the means of subsistence, mainly the food supply, tends to increase at most by arithmetic progression, so that a wheat production of 1000 lb per acre might increase to 2000 lb over 30 years, to 3000 over the next 30 years and to 4000 by 90 years on. The ensuing gap between growth in population and food supplies is illustrated by Fig. 4.2.4.

Whether the minute details are accurate or not, the basic Malthusian principle that subsistence only increases at best by regular additions whereas populations multiply implies a gloomy prospect for mankind: starvation. It is this spectre

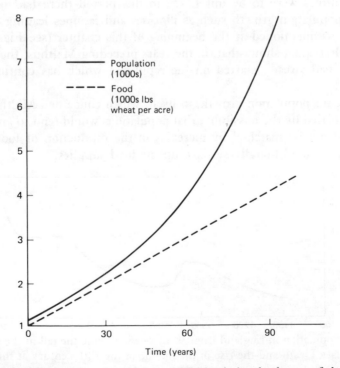

Fig. 4.2.4 Diagrammatic representation of Malthus' simple theory of the relationship between population growth and food supply.

which today still haunts many people who share similar concerns to those of Malthus.

In the event, for 150 years after Malthus' bleak predictions, a population crisis on a global scale did not occur. There were many reasons for this. For example, the Industrial Revolution was also associated with improved agricultural techniques such as the use of more scientific systems of crop rotation, of improved strains of seed and of fertilisers. Also, huge new areas of land became available for growing food, in Russia, the Americas and Australia, and there were great improvements in transport facilities for the bulk carriage of produce. The growth of a world economy combined with new resources, improved technology and better transport and communications have allowed the world population to expand without causing a population crisis on a world scale. However, the population problems now being experienced by many third world nations are in many ways reminiscent of Malthus' predictions.

So we can see that, with amazing prescience, Malthus in the 18th century had identified some major human problems which are still important and unsolved today. This is only one example of how the social sciences were beginning to generate ideas which had to be taken seriously. Malthus' ideas were also taken seriously by natural scientists in the 19th century and had an influence on the way that Charles Darwin and Alfred Wallace developed their ideas on evolution (see Chapter 4.8)

IDEAS HAVE CONSEQUENCES

This brief survey of the origins of the social sciences would be incomplete without some discussion of the relationship between ideas and their consequences. Clearly, ideas do not occur in a vacuum and they will reflect the social, political and economic situation; and ideas may, in turn, influence social, political and economic developments. The attempts in the 18th century to apply a scientific analysis to human societies reflected the prevailing theories of the nature of man. For example, the themes and the concepts which we have seen recurring again and again in the writing of men such as Locke, Montesquieu and Adam Smith, influenced the political credos of the American Declaration of Independence (see Fig. 4.2.5) and the French Revolution (see Fig. 4.3.3, page 121 and Chapter 4.3).

This excerpt from the Declaration of Independence resonates with those themes we have been considering: Law; Nature's God; unalienable Rights; Liberty; Equality and the pursuit of Happiness:

> When in the Course of human events, it becomes necessary for one people to dissolve the political bands which have connected them with another, and to assume among the Powers of the earth, the separate and equal station to which the Laws of Nature and of Nature's God entitle them, a decent respect to the opinions of mankind requires that they should declare the causes which impel them to separation.
>
> We hold these truths to be self-evident, that all men are created equal, and

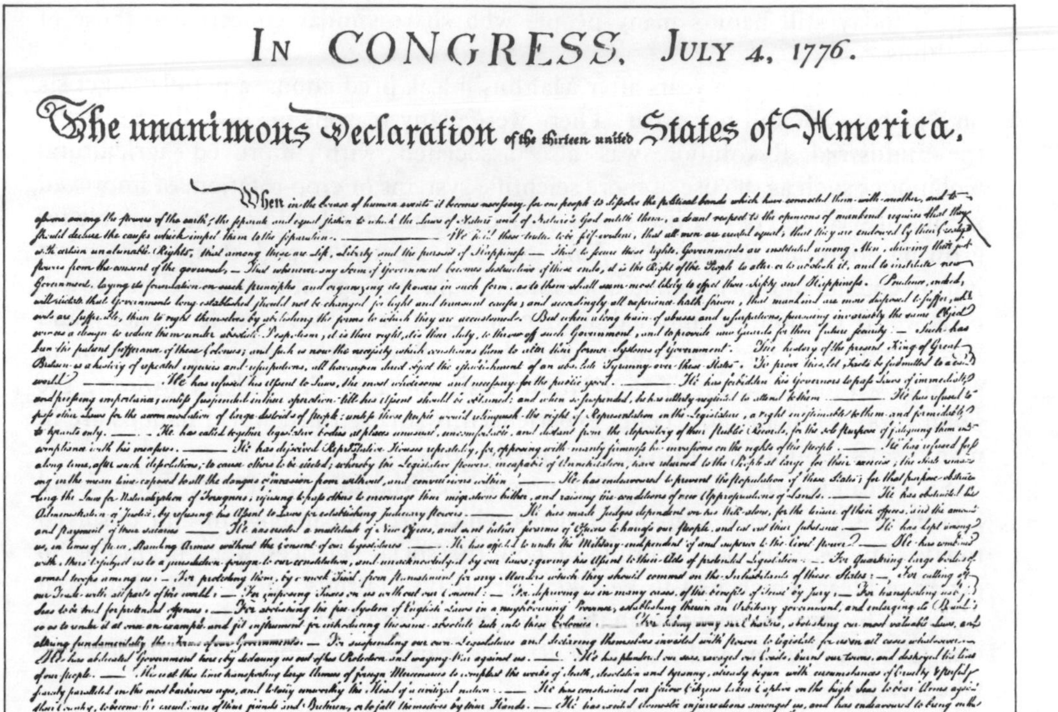

Fig. 4.2.5   The American Declaration of Independence of 1776.

that they are endowed by their Creator with certain unalienable Rights, that among these are Life, Liberty and the pursuit of Happiness.[9]

Or, turning from America to France, listen to this account of the relationship between ideas and their consequences in the French Revolution. Here the author is writing about the *philosophes*, who had so passionately embraced the ideas of the

Enlightenment, and about the Festival of the Supreme Being organised during the French Revolution (see Chapter 4.3):

> Emancipated themselves, they were conscious of a mission to perform, a message to deliver to mankind; and to this messianic enterprise they brought an extraordinary amount of earnest conviction, of devotion, of enthusiasm. We can watch this enthusiasm, this passion for liberty and justice, for truth and humanity, rise and rise throughout the century until it culminates, in some symbolic sense, in that half admirable, half pathetic spectacle of June 8, 1794, when Citizen Robespierre, with a bouquet in one hand and a torch in the other, inaugurated the new religion of humanity by lighting the conflagration that was to purge the world of ignorance, vice and folly.[10]

This idea that people can shape society into a form which will achieve Utopia or heaven on earth, and which will create a perfect environment in which people's good nature can achieve perfection, is one which we will encounter again in the work of later social scientists. The Enlightenment initiated a deep divide between two schools of thought about society. There are those who believe in the ability of people to understand human behaviour so well that they are justified in developing a blueprint for society and constructing it according to a master plan, a view held by many contemporary Marxists and socialists. And then there are those who argue that natural and social reality are so complex that it is impossible to grasp their diversity and richness. These thinkers, of whom Karl Popper is a contemporary example, emphasise the dangers of trying to build a heavenly city on earth. Because human knowledge must inevitably be limited and inadequate, great humility is required, and a piecemeal approach may be more appropriate than the complete reconstruction of an entire society. These thinkers argue that without such humility there is a danger that naïve policies and new dogmatisms may generate developments which are more dangerous, and oppressive of the human spirit, than those which characterised the era which preceded the Enlightenment—the era of the witchcraft trials and the suppression of heretics and dissidents.

Social scientists, like natural scientists, therefore, have great responsibilities because their theories and ideas may promote social and political changes which can influence the lives of countless people for good or for ill. We will see, as we come to consider the 19th and 20th centuries, just how some of these ideas have worked out in practice. And when we do this we will find indispensable one of the methods of analysis pioneered by men like Locke, Montesquieu, Adam Smith and Malthus—the method of comparing the institutions and practices in as many different societies as possible.

REFERENCES

1 Wrigley, E. A., *Population and History*, (London: Weidenfeld and Nicolson, 1969), Table 4.1, p. 114.
2 Becker, C., *The Heavenly City of the Eighteenth Century Philosophers*, (New Haven: Yale University Press, 1932), p. 58.

3   Montesquieu, *The Spirit of the Laws*, (New York: Hofner Press, 1949), p. lxvii.
4   Ibid., p. 8.
5   Ibid., p. 231.
6   Aron, R., *Main Currents of Sociological Thought 1*, (Harmondsworth: Penguin, 1968), p. 60.
7   Smith, A., *The Wealth of Nations*, (London: Methuen, 1930), p. 421.
8   Friedman, M., *Adam Smith's Relevance for 1976*, (Los Angeles: International Institute for Economic Research, 1976), pp. 15–16.
9   Declaration of American Independence.
10 Becker, C., op. cit., p. 43.

SUMMARY

1   Before the 18th century life for most people was harsh with high death rates and widespread disease and poverty. Ideas were dominated by religious beliefs which did not encourage people to think independently or to try to shape their social environment.
2   Newton's work—both his mathematical theories and his empirical methods—was influential in the early development of social science and so were the ideas of the Enlightenment.
3   *The origins of the social sciences:* Psychology had its principal origins in John Locke's *An Essay on Human Understanding*, sociology in Montesquieu's *The Spirit of the Laws*, economics in Adam Smith's *The Wealth of Nations*, and demography in Malthus' *Essay on Population*.
4   Together, these developments represent a turning point in the history of man's attitude to himself, to society and to the world. He now believed that, using the methods of science, he could develop sufficient knowledge to create an environment which could achieve human happiness and well-being.

4.3   The American and French revolutions: ideas and consequences

Two major political revolutions took place in the 18th century–the American Revolution which culminated in the Declaration of Independence in 1776 (see Fig. 4.2.5) and the French Revolution, which started in 1789 with the fall of the Bastille and the Declaration of the Rights of Man and of Citizens. Over the next 200 years these revolutions, and the ideas of democracy and liberty on which they were based, inspired the political institutions of many countries.

Both these revolutions drew much of their inspiration from the ideas of the Enlightenment which was itself greatly impressed and influenced by the development of science in the 17th century and by the work of Newton in particular (see Chapters 4.1 and 4.2). Both these revolutions also had a significant impact on the subsequent development of the natural and social sciences. So in this chapter we will first look at the roots of these revolutions in the ideas of the Enlightenment and then consider some of the political and scientific events which were, at least in part, a consequence of these ideas.

THE INFLUENCE OF THE ENLIGHTENMENT

The Enlightenment was largely a French movement. Its writers developed their ideas against the background of a France in which the Catholic Church was powerful and which was governed by a nearly absolute monarchy that made a deliberate display of its wealth. Louis XIV (1638–1715), who reigned for more than 50 years, exemplified this style of government; he was known as the Sun King. It was Louis XIV (see Fig. 4.3.1) who said '*L'état, c'est moi*' and who built the enormous palace at Versailles, just outside Paris.

This powerful combination of Church and King made up what we now call the Ancien Régime, or old order, in France. Fig. 4.3.2 shows a cartoon from the French Revolution which indicates how such a society appeared from below; a peasant is carrying a member of the clergy and of the nobility on her back.

It was in criticism of such a society that *philosophes* such as Voltaire and Diderot wrote their books and pamphlets. The *philosophes* believed that happiness was the main aim of society and that reason was the instrument by which this aim should be achieved. As Diderot put it:

> Our aim is to gather all knowledge together, so that our descendants, being better instructed, may become at the same time happier and more virtuous.[1]

In the words of another *philosophe*, Condorcet, they tried to track down 'prejudice in those sheltered places where the Clergy, the Government and the ancient corporations are protecting it.'[2]

To sum up the influence of the *philosophes* without being over-simple or misleading is not easy. According to the historian C. B. A. Behrens, the

Fig. 4.3.1 Louis XIV, the Sun King.

Fig. 4.3.2 A cartoon from the French Revolution: a peasant woman carrying a member of the nobility and of the clergy on her back.

philosophes 'always argued on the assumption that man was the creature of his environment, and that the propensities for evil which he only too obviously displayed could be diminished or eradicated if his environment could be changed'. 'They attacked all the assumptions on which the old order rested' and 'held all existing institutions up to ridicule'. So their teaching 'on social and political questions was profoundly revolutionary'. Yet 'none of them . . . ever concerned himself with the means by which power could be seized' and '. . . nothing was further from their minds than the idea of revolution.'[3]

In summary, Behrens states that:

> . . . this great intellectual movement . . . laid the foundations of the social sciences of psychology, anthropology, sociology and economics, and provided Liberals, Socialists and finally Communists with their inspiration—Marx, for example, said that Diderot was his favourite reading, and Trotsky was greatly influenced by Condorcet.

The Enlightenment was thus a potent 'solvent of the old order . . . because it provided all the discontented with principles by which to justify their discontents and with visions of a better future to inspire the attempts at change'.[4]

THE AMERICAN REVOLUTION

France was the major centre of ideas in the 18th century, but the first major political revolution of the century took place not there but across the Atlantic in America. The American Revolution grew out of the revolt of the British colonies there against British rule. The desire for more independence from Britain had been developing for some time, and in 1775 war broke out between Britain and the colonists. This led on 4 July 1776 to the famous *Unanimous Declaration of the Thirteen United States of America*. Fig. 4.2.5 (page 114) shows the original document and the first part of the text was quoted in Chapter 4.2. The signatories of the Declaration then go on to justify their revolution in these words:

> That to secure these rights, Governments are instituted among Men, deriving their just powers from the consent of the governed, That whenever any Form of Government becomes destructive of these ends, it is the Right of the People to alter or to abolish it, and to institute new Government, laying its foundation on such principles and organizing its powers in such form, as to them shall seem most likely to effect their Safety and Happiness.
>
> Prudence, indeed, will dictate that Governments long established should not be changed for light and transient causes; and accordingly all experience hath shown, that mankind are more disposed to suffer, while evils are sufferable, than to right themselves by abolishing the forms to which they are accustomed. But when a long train of abuses and usurpations, pursuing invariable the same Object evinces a design to reduce them under absolute Despotism, it is their right, it is their duty, to throw off such Government, and to provide new Guards for their future security.[5]

Many of the ideas expressed in this declaration are very similar to those of the French Enlightenment. This is not surprising since two of the people who drafted it, and who also were amongst those who signed it, were Benjamin Franklin and Thomas Jefferson, both of whom were strongly influenced by the Enlightenment and spent considerable periods of their lives in France. Franklin's is the third signature in column 4 and Jefferson's the seventh in column 3 in Fig. 4.2.5.

Franklin (1706–90) is probably best known to us for his scientific work on the electrical nature of thunderstorms. He was elected a foreign member of the *Académie des Sciences* in 1773 and was an ambassador to France for the new United States government from 1776 to 1784. In 1784 he represented the *Académie des Sciences* on a royal commission on mesmerism. Jefferson (1743–1826) succeeded Franklin as American ambassador to France from 1784 until just after the French Revolution started in 1789. His career was almost the epitome of the ideas of the Enlightenment. At various times he was an architect, an educator, a scientist, a scholar and a political philosopher. He was President of the United States from 1801 to 1809.

THE FRENCH REVOLUTION

Jefferson was in France for the event which is generally taken to be the start of the Revolution—the storming by Parisians of the Bastille, the fortress prison of Louis XVI, on 14 July 1789. And, like many others at the time, Jefferson was sympathetic to the Revolution in its early stages.

Almost at once the French revolutionaries set up a National Assembly which debated and proclaimed a Declaration of the Rights of Man and of Citizens. This declaration has many similarities with the American Declaration of Independence; Fig. 4.3.3 shows the first page of the original document. This is how it starts in a translation given by Tom Paine in his *Rights of Man* of 1791:

Declaration of the Rights of Man and of Citizens, 1789
The representatives of the people of France, formed into a National Assembly, considering that ignorance, neglect, or contempt of human rights are the sole causes of public misfortunes and corruptions of Government, have resolved to set forth in a solemn declaration, these natural, imprescriptible, and inalienable rights; that this declaration being constantly present to the minds of the members of the body social, they may be ever kept attentive to their rights and their duties; that the acts of the legislative and executive powers of Government, being capable of being every moment compared with the end of political institutions, may be more respected; and also, that the future claims of the citizens, being directed by simple and incontestable principles, may always tend to the maintenance of the Constitution and the general happiness.

For these reasons the National Assembly doth recognise and declare *in the presence of the Supreme Being, and with the hope of his blessing and favour*, the following *sacred* rights of men and of citizens:

I   Men are born, and always continue, free and equal in respect of their rights.

Du 3 Novembre 1789.

LOUIS, PAR LA GRÂCE DE DIEU, ROI DE FRANCE ET DE NAVARRE: A tous ceux qui ces présentes Lettres verront; SALUT.

L'Assemblée Nationale nous a fait présenter le Décret dont la teneur suit :

EXTRAIT du Procès-verbal de l'Assemblée Nationale.

Du Mardi 20 Octobre 1789.

L'ASSEMBLÉE NATIONALE a décrété que les arrêtés du 4 août & jours suivans dont le Roi a ordonné la publication, ainsi que tous les arrêtés & Décrets qui ont été acceptés ou sanctionnés par Sa Majesté, soient, sans aucune addition, changement, ni observations envoyés aux Tribunaux, Municipalités & autres Corps administratifs, pour y être transcrits sur leurs registres, sans modification ni délai, & être lûs, publiés & affichés. *Signé* FRETEAU, Président.

Collationné conforme à l'original, par nous Président & Secrétaires de l'Assemblée Nationale. A Paris, le vingt trois Octobre mil sept cent quatre-vingt-neuf. *Signé* FRETEAU, Président; FAYDEL, THIBAULT, Curé de Souppes, & ALEXANDRE DE LAMETH, Secrétaires.

Suit la teneur desdits Décrets.

EXTRAIT des Procès-verbaux de l'Assemblée Nationale.

DÉCLARATION DES DROITS DE L'HOMME ET DU CITOYEN.

PRÉAMBULE.

Séance du 20 août 1789.

LES Représentans du peuple François, constitués en Assemblée Nationale, considérant que l'ignorance, l'oubli ou le mépris des droits de l'homme, sont les seules causes des malheurs publics & de la corruption des Gouvernemens, ont résolu d'exposer dans une Déclaration solennelle, les droits naturels, inaliénables & sacrés de l'homme, afin que cette Déclaration constamment présente à tous les membres du corps social, leur rappelle sans cesse leurs droits & leurs devoirs; afin que les actes du pouvoir législatif & ceux du pouvoir exécutif, pouvant être à chaque instant comparés avec le but de toute institution politique, en soient plus respectés; afin que les réclamations des citoyens, fondées désormais sur des principes simples & incontestables, tournent toujours au maintien de la Constitution & au bonheur de tous.

En conséquence, l'Assemblée Nationale reconnoît & déclare, en présence & sous les auspices de l'Etre Suprême, les droits suivans de l'homme & du citoyen.

ARTICLE PREMIER.

LES hommes naissent & demeurent libres & égaux en droits; les distinctions sociales ne peuvent être fondées que sur l'utilité commune.

Fig. 4.3.3 Declaration of the Rights of Man and of Citizens, Paris, 1789.

II   The end of all political associations is the preservation of the natural and imprescriptible rights of man; and these rights are Liberty, Property, Security, and Resistance of Oppression. . . .[6]

The Declaration goes on to uphold freedom of opinion, freedom of speech and publication, the separation of powers and the necessity for due processes of law which include freedom from arbitrary arrest. The Declaration was drawn up by the National Assembly in August 1789, and a month or so later the King, Louis XVI, reluctantly agreed to it.

The early days of the Revolution were a time of hope. The old order was to be permanently changed. They even dismantled the Bastille brick by brick. The world was to be made anew. The hopes of the Revolution were also welcomed enthusiastically by many people outside France. For example, the English poet Wordsworth wrote his famous lines:

> Bliss was it in that dawn to be alive
> But to be young was very heaven.

The National Assembly soon went on to legislate in earnest, and in 1790 asked the *Académie des Sciences*, founded by Louis XIV in 1666, to advise on the best method of reforming the traditional and somewhat chaotic French system of weights and measures. The Academy did this with enthusiasm. They decided that the new unit of length should be one-ten-millionth part of the distance from the equator to the North Pole along the meridian through Paris. And they sent an expedition to survey this meridian. They named the new unit the *metre* and the system they recommended was based on the metre and two other new units, the gram and the litre, and on decimal multiples and sub-multiples of these basic units. Thus the metric system of units, with which we are so familiar, was first adopted in revolutionary France.

Another legacy to the modern world from the French Revolution are the political terms Left and Right. In 1790 and 1791 the National Assembly continued to legislate and tried to draw up a detailed constitution to follow the Declaration of the Rights of Man of 1789. In these debates factions emerged. Some wanted radical changes straight away and others wanted to move more slowly. The more revolutionary faction sat on the left of the Assembly while the more traditional faction sat on the right. It is from this accident that our political terms Left and Right derive.

In 1791 the Assembly passed a law which severely limited the powers and rights of the Church. This was denounced by the Pope and caused Louis XVI to try, unsuccessfully, to escape from the country. The Revolution became more openly anti-religious and in July, 1791, organised a massive festival—the Festival of Voltaire. Fig. 4.3.4 shows one of the processions in the festival which was planned and organised by the artist Jacques-Louis David. Voltaire, when he died in 1778, had been refused a Christian burial and had been buried outside Paris in unconsecrated ground. Now his body was exhumed and given a ceremonial burial in a church in Paris which had been transformed by the Revolution into a

Fig. 4.3.4 Festival of Voltaire, 11 July 1791; the carriage carrying Voltaire's remains passing the Royal Palace on its way to the Pantheon. *From Open University Course A202, The Age of Revolution, Units 29–30 Art and Politics in France.*

Pantheon or place for the burial of the great. Fig. 4.3.4 shows the carriage bearing Voltaire's remains as it passed the Royal Palace on its way to the Pantheon. The whole festival was anti-religious in tone, and was also anti-royalist, since Louis' unsuccessful bid to escape had taken place only a month earlier in June 1791. This was one of several propaganda festivals for the Revolution which were organised by the artist David.

In 1792 a great crisis occurred. France went to war with Catholic Austria and many French people believed that foreign powers wished to intervene in France to bring the Revolution to an end. The war went badly and the King was suspected of sympathising with the Austrians. So, in August 1792, an insurrection overthrew the King and a new governing body, the Convention, was set up which formally abolished the monarchy and declared a republic in September 1792. A few months later the Convention decided to try the King and in January, 1793, voted narrowly, by 387 to 310, for his execution by guillotine, which was a new and more humane method of execution introduced by the Revolution. The execution took place within a few days on 21 January 1793, and led almost at once to an extension of the war. In February 1793, France was at war with England as well as with Austria and Prussia. Fig. 4.3.5 shows the execution as it was depicted on an English poster of the time.

Now that the war had spread, conditions in France again worsened and the Convention adopted new measures to try to assure the survival of the Revolution

MASSACRE of the Unfortunate FRENCH KING,
with a View of LA GUILLOTINE, or the Modern French
BEHEADING MACHINE.

1 The Monarch.   3 General Santerre.
2 His Confessor. 4 The Mayor of Paris.

Published by Alex.r Hogg., April 1.17.93.

Fig. 4.3.5   The execution of Louis XVI by guillotine on 21 January 1793 as depicted in an English poster of the time.

against enemies at home and abroad. A Committee of Public Safety was set up which, amongst other things, tried to mobilise scientists to use their scientific knowledge in defence of the republic. Month by month, the atmosphere grew more tense and the Convention became more extreme. In August 1793, the Convention abolished the Academy of Sciences, along with other learned academies of France, as being incompatible with a republic. This was the culmination of a long campaign against the Academy and certain of its members, such as the chemist Lavoisier, since the early days of the Revolution. But, according to the historian of science, Charles Gillispie, another factor was the nature of science itself. Gillispie sums up the Convention's attitude thus:

> Science was undemocratic in principle, not a liberating force of enlightenment, but a stubborn bastion of aristocracy, a tyranny of intellectual pretension stifling civic virtue and true productivity.[7]

This hostility to science, however, did not extend to natural history. This was seen as less arrogant and more popular than the other sciences—as essentially democratic in nature. The Museum of Natural History in Paris was one of the few learned institutions of the old order to prosper under the Revolution.

Having abolished organised science almost in passing, the Convention turned to wider matters. On 17 September 1793, they decreed the Law of Suspects of which the key part states: 'The National Convention . . . decrees that all suspected persons be arrested'. This effectively removed the rights of individuals, set out in the Declaration of 1789, and marked the start of the period of ten months or so which is known as the Terror. During this time probably more than 10,000 people died by guillotine.

Two distinguished scientists were victims of the Terror. One was the *philosophe* Condorcet, a good mathematician who had been permanent secretary of the Academy of Sciences since 1776 and who was, like a true *philosophe*, a persistent advocate of the rational reorganisation of science and of scientific education. Condorcet never saw this reorganisation in his life-time but, a few years later when the Terror was over, French science and education were transformed by the new government in the way that he had proposed. He died in prison in March 1794.

The second scientific victim of the Terror was Antoine Lavoisier—probably the greatest chemist France has ever produced (see Chapter 4.5). The story is told that at his trial Lavoisier appealed for time to complete some scientific experiments and that the judge replied: 'The Republic has no need of scientists'. So, at the age of 50, Lavoisier went to the guillotine on 8 May 1794.

The Terror aimed to intimidate opponents of the Revolution but other methods were also tried in the attempt to save it. One development was a Cult of Reason which tried to provide an organised non-religious alternative to the Catholic Church. Many churches were turned into 'Temples of Reason' and in November 1793, a Festival of Reason took place in the Church of Notre Dame in Paris; another Feast of Reason was held in the Cathedral at Chartres.

Another development was a new Revolutionary Calendar which was seen as a clear break with the old Christian calendar. It was backdated to start on 22 September 1792—the date when the monarchy was abolished and the Republic set up. The period from 22 September 1792 to 21 September 1793 was Year I of the Republic. The new year had twelve months of 30 days, all of which were given new names taken from nature; Fructidor—the month of fruit, Floreal—the month of flowers, Ventose—the month of wind and so on. This left five days (365−[12 × 30]) which were declared holidays to celebrate the Revolution. The seven-day week was abolished, and each month was divided into three periods of ten days with every tenth day being a day of rest. Even with the five extra days, this still left the Revolutionary Calendar with fewer rest days than the Christian one, and it was not popular. So extra feast days had to be introduced. The new calendar lasted about 12 years until it was abandoned by Napoleon in 1806.

Meanwhile in 1794 the Terror continued and the Cult of Reason was spreading and proving in many ways a disruptive force. The principal leader at this time in the Convention was Robespierre who spoke out against the atheism of the Cult of Reason and in favour of a new Cult of the Supreme Being. Jacques-Louis David, the man who had organised the Festival of Voltaire, was commissioned to organise a Festival of the Supreme Being, which took place on 8 June 1794. This festival culminated with Robespierre inaugurating the new religion by lighting an enormous bonfire which was to purge the world of all its ills (see Chapter 4.2, p. 114). At this time Robespierre was also responsible for increasing the severity of the Terror until in the end on 28 July 1794, he himself was overthrown and sent to the guillotine and the Terror ended.

SCIENTIFIC INSTITUTIONS AFTER THE FRENCH AND AMERICAN REVOLUTIONS

After the Terror the new government in France soon made three innovations which had long-lasting effects on French science and the way it was organised. In 1795 they set up a national system of secondary schools in which the curriculum emphasised mathematics and science. These schools had chemical and physical laboratories and strongly resembled the schools advocated in Condorcet's Scheme for National Education which had been turned down by the Convention in 1792.

And in 1795 the government also reconstituted the *Académie des Sciences* under a new name the *Institut National de France*. This new *Institut* had divisions dealing with the arts and with the natural and social sciences.

But probably the best known new scientific institution of this period was the *École polytechnique*, which was established in Paris in 1794 in order to train engineers for the defence of the Republic. The aim was to make the *École polytechnique* an institution 'without parallel in Europe', and it was equipped with a library, workshops and laboratories for both teaching and research. Its staff included some of the best mathematicians and scientists of the time—for example, the mathematicians Laplace and Lagrange and the chemist Berthollet. Its students became the best trained engineers in the world and some of them, such as

Fresnel in optics and Carnot in thermodynamics (see Chapter 4.9), went on to become distinguished scientists of the first rank.

Under Napoleon many of these institutions were modified and brought under more centralised control. Napoleon created the Imperial University—a unified system of secondary and higher education for the whole of France which was directed from Paris. By 1804 Napoleon had also turned the *École polytechnique* into a military academy producing officers for the army; the curriculum became much more practical and the budgets for research and for laboratories were severely cut. And Napoleon even abolished the Second Class of the *Institut National* which was responsible for the social sciences. These new centralised institutions, deriving from the Revolution, are still of major importance in France right down to the present day.

But in America the legacy of revolution was very different. The American Constitution was greatly influenced by Montesquieu's principle of the separation of powers as a check on centralised government. The result was a system of checks and balances on the exercise of power, and many powers were retained by the individual states of the republic. This meant that George Washington's proposals for an American National University were never accepted, and it was not until nearly a hundred years later that the National Academy of Sciences was set up in 1863 during the crisis brought about by the American Civil War.

However, universities were set up in some of the individual states. For example, Thomas Jefferson was influential in setting up the University of Virginia, which strongly emphasised science and mathematics in its curriculum. Like a true man of the Enlightenment, Jefferson even designed all the buildings for the new university. When Jefferson died, he left instructions that on his tomb was to be inscribed this sentence, 'and not a word more': 'Here was buried Thomas Jefferson, author of the Declaration of American Independence, of the Statute of Virginia for Religious Freedom, and Father of the University of Virginia.' But on the Jefferson Memorial in Washington, as well as some noble phrases from the Declaration of Independence, are inscribed these less familiar words: 'I tremble for my country when I reflect that God is just, that his justice cannot sleep forever. Commerce between master and slave is despotism. Nothing is more certainly written in the book of fate than that these people are to be free'.

Jefferson's prophetic words both here and in the Declaration of Independence still speak to us today, nearly 200 years later. They show that ideas have consequences and that the ideas of the 18th-century *philosophes* are still a potent force in shaping the modern world.

REFERENCES

1 Diderot, D., in the article *Encyclopedia* in Diderot, D. and D'Alembert, J., *Encyclopédie*, (Paris: 1755–67).
2 Condorcet, quoted in Behrens, C. B. A., *The Ancien Régime*, (London: Thames & Hudson, 1967), p. 124.
3 Behrens, C. B. A., op. cit., pp. 126–7.

4   Ibid. p. 124.
5   See Reference 9, Chapter 4.2 (page 116).
6   Translation in Paine, T., *Rights of Man* (Harmondsworth: Penguin, 1969), p. 132.
7   Gillispie, C. C., 'The Encyclopédie and the Jacobin Philosophy of Science: A Study in Ideas and Consequences' in Clagett, M. C. (ed.), *Critical Problems in the History of Science*, (Madison: University of Wisconsin Press, 1959), p. 257.

SUMMARY

1   The American and French revolutions drew much of their inspiration from the *philosophes* of the Enlightenment and in particular from their emphasis on the role of reason in human affairs and their vision of a better future.

2   The American Declaration of Independence (1776) emphasised 'that all men are created equal', that their 'unalienable rights' include 'Life, Liberty and the pursuit of Happiness' and that governments derive 'their just powers from the consent of the governed'.

3   The French Declaration of the Rights of Man and of Citizens (1789) stated 'that ignorance, neglect or contempt of human rights are the sole causes of public misfortunes and corruptions of Governments', and declared that 'Men are born, and always continue, free and equal in respect of their rights' which are 'Liberty, Property, Security and Resistance against Oppression'.

4   As the Revolution developed, conditions in France got worse and many of these rights disappeared until, during the Terror of 1793–4, no-one was safe.

5   Initially the French Revolution made use of science and scientists, for example in the development of the new metric system. Later a more hostile atmosphere developed, the Academy of Sciences was abolished and Lavoisier went to the guillotine. But when the Terror was over, the new government encouraged science and technology and created many new centralised scientific institutions and colleges.

6   The new American constitution laid great emphasis on the balance of powers and the rights of individual states. This emphasis was reflected in the development of American universities and scientific institutions.

4.4   The early Industrial Revolution

In this chapter we are going to look at the early stages of the Industrial Revolution which took place in England during the 18th century.

Before the Industrial Revolution, most people lived in the countryside and worked in agriculture. After it, many more people lived in towns and worked in the manufacturing industry. These changes are illustrated in Fig. 4.4.1 which shows the start of the Industrial Revolution in various countries together with the proportion of the population engaged in agriculture. Even before 1800 less than 40 per cent of the population in England were working on the land and by soon after 1850 this proportion had fallen below 20 per cent.

How did this dramatic shift to the towns come about? And how far was it due to the increase in scientific knowledge arising from the scientific revolution of the 17th century? We will try to answer these questions later. But first we need to describe some of the changes which took place during the 18th century in Britain and, in particular, to discuss those three pillars of the Industrial Revolution—Cotton, Iron and Steam.

Fig. 4.4.1   The Industrial Revolution and the proportion of the population employed in agriculture. *Redrawn from Cipolla, C. M., The Economic History of World Population, figure 4, page 27, copyright © Carlo M. Cipolla, 1962, 1964, 1965, 1967, 1970, 1974, 1978, reprinted by permission of Penguin Books Ltd.*

MACHINES IN THE ENGLISH COTTON INDUSTRY

Around 1720 cotton cloth was made in England in much the same way as it had been for generations. Raw cotton was spun into yarn on hand-operated spinning wheels (see Fig. 4.4.2) and then woven, again by hand, into cloth. The work could be done in people's homes because the machines used were simple and small enough to fit easily into a rural cottage.

By 1820, all this had changed. Machines for both spinning and weaving (see Figs. 4.4.5 and 4.4.6) were now much larger and more complex and they had to be driven by external sources of power like water wheels or steam engines. Numbers of spinning and weaving machines were now brought together around a single central source of power. The pattern of the classical Victorian cotton factory had started to emerge. And it was the growth of these factories which increased the consumption of cotton in Britain from about 10 million pounds a year in 1780 to nearly 500 million pounds in 1840.

By what intermediate stages did this dramatic change from cottage industry to factory take place? The conversion of raw cotton into woven cloth is a complex process involving a number of separate operations. Spinning is particularly difficult to mechanise as it involves three separate stages. First the tangled fibres in the raw cotton are straightened in a process known as carding. Then the fibres are put together into coarse ropes known as rovings. Finally they are spun into thread suitable for weaving. Fig. 4.4.2 shows a traditional spinning wheel in operation in a cottage; note that only one bobbin of thread is being spun. And Fig. 4.4.3 shows weaving on a narrow handloom. The shuttle carrying the cross-threads or weft had to be thrown by hand which limited the width of the cloth to about three feet.

It was weaving that changed first, with the invention of Kay's Flying Shuttle in 1733. This shuttle could be propelled across the loom by jerking on a cord and it greatly increased the efficiency of weaving. The widespread use of the flying shuttle from about 1760 disturbed the equilibrium between spinners and weavers and thus stimulated attempts to improve the productivity of spinning machines. Many different machines were tried and perhaps the most important were those like Hargreave's Spinning Jenny of 1768 in which several bobbins of thread could be spun simultaneously. As the number of bobbins increased it became necessary to drive the machines by water power rather than by hand. By about 1771 Richard Arkwright (Fig. 4.4.4) had succeeded in mechanising the whole spinning process in a single factory. Arkwright brought together numbers of carding, roving and spinning machines all driven by gears and belts from a common water wheel. Fig. 4.4.5 shows spinning machines in a typical cotton factory; note the number of bobbins which are being spun simultaneously. Power driven looms for weaving took a little longer to develop but by about 1800 looms like those shown in Fig. 4.4.6 were in use; note the driving belts above the machines. So, by a process of piecemeal development, the machinery used in the English cotton industry gradually grew in complexity, size and precision, and as they developed, the new machines irrevocably changed the rural pattern of 18th-century life.

Fig. 4.4.2 Traditional hand-operated spinning wheel on which only one bobbin at a time can be spun.

Fig. 4.4.3 Weaving on a hand-operated narrow loom. *From the Rickitt Encyclopaedia of Slides* (*The Industrial Revolution, 1179/8*).

Fig. 4.4.4   Richard Arkwright.

THE GROWTH IN IRON PRODUCTION

Drastic changes also took place in the manufacture of iron in the 18th century. As with cotton manufacture, there were many new techniques, an increase in the scale of operation and new sources of power to operate the larger machines.

Traditionally in Europe iron had been produced in charcoal-fired blast furnaces fanned by bellows worked by water-power (see Fig. 4.1.9, page 100). This traditional method gave good quality iron but had several disadvantages. Charcoal crumbles easily, which limits the size of furnace. And, if the supply of water dried up, the bellows stopped and the fire went out. So the production of iron was limited and intermittent. Also, timber for charcoal was becoming scarcer—there was an energy crisis in the 18th century too. Attempts had been made to use coal instead but the iron produced was too brittle to be of any use.

Then in 1709, at Coalbrookdale in Shropshire, Abraham Darby opened a new kind of blast furnace which used coke as a fuel. Since coke is considerably stronger than charcoal, much larger furnaces could be used. But the iron from Darby's furnaces was still too brittle to be easily forged, and so he developed the technique of casting iron in moulds—initially on a small scale for domestic pots and pans and later on a larger scale for the boilers and other metal parts of the early steam engines.

Fig. 4.4.5 Spinning machines in an early spinning factory; note the belt-driven machines and the number of bobbins which are being spun simultaneously (cf. Fig. 4.4.2).

Fig. 4.4.6 Power-driven looms from around 1800; note the driving belts above the machines.

Fig. 4.4.7 Iron road bridge at Coalbrookdale. 1779.

Fig. 4.4.8 Coalbrookdale ironworks by night; painting by De Loutherbourg.

The technique of cast iron production was continually improved during the 18th century and the scale of production increased. This led to completely new uses for iron as in the construction of the famous iron bridge across the Severn at Coalbrookdale (see Fig. 4.4.7). This bridge took four years to build and was completed in 1779. Its span was 100 feet and it contained nearly 400 tons of iron castings. To build it, the Coalbrookdale iron works were considerably expanded and must have been a startling and extraordinary sight in the surrounding rural countryside, particularly at night (see Fig. 4.4.8).

The Coalbrookdale iron bridge was a road bridge, but large-scale iron castings were also used in aqueducts for the extensive network of canals which was built in Britain from 1750 to 1800 (see Fig. 4.4.9). Later, during the railway boom of the 19th century—from 1830 onwards—iron was a major material, being used for bridges, rails and the engines themselves (see Chapter 4.9 and Fig. 4.9.3, page 189).

THE STEAM ENGINE—A NEW SOURCE OF POWER

The increase in the scale of production of both iron and cotton would not have been possible without the introduction of a new source of power to replace the water-wheel. This brings us to the third major development, after cotton and iron, in the early Industrial Revolution—steam power.

A steam engine of a sort had been devised by a Greek, Hero of Alexandria, in the 1st century AD, just before Ptolemy. But Hero's engine couldn't do significant amounts of work and was little more than a toy. The modern steam engine grew from two roots. The first root was the recognition of the large forces exerted by the atmosphere, as for example in von Guericke's experiments in the 17th century (see Chapter 3.2 and Fig. 3.2.7, page 52). The problem was how to harness these forces.

The second root was the understanding of air pressure which came from Torricelli's work on early barometers (see Chapter 3.2 and Fig. 3.2.6, page 51). Torricelli's work on water barometers showed that the pressure of the atmosphere could balance a column of water just over 30 feet in height. This finding explained a fact which had been known since antiquity—that traditional water pumps, which in effect make use of the pressure of the atmosphere, can only raise water through a height of about 30 feet.

One way to overcome this limitation was to use pressures greater than one atmosphere, and this was done by Thomas Savery in 1698 in a steam engine designed to pump water from mines. Savery's engine worked in two stages. First of all it raised water through 30 feet by utilising the pressure of the atmosphere. Then it pumped the water through a considerably greater height by means of high pressure steam generated in an iron boiler.

In principle, the higher the steam pressure, the higher the water could be pumped. In practice Savery's engine could pump water to about 150 feet, five times higher than an atmospheric pump, at a rate of about 1 horse-power. Even then the boilers often burst or came apart at the seams. The iron technology of the

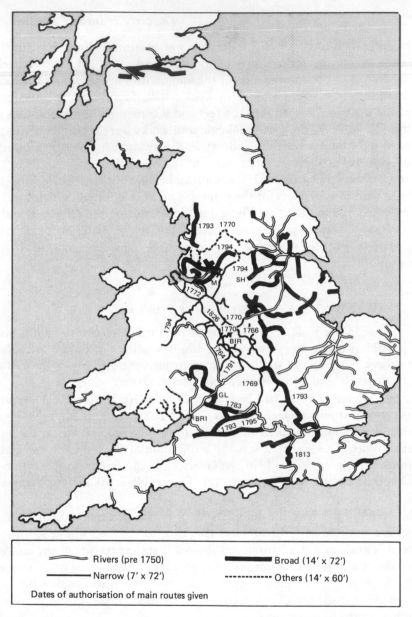

Fig. 4.4.9 Canals and navigable rivers in 1830 in England and Wales, with dates of building. *Redrawn from Open University Course A202, The Age of Revolution, Units 5–6 The Industrial Revolution, copyright © The Open University Press.*

time wasn't able to do any better. For this reason it is unlikely that many such engines were built!

A safer and better steam engine for pumping water was developed by Newcomen in 1712. Fig. 4.4.10 shows an 18th-century engraving of a Newcomen engine in operation. Its most prominent feature is the massive beam. The left-hand end of this beam is connected to the underground water pump. The other

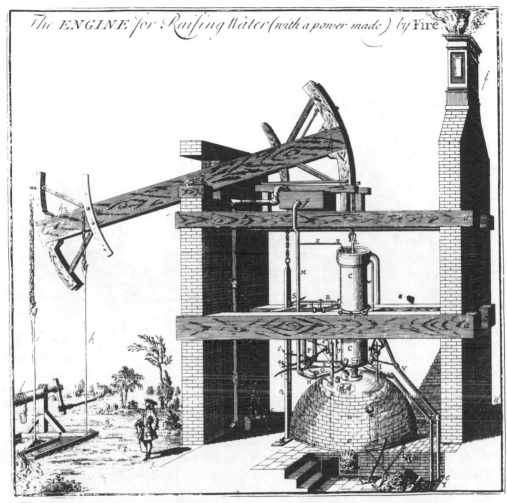

Fig. 4.4.10 18th-century engraving of a Newcomen water pumping engine in operation. Note the massive beam which was such a prominent feature of nearly all the early steam engines.

end of the beam is driven by the steam engine on the right, which mainly consists of a single large cylinder and a steam-raising boiler at the bottom. The piston in the cylinder is connected to the beam. At rest the beam is in the position shown and the piston is at the top of the cylinder. In the first stage of a pumping cycle the cylinder is filled with steam from the boiler. Then cold water is sprayed into the cylinder causing the steam to condense and creating a partial vacuum inside the cylinder. Air pressure then forces the piston down to the bottom of the cylinder and this movement in turn brings the right-hand end of the beam down. The beam then returns to its original position under its own weight and operates the underground water pump. This engine was safer than Savery's engine because it didn't use high pressure steam—the piston is brought up by the beam and not by steam pressure. And, by making the cylinder larger, the power of the engine could

STEAM ENGINE.   *PLATE III*

M? *WATT'S ENGINE.*

Fig. 4.4.11   Steam pumping engine designed by James Watt for the Chelsea Water
Works.

be increased considerably. Large cast-iron cylinders from the Coalbrookdale
Ironworks were used in Newcomen engines all over England in the 18th century.
Their power was also increased by introducing a control rod which moved up and
down with the beam. Projections on the control rod operated the valves which
turned the water and steam on or off as required. This early form of automatic
control enabled a whole pumping cycle to be completed every five seconds, giving
a typical power of about five or six horse-power. These water-pumping engines
were very successful but, as can be seen from Fig. 4.4.10, they were very bulky
machines and this limited their application to other tasks.

   The main source of inefficiency in Newcomen's engines was the repeated
alternate heating and cooling of the main cylinder, and it was James Watt (1736–
1819) who devised an engine which avoided this wasteful procedure. Watt was
originally an instrument maker who realised the inefficiency of Newcomen's
engine when he was asked to repair a model of one at the University of Glasgow.
Watt's solution, which he patented in 1769, was to use two cylinders, not one. One
cylinder was always kept hot and the other, which he called the condenser, always
cold. Fig. 4.4.11 shows a Watt pumping engine built for the Chelsea Water
Works. This engine has many features in common with Newcomen's engine
(Fig. 4.4.10): the boiler G raising steam, the main cylinder Q with its piston
connected to the end C of the massive beam C B A, the other end of which, A,
drives the pump. But in Watt's engine the partial vacuum in the main cylinder
Q is created not by cooling the cylinder but by connecting it to the cold condenser

cylinder N, which in turn is pumped out by an air pump. Watt's engine also used hot steam, not cold air, to force down the piston. It was about three times as efficient as Newcomen's engine.

Watt's steam engine attracted the attention of a man called Matthew Boulton (1728–1809) who manufactured metal objects such as buttons and ornaments in a small factory in Birmingham. Boulton's factory was powered by water-wheels but these stopped turning whenever the streams dried up. So Watt's new and efficient source of continuous power interested him, and he wrote to Watt to suggest that they work together:

> I was excited by two motives to offer you my assistance which were love of you and love of a money-getting ingenious project. I presumed that your engine would require money, very accurate workmanship and extensive correspondence to make it turn out to the best advantage and that the best means of keeping up the reputation and doing the invention justice would be to keep the executive part out of the hands of the multitude of empirical engineers, who from ignorance, want of experience and want of necessary convenience, would be very liable to produce bad and inaccurate workmanship; all of which deficiencies would affect the reputation of the invention. To remedy which and to produce the most profit, my idea was to settle a manufactory near to my own by the side of our canal where I would erect all the conveniences necessary for the completion of engines, and from which manufactory we would serve all the world with engines of all sizes. By these means and your assistance we could engage and instruct some excellent workmen (with more excellent tools than would be worth any man's while to procure for one single engine), could execute the invention 20 per cent cheaper than it would be otherwise executed, and with as great a difference of accuracy as there is between the blacksmith and the mathematical instrument maker. It would not be worth my while to make for three counties only, but I find it very well worth my while to make for all the world.[2]

So in 1774 Watt moved to Birmingham, and the next year he and Boulton became partners. Together they set up a factory which produced all kinds of steam engines at the rate of about 20 a year for the next 30 years or so. It was the first great engineering works in the world. The iron parts of these engines may well have been cast at the Coalbrookdale iron works, which was much used by Boulton and Watt. And the cylinders were usually made by the Staffordshire ironmaster, John Wilkinson, who had devised a much improved method of accurate boring. Fig. 4.4.12 shows a later Boulton and Watt engine, dating from 1788, in which the up and down motion of the beam is converted into a rotary motion. It was developments like this that led to the widespread use of steam engines in iron works and in the early textile factories. Instead of pumping water into a reservoir to turn a water-wheel, rotatory motion could be obtained directly from the steam engine.

These engines were not really very powerful or efficient, as can be seen from Fig. 4.4.13, which shows the fuel consumptions and efficiencies of many different

Fig. 4.4.12 Boulton and Watt rotative engine of 1788 now in the Science Museum, London.

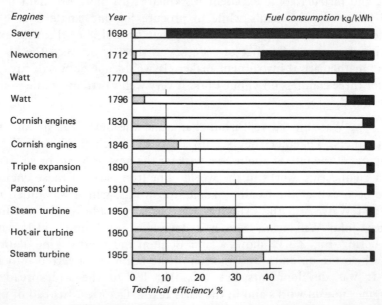

Fig. 4.4.13 Improvements in the efficiency and fuel consumption of steam engines, 1698–1955. *Redrawn from Cipolla, C. M., The Economic History of World Population, figure 7, page 61, copyright © Carlo M. Cipolla, 1962, 1964, 1965, 1967, 1970, 1974, 1978, reprinted by permission of Penguin Books Ltd.*

kinds of steam engine. Watt's engines were clearly much more economical than those of Savery and Newcomen but were relatively inefficient by modern standards. But, above all, what the new steam engines did offer was a continuous source of power—not subject to the changes of the weather and seasons, like windmills and water-wheels, nor to fatigue like human or animal muscle-power. As Boulton was reputed to say to visitors to his factory: 'I sell here, sir, what all the world desires to have—power'. And it was an increasingly versatile source of power which could be used almost anywhere.

THE MEN OF THE INDUSTRIAL REVOLUTION

Most of the dramatic new developments in the early Industrial Revolution were pioneered by a number of remarkable men. These men were all adventurous, determined, adaptable and confident. But above all they were supremely practical. They were all very highly skilled technically and they had a kind of genius for actually making things work. The success of Richard Arkwright (Fig. 4.4.4) in simultaneously mechanising all the diverse processes involved in cotton spinning is an example of this practical genius. The confidence of these men is well illustrated by John Wilkinson, the famous ironmaster who made many of the precisely bored cylinders for Boulton and Watt's steam engines. Fig. 4.4.14 shows a coin, minted by Wilkinson himself in 1788, which he used to pay his workers. On it is imprinted not a royal head but Wilkinson's own face and the legend 'John Wilkinson, Ironmaster'! And consider the confidence of Matthew Boulton when he wrote to Watt—'It would not be worth my while to make for three counties only, but I find it very well worth my while to make for all the world'.

Many of these men were members of an informal society which met every month in Birmingham to discuss technical and scientific matters. The society was

Fig. 4.4.14 Wage token of 1788 showing the head of John Wilkinson, Ironmaster. Wilkinson minted these tokens himself and paid his workers with them.

called the Lunar Society—not because the members were lunatics but because some of them came a considerable distance to the meetings and it was safer going home on a moonlit night. So they always met near the full moon. The society was not large. Over the years it had little more than a dozen members. But it had contacts with nearly all the important new technologies throughout the country. The driving spirit of the Lunar Society was Matthew Boulton, and other members included James Watt, the famous pottery manufacturer Josiah Wedgwood, Erasmus Darwin, the grandfather of Charles Darwin (see Chapter 4.8) and the chemist Joseph Priestley (see Chapter 4.5). Priestley, as well as being John Wilkinson's brother-in-law, was scientific adviser to Wedgwood, who in turn was no mean experimentalist. Wedgwood was made a Fellow of the Royal Society for his work on the measurement of high temperatures which he did in trying to improve the furnaces used in his potteries. And it is worth emphasising that it was not his highly decorated ornamental pottery which made Wedgwood prosperous. He mainly manufactured white earthenware pottery, called creamware, which was produced on a large scale for everyday common use.

It was mass-produced objects like these, together with new and cheaper textiles and products made of cast iron for the home and kitchen, which eventually were to transform the homes and lives of millions of people during the Industrial Revolution. Much of the inspiration for these changes came from the vision and determination of men like those who made up the Lunar Society of Birmingham in the second half of the 18th century.

CONCLUSION

Having described some of the events of the early Industrial Revolution, let us now try to answer the questions we posed at the beginning of this chapter.

First, how did the shift to the towns come about? It happened primarily because of a change of scale in all the techniques we have looked at. Local crafts began to change into larger scale industries and the first factories started to emerge. New techniques led to an increase in specialisation and hence to the division of labour and also to the beginnings of mass production. All these changes helped the towns to grow but it is important to realise that the growing towns depended on the countryside for food. Without gradual but substantial improvement in agriculture, the Industrial Revolution could never have taken place.

Secondly, how much were the changes of the early Industrial Revolution due to increased scientific knowledge? The answer here is—not very much. In the cotton industry, the advances were largely empirical. In iron manufacturing, little was known or understood of the underlying chemistry involved until the second half of the 19th century. And the fundamental physical principles governing the efficiency of steam engines were only just beginning to be understood more than 50 years after Watt's first patent of 1769—in the work of Carnot in France, who laid the foundations of thermodynamics in the 1820s (see Chapter 4.7).

What was important in the early Industrial Revolution was not scientific knowledge, particularly theoretical knowledge, but the empirical and experimental scientific attitude—a combination of inspired trial-and-error and increasing control of practical techniques. The key factor was the sort of experimental science aimed at practical usefulness which had been strongly advocated by Francis Bacon early in the 17th century and which was later taken up, with due acknowledgement to Bacon, by the French *philosophes* and encyclopedists in the 18th century. But it was not in Paris, the centre of civilised Europe, that these developments actually took place. In France they were talking and writing about the beneficial changes which they thought empirical advances would bring. Yet, at the very same time, the dramatic changes we have looked at were actually happening in England. And they were happening not because the government in London intended them to happen but as a result of spontaneous initiatives by many ingenious individuals in the provinces of England.

REFERENCES

1 Cardwell, D. S. L., *From Watt to Clausius*, (London: Heinemann, 1971), pp. 13–15.
2 Letter from Matthew Boulton to James Watt; quoted in *James Watt and Steam Power*, Jackdaw No. 13, Jackdaw Publications Ltd.

SUMMARY

1 The Industrial Revolution started in England during the 18th century as a result of major changes in the manufacture of cotton and iron, and the development of steam power.
2 The changes in cotton manufacture were due to a series of technical inventions which increased the scale of operation and improved the quality of the product.
3 In iron manufacture the use of a new fuel, coke, and the development of the technique of casting led to a significant increase in the scale of iron production, as was shown most dramatically in the new iron bridges and aqueducts.
4 The development and large-scale production of more efficient steam engines by Watt and Boulton gave, for the first time, a versatile and continuous source of power which hastened the emergence of the early factories.
5 All these changes were largely empirical; they owed little to previously existing scientific knowledge but much to enterprising, even inspired, trial-and-error, and to continually improving levels of skill and craftsmanship.

During the late 18th and early 19th centuries—from about 1770 to 1840—the broad outlines of many of our modern sciences began to emerge. Before this time natural philosophers studied and wrote about any aspect of the natural world which caught their interest. The same person might range from mathematics to chemistry or natural history and from the most speculative and theoretical work to the most practical and applied. Galileo and Newton are obvious examples, but they are not isolated cases. The scientific and technical interests of many other less gifted people were just as wide. This can be seen in the very broad activities of the early scientific societies like the Royal Society and the French Academy of Science (see Chapter 3.5)—or in the pages of the scientific journals of the time. In Fig. 3.5.4 (page 84) we saw that the contents of the *Philosophical Transactions of the Royal Society* in 1685 ranged far and wide. A hundred years later, the spread of subjects was nearly as great.

But after the period from about 1770 to 1840 this wide range of interests became much less common. Scientists started to specialise more—indeed, the very word 'scientist' in its modern sense only dates from the 1830s (see Chapter 3.5, Note 1). And many of our modern sciences first began to be recognisably distinct fields of investigation at about this time.

In the next three chapters we are going to describe, in turn, some of the origins and roots of modern chemistry, of geology, and of much of physics—in particular the study of energy and of electricity and magnetism. Then we will discuss what is probably the most important scientific event of the 19th century—the emergence of the theory of evolution in the work of Charles Darwin and Alfred Wallace.

4.5   The origins of modern chemistry

Modern chemistry has many roots. Some of the most important were empirical recipes for dyes and paints, the use of chemicals in medicine as in the iatro-chemistry of Paracelsus (see Chapter 6.4), the smelting and working of metals and, of course, alchemy (see Chapter 4.1).

All these activities were very ancient, as was Aristotle's theory that everything was made of four elements—Earth, Air, Fire and Water. Fire in particular was regarded as the transforming element, and it played a large part in people's imagination and myths, as well as in the practical techniques of metal working and alchemy. So it was oddly appropriate that it was a series of remarkable discoveries about the nature of Fire—or combustion—between 1770 and 1800 which laid the foundations of modern chemistry. And these discoveries also established the nature of two more of Aristotle's elements, Air and Water.

These new discoveries were partly due to improved practical techniques of gas collection and identification, and of quantitative analysis, both by weight and by volume. But they were also due to the individual contributions of three remarkable men—Joseph Priestley (1733–1804), Antoine Lavoisier (1743–94) and John Dalton (1766–1844).

PRIESTLEY AND THE NATURE OF AIR

Priestley came from the north of England and was a Dissenter, that is, a member of one of the Christian groups—Methodists, Presbyterians, Quakers or Unitarians—which had split away from the Church of England.

These groups were not permitted to attend Church schools or the ancient universities of Oxford and Cambridge. So they set up their own schools and colleges, called Dissenting Academies, and Priestley became a successful lecturer at such an academy. He was also an ordained minister and a well-known preacher. Priestley wrote on theology, history, education, language and politics as well as on scientific subjects, and in his own day he was as well known for his controversial religious and political views as he was for his science. In the 1780s Priestley settled in Birmingham at the suggestion of his brother-in-law, John Wilkinson, the ironmaster, and became a member of the Lunar Society, that group of intellectuals, scientists and industrialists which included Matthew Boulton, James Watt and Josiah Wedgwood amongst its members (see Chapter 4.4). In Birmingham Priestley continued to voice his criticisms of the established Church and he was an enthusiastic supporter of the early stages of the French Revolution. Partly as a result of this, his house and laboratory in Birmingham were destroyed by a mob in 1791 (see Fig. 4.5.1) and continuing hostility led, in 1794, to his emigrating to the newly independent United States of America. That is why Priestley's chemical apparatus is now in the Smithsonian Museum in Washington.

Priestley's main scientific work was in the study of gases or 'airs' as they were usually known at the time. In the early 18th century it began to be realised that the ordinary 'air' of the atmosphere was not the only possible kind of air. 'Fixed Air', or carbon dioxide as we call it, was the first such air to be identified. This was done by two techniques. The first was the collection of a gas by displacement of water in a pneumatic trough (see Fig. 4.5.2). No more would elusive gases necessarily vanish in an experiment. They could be caught, their properties tested and their volumes measured. The second technique was careful weighing, as first adopted by Joseph Black of Edinburgh and reported in his famous book of 1758, *Experiments upon Magnesia Alba, Quicklime and some other Alcaline Substances*. In a typical experiment Black dissolved 120 grains of chalk in 421 grains of what we now call hydrochloric acid and found that the weight of 'fixed air' lost by effervescence was 40 per cent. Then Black took another 120 grains of chalk and found that it lost 43 per cent of its weight when heated to form quicklime. And

Fig. 4.5.1　The looting and burning of Priestley's house in Birmingham by a mob in 1791.

Fig. 4.5.2　Priestley's pneumatic trough for the collection of 'airs'.

this quicklime, in turn, required 414 grains of hydrochloric acid to neutralise it. Within the limits of Black's errors, the weights balanced. As one historian put it, '. . . with Black . . . the balance as the symbol of the chemist's science displaced the still and retort as symbols of the chemist's craft'.[1]

Black seldom collected fixed air or studied its properties directly. But later workers did, and they also collected other 'airs' such as the 'inflammable air' (modern hydrogen) which is produced when oil of vitriol (sulphuric acid) reacts with iron or tin or zinc. The pneumatic trough was also made more versatile by using mercury instead of water.

But more than anyone else it was Priestley who seized on this new area of study and made it his own. He set out to discover the nature of all the 'airs' that might be released from substances. Between 1774 and 1786 Priestley published various volumes of his book *Experiments and Observations on Different Kinds of Air*, in which he described the preparation and properties of a whole range of gases, most of them identified for the first time. These gases included what we call carbon dioxide, hydrogen, nitric oxide, nitrous oxide, nitrogen dioxide, hydrochloric acid, sulphur dioxide, ammonia and silicon tetrafluoride, or, as Priestley called them, fixed air, inflammable air, nitrous air, dephlogisticated nitrous air, red nitrous vapour, marine acid air, vitriolic acid air, alkaline air and fluor acid air.

Priestley's work was mainly descriptive. He sometimes measured volumes carefully but he seldom used the balance. Usually he prepared the gases, collected them over water or mercury and then identified them by simple tests—smell, taste, would they burn or support combustion, did they turn lime-water milky, how long would a mouse live in them? Fig. 4.5.2 shows some of Priestley's apparatus with a mouse in one of the jars of gas.

Priestley's most famous discovery was oxygen, which he made by heating the red calx of mercury, or mercuric oxide as we call it. Similar experiments had been done many times by the alchemists, although they never collected the emitted oxygen as Priestley did. But Priestley didn't call it oxygen. His name for it was 'dephlogisticated air', a name which derives from an earlier theory of combustion, the phlogiston theory. It was Lavoisier who named the new air oxygen and made it the basis of his new chemical system.

LAVOISIER AND THE NEW CHEMISTRY

Lavoisier's first interest was in geology, and he made some interesting suggestions about the action of water in the deposition of different rock layers. Water fascinated Lavoisier—he called it 'the favourite agent of nature'. Later in life, when he was a member of the Academy of Sciences, he worked on the problem of supplying Paris with water. But it is his chemistry that made Lavoisier famous and here too the properties and nature of water played an important part. Fig. 4.5.3 shows Lavoisier with his wife and some chemical apparatus in a portrait by Jacques-Louis David.

The key to Lavoisier's chemistry was the balance. In his *Elements of Chemistry*, published in 1789, he writes:

Fig. 4.5.3   Lavoisier with his wife and some chemical apparatus in a portrait by Jacques-Louis David.

We must lay it down as an incontestable axiom that in all the operations of art and nature, nothing is created; an equal quantity of matter exists both before and after the experiment.[2]

In other words Lavoisier assumes the conservation of matter.

This quantitative emphasis is shown in Lavoisier's careful work on the increase in weight of metals and other substances, such as sulphur and phosphorus, when they are heated in air. He noted that this increase seemed to be associated with absorption of air and he determined to study these matters further. In 1773, at the beginning of a new research notebook, Lavoisier wrote that he was going to carry out a 'long series of experiments . . . on the elastic fluid that is set free from substances, either by fermentation, or distillation or in every kind of chemical change, and also on the air absorbed in the combustion of a great many substances'. He added that 'this work . . . seemed to me destined to bring about a revolution in physics and chemistry'. It sounds like Newton speaking. Like Newton, Lavoisier was not modest about his abilities but, also like Newton, his achievements over the next 20 years were equal to his claims.

These experiments on combustion soon led Lavoisier to work with mercury and its red calx. Just as Priestley had done, he heated red mercuric oxide and studied the gas which it gave off. There has been much dispute as to whether Lavoisier was already working on these experiments when Priestley met Lavoisier in Paris in October 1774 and talked about the new 'air' he had prepared. In any event, although he probably learnt much from Priestley, there is no doubt that Lavoisier took the experiment much further in ways that were typical of his methods of chemical experiment. As always Lavoisier's first principle was to make the experiment quantitative. He weighed the red calx, drove off the new 'air', collected it and measured its volume, and then weighed the mercury that was left. Lavoisier's second principle was what he called 'decomposition and reconstitution', or what we might call analysis and synthesis. So, in this case, he went on to heat the residual mercury in air, to measure the volume of air it absorbed and to weigh the resulting red calx. The volumes and weights agreed precisely. Lavoisier had clearly taken mercury and the new 'air' apart and put them together again. He went on to demonstrate in the same precise, quantitative and convincing way that the new 'air', which he called oxygen, was involved in all the many combustions that he studied, and in respiration too. Fig. 4.5.4 from the *Elements of Chemistry* shows an experiment on respiration in progress in Lavoisier's laboratory.

Lavoisier's chemical revolution had thus dealt with two of Aristotle's elements, Air and Fire, and he now applied his techniques to Water. He carried out extensive experiments with 'inflammable air' and showed conclusively that, when burned in oxygen, it produced nothing but water. So Lavoisier changed the name of 'inflammable air' to hydrogen, which means water-generator. In this case the synthesis came first, but Lavoisier later decomposed water by passing it through a red-hot gun-barrel and collected the hydrogen produced. Lavoisier made large amounts of hydrogen in this way and some of it was used to fill the balloons used in the earliest balloon flights which took place during the 1780s.

Lavoisier's work had changed chemistry so much that he and some of his collaborators now saw the need for a new system of chemical names or nomenclature. In 1787 they published a book called *The Method of Chemical Nomenclature* and Lavoisier wrote in the foreword:

That method which it is so important to introduce into the study and teaching

Fig. 4.5.4   An experiment on respiration in Lavoisier's laboratory; Lavoisier is directing operations and his wife is making notes.

of chemistry is closely linked to the reform of its nomenclature. A well-made language, a language which seizes on the natural order in the succession of ideas, will entail a necessary and even a prompt revolution in the manner of teaching. It will not permit professors of chemistry to deviate from the course of nature. Either they will have to reject the nomenclature, or else follow irresistibly the road it marks out. Thus it is that the logic of a science is related essentially to its language.

What enormous confidence! But Lavoisier was justified by posterity. For the system of chemical names he devised is largely the same as the one we still use 200 years later. The names of the simpler elements and their compounds, oxides, sulphates, carbonates, nitrates and so on, all date from Lavoisier, as does the principle that chemical names should reflect composition and structure.

In 1789 Lavoisier and his co-workers established a new journal, the *Annales de Chimie*, and in the same year Lavoisier summed up his discoveries and methods in the *Traité Elémentaire de Chimie* or *Elements of Chemistry*. In this book Lavoisier published a new table of chemical elements or *substances simples* (see Fig. 4.5.5). This table contains many substances which we still regard as elements—oxygen, hydrogen, sulphur, phosphorus, carbon, silver, arsenic, bismuth and so on. But light and caloric (see Chapter 4.7) as well as substances we now know to be compounds—chalk, magnesia, barytes, alumina and silica—were also included.

	Noms nouveaux.	Noms anciens correspondans.
Substances simples qui appartiennent aux trois règnes & qu'on peut regarder comme les élémens des corps.	Lumière.........	Lumière.
	Calorique........	Chaleur.
		Principe de la chaleur.
		Fluide igné.
		Feu.
		Matière du feu & de la chaleur.
	Oxygène.........	Air déphlogistiqué.
		Air empiréal.
		Air vital.
		Base de l'air vital.
	Azote...........	Gaz phlogistiqué.
		Mofete.
		Base de la mofete.
	Hydrogène.......	Gaz inflammable.
		Base du gaz inflammable.
Substances simples non métalliques oxidables & acidifiables.	Soufre...........	Soufre.
	Phosphore........	Phosphore.
	Carbone..........	Charbon pur.
	Radical muriatique.	Inconnu.
	Radical fluorique	Inconnu.
	Radical boracique,.	Inconnu.
Substances simples métalliques oxidables & acidifiables.	Antimoine........	Antimoine.
	Argent...........	Argent.
	Arsenic..........	Arsenic.
	Bismuth..........	Bismuth.
	Cobolt...........	Cobolt.
	Cuivre...........	Cuivre.
	Etain............	Etain.
	Fer..............	Fer.
	Manganèse........	Manganèse.
	Mercure..........	Mercure.
	Molybdène........	Molybdène.
	Nickel...........	Nickel.
	Or...............	Or.
	Platine..........	Platine.
	Plomb............	Plomb.
	Tungstène........	Tungstene.
	Zinc.............	Zinc.
Substances simples salifiables terreuses.	Chaux...........	Terre calcaire, chaux.
	Magnésie.........	Magnésie, base du sel d'Epsom.
	Baryte...........	Barote, terre pesante.
	Alumine..........	Argile, terre de l'alun, base de l'alun.
	Silice...........	Terre siliceuse, terre vitrifiable.

Fig. 4.5.5 Lavoisier's table of elements from the *Traité Elémentaire de Chimie*, 1789.

But, despite his occasional arrogance, Lavoisier clearly regarded his list of *substances simples* as provisional, as this comment shows:

> We cannot be certain that what we regard today as simple really is so; all that we can say is that such a substance is the actual limit reached by chemical analysis, and that, in the present state of our knowledge, it cannot be further subdivided.[3]

Lavoisier was a remarkable man. As the mathematician Lagrange said on the day after Lavoisier died on the guillotine, 'It took them only an instant to cut off that head, and a hundred years may not produce another like it'.

DALTON'S ATOMIC THEORY

Lavoisier's revolution in chemical ideas was carried further by John Dalton, a Quaker who lived in Manchester and who, like Priestley, taught in a Dissenting Academy.

Dalton took all Lavoisier's hard-won discoveries for granted, including his new nomenclature. Partly inspired by a suggestion of Newton's in the *Principia* and partly from his own extensive work on the solubility of gases in water and other liquids, Dalton put forward his atomic theory. He supposed that matter was composed of 'a considerable number of what may properly be called *elementary* particles, which can never be *metamorphosed* one into another.'[4]

In his book *New System of Chemical Philosophy*, published in 1808, Dalton emphasised the importance of atomic weights and proportions:

> . . . it is one great object of this work, to show the importance and advantage of ascertaining the relative weights of the ultimate particles both of simple and compound bodies, the number of simple elementary particles which constitute one compound particle, and the number of less compound particles which enter into the formation of one more compound particle.[5]

Fig. 4.5.6 shows the symbols which Dalton devised to represent atoms of the elements and some simple compounds. These symbols were accompanied by the following key:

> This plate contains the arbitrary marks or signs chosen to represent the several chemical elements or ultimate particles.

Fig.			Fig.		
1	Hydrog. its rel. weight . . .	1	11	Strontites	46
2	Azote	5	12	Barytes	68
3	Carbone or charcoal	5	13	Iron	38
4	Oxygen.	7	14	Zinc	56
5	Phosphorus	9	15	Copper	56
6	Sulphur	13	16	Lead	95
7	Magnesia	20	17	Silver	100
8	Lime	23	18	Platina	100
9	Soda	28	19	Gold	140
10	Potash	42	20	Mercury	167

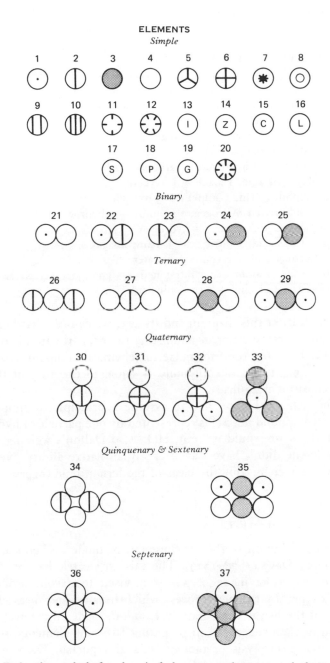

Fig. 4.5.6 Dalton's symbols for chemical elements and compounds from *New System of Chemical Philosophy*, 1808.

Fig.

21 An atom of water or steam, composed of 1 of oxygen and 1 of hydrogen, retained in physical contact by a strong affinity, and supposed to be surrounded by a common atmosphere of heat; its relative weight = ... 8

22 An atom of ammonia, composed of 1 of azote and 1 of hydrogen 6

23 An atom of nitrous gas, composed of 1 of azote and 1 of oxygen 12

24 An atom of olefiant gas, composed of 1 of carbone and 1 of hydrogen . . . 6

25 An atom of carbonic oxide composed of 1 of carbone and 1 of oxygen . . . 12

26 An atom of nitrous oxide, 2 azote + 1 oxygen 17

27 An atom of nitric acid, 1 azote + 2 oxygen 19

28 An atom of carbonic acid, 1 carbone + 2 oxygen 19

29 An atom of carburetted hydrogen, 1 carbone + 2 hydrogen 7

30 An atom of oxynitric acid, 1 azote + 3 oxygen 26

31 An atom of sulphuric acid, 1 sulphur + 3 oxygen 34

32 An atom of sulphuretted hydrogen, 1 sulphur + 3 hydrogen 16

33 An atom of alcohol, 3 carbone + 1 hydrogen 16

34 An atom of nitrous acid, 1 nitric acid + 1 nitrous gas 31

35 An atom of acetous acid, 2 carbone + 2 water 26

36 An atom of nitrate of ammonia, 1 nitric acid + 1 ammonia + 1 water . . . 33

37 An atom of sugar, 1 alcohol + 1 carbonic acid 35

If we look carefully at this diagram and its key, we can see that Dalton is not only giving symbols for the elements and listing their relative atomic weights. He is also giving the first structural formulae for chemical compounds and stating that the relative weight of any compound is equal to the sum of the relative weights of its constituent atoms.

To derive his scheme Dalton had to assume a principle of simplicity. For example, water (compound no. 21) is represented by one particle of hydrogen and one of oxygen or, as we would write it, HO. Also Dalton's weights were often much in error for he didn't have Lavoisier's quantitative ability. Nevertheless, Dalton's scheme is recognisably the basis of modern atomic theory.

DAVY AND ELECTROCHEMISTRY

Another major contribution to the development of modern chemical ideas was made by Humphry Davy (1778–1829). The earliest electric battery, the voltaic pile constructed by Volta in 1799, was soon used to decompose water into hydrogen and oxygen by the new process which became known as electrolysis. Davy, working at the Royal Institution in London, made an extensive study of electrolysis and in 1806 succeeded in preparing two new elements, sodium and potassium, by the electrolysis of molten soda and potash. Davy went on to discover a number of other new elements—magnesium, calcium, strontium, barium, boron and silicon—all by electrolysis using Volta's new battery. Davy's work considerably extended Lavoisier's table of elements and showed that some of the materials Lavoisier had classified as elements, like magnesia or barytes, were in fact compounds which could be decomposed by electrolysis.

But perhaps the most important aspect of Davy's work was the support it gave to the idea that chemical compounds are held together by forces that are ultimately electrical in nature.

REFERENCES

1   Gillispie, C. C., *The Edge of Objectivity*, (New Haven: Princeton University Press, 1960), pp. 206–7.
2   Lavoisier, A., *Elements of Chemistry*, (Paris: 1789).
3   Ibid.; quoted in Gillispie, C. C. (ed.), *Dictionary of Scientific Biography*, (New York: Charles Scribner, 1973), Vol. VIII, p. 82.
4   Quoted in Gillispie, C. C., op. cit., p. 259.
5   Dalton, J., *New System of Chemical Philosophy* (1808); quoted in Greenaway, F., *John Dalton and the Atom*, (1966: Heinemann, London), p. 132.

SUMMARY

1   Between 1770 and 1800 chemical knowledge was transformed and chemistry had begun to emerge as a separate science in something like its modern form.
2   Priestley's work on the properties of gases and Lavoisier's emphasis on quantitative experiments were major factors in this development of chemical knowledge.
3   Lavoisier established that combustion and respiration were similar processes, both involving oxygen; he also introduced a new chemical nomenclature which is still used today.
4   Dalton devised an atomic theory for chemistry and gave the first table of atomic weights and the first structural formulae.
5   Davy used electrolysis to discover many new chemical elements; his work suggested that chemical forces were probably electrical in nature.

4.6   The making of geology

In the late 18th and early 19th centuries the main outlines of modern geology began to emerge. Two kinds of evidence were involved in this process—observations of the rocks themselves and of the fossils they contained. These observations laid the foundations not only for geology but also for one of the central ideas of modern biology—Darwin's theory of evolution (see Chapter 4.8).

FORCES SHAPING THE ROCKS

In the late 18th century the major questions which were being asked about the earth concerned the force or forces which had shaped the rocks into their present forms. Two of Aristotle's elements were the prime candidates. Was Water—rivers, the sea, Noah's Flood—the major shaping force? Or was it Fire—acting via volcanoes and pressures from within the earth? How long had the earth existed for these forces to act? Had the forces we now see, shaping and reshaping the earth's surface, always been the same? Or had there been some dramatic creative event, possibly of divine origin?

The prevailing view about the origin of rocks towards the end of the 18th century was that of Abraham Werner (1749–1817) who was Professor of Mineralogy in a well-known German mining school at Freiburg in Saxony. Werner was probably the first person to work out a complete and systematic account of the origin of all the rocks which make up the earth. His account was basically an historical one including several stages of development:

> The two basic postulates of the Wernerian theory were that the earth was once enveloped by a universal ocean and that all the important rocks that make up the crust of the earth were either precipitates or sediments from that ocean. Werner placed the rocks in four (later five) classes according to the period in which they were formed, believing that characteristics of the rocks were the result of the depth, content, and conditions of the universal ocean at the time when they were formed.[1]

> The first, or primitive, period was the most important.

> At the beginning of the primitive period, according to Werner's theory, the universal ocean was very deep and calm; and the first rocks were chemical precipitates which adhered to an originally uneven surface, granite being the first rock formed. Gradually the water became less calm, so that later rocks of the primitive period are not as crystalline as the older ones; and toward the end of the period there was a general rising of the waters, followed by a comparatively rapid recession, which explains the position of some of the later primitive rocks relative to the older ones.[2]

On Werner's view, living species appeared after the formation of the primary rocks and in the order—fish, mammals, man—in which they are mentioned in the Bible. But Werner did write of a time 'when the waters, perhaps a million years ago, completely covered the earth . . .'. So the time-scale which he contemplated was much greater than the 6000 years or so which was sometimes claimed to be the Biblical estimate of the age of the earth.

Werner was an inspiring teacher and he trained field-workers who went out and collected evidence from many regions—with the ultimate aim of confirming the Wernerian system. So it is to Werner's credit that many of these students were often in the forefront of investigations which ultimately proved him wrong!

The main challenge to Werner's ideas came from those who were sometimes called Vulcanists—the most prominent of whom was James Hutton (1726–97). Hutton was from Edinburgh and he became a prominent figure in a movement which has been called the Scottish Enlightenment. Hutton was closely associated with Joseph Black, Adam Smith and James Watt and corresponded with Erasmus Darwin and Matthew Boulton who were leading members of the Lunar Society of Birmingham.

Hutton qualified as a doctor but then took up farming, and it was from his studies of the latest agricultural methods that his interest in rocks developed. He studied geology as systematically as he could for more than 30 years and towards the end of his life published the ideas that he had gradually developed. His *Abstracts of a Dissertation . . . Concerning the System of the Earth, Its Duration and Stability*—a 30-page pamphlet—was circulated privately in 1785. Then, in 1795, two years before he died, Hutton published the detailed two-volume work *Theory of the Earth: With Proofs and Illustrations*.

Hutton's ideas have been summed up like this:

> Past events can be described only by inductive analogy to processes which we observe working in the present, and by the evidence of the rocks. There are two sorts in the crust of the earth, one of igneous and the other of aqueous origin. The primary igneous rocks (granite, porphyry, basalt, etc.) usually occur under the aqueous, except where formations are overthrust, or where molten dikes and plugs have intruded through the limestones. Weathering and erosion are constantly carrying a fine silt of sandstones, clay, and topsoils down the rivers and out to settle in the ocean bottoms. Some agency must have transformed, and must still transform, these loose deposits into the rocks around us. It cannot have been water. They are utterly insoluble. That agency must, therefore have been heat. The intense heat of the earth's heart, acting under enormous pressure, consolidates the rocks, and its expansive force uplifts the continents thus born in the bed of the sea.[3]

Throughout his book Hutton laid great emphasis on the principle which later came to be called uniformitarianism. His model of the earth was dynamic. He assumed that the same processes which had shaped the rocks in the past are still shaping them today. And since the current processes of change are so slow as to be almost imperceptible, Hutton had to assume that the earth had been in existence for vast amounts of time. His conclusions were:

... That it had required an indefinite space of time to have produced the land which now appears; .... That an equal space had been employed upon the construction of that former land from whence the materials of the present came; .... That there is presently laying at the bottom of the ocean the foundation of a future land, which is to appear after an indefinite space of time. ... so that, with respect to human observation, this world has neither a beginning nor an end.[4]

Both Werner and Hutton recognised the existence of rocks formed by sedimentation and by igneous activity. The questions which divided them were: which rocks were primarily sedimentary and which igneous? and, were past processes of rock formation different from those now in operation?

The views which eventually prevailed were mainly those of Hutton. But his division of sedimentary and igneous rocks, and his assertion of the principle of uniformitarianism, were only accepted after a lengthy period in which many conflicting views were advanced and much new, and often confusing, observational evidence had been acquired. A major factor in this process was the increased understanding of the nature of fossils.

EVIDENCE FROM FOSSILS

Fossils had been known and described since antiquity but the earliest drawings which we have date only from the 16th century—and these are mainly crude and simple sketches. By the 17th century records of fossils had become both more extensive and more detailed. For example, Fig. 4.6.1 is from a book *Discourse on Earthquakes* by Robert Hooke—the same Hooke who had been in dispute after

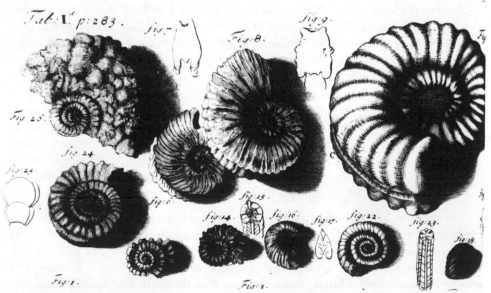

Fig. 4.6.1   Fossil ammonites from Robert Hooke's *Discourse on Earthquakes*.

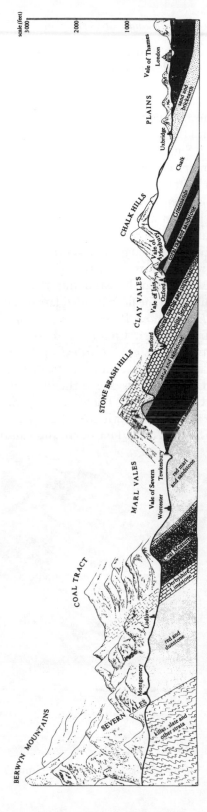

Fig. 4.6.2 Geological section from London to Snowdon in Wales as drawn up by William Smith in 1817. *From Open University Science Foundation Course S101, Unit 26, copyright © The Open University Press.*

dispute with Isaac Newton. Hooke studied a variety of fossils in the course of his work with his improved compound microscope, and these are some of the careful sketches that he made. In this book Hooke also speculated that fossils might one day be used to date rocks.

However, the close connection between fossils and the rocks which contain them was not established until the work of William Smith (1769–1839) in the late 18th and early 19th centuries. Smith was a practical man. His interest in rocks was initially as a surveyor and engineer concerned with mining and canal building. A whole network of canals was built in England in the late 18th century (see Chapter 4.4) and the cutting of canals often revealed details of the underlying rock strata. It is interesting that both Hutton and Smith, of the pioneer geologists, were closely involved in canal building.

William Smith gained a great knowledge of strata from his work on canals, and by 1796 his notebooks show that he realised that geologically similar strata in different regions could be identified by studying the types of fossils they contained. In 1815 Smith published the first detailed large-scale geological map *A Definition of the Strata of England and Wales, with part of Scotland*. Fig. 4.6.2 shows the strata he identified between London and Snowdon. The next year Smith published his famous book *Strata Identified by Organised Fossils* (1816) which contained coloured plates of fossils characteristic of 19 different strata from the London Clay downwards. Fig. 4.6.3 is one of the illustrations from that book showing fossils from the Upper Chalk. Then in 1817 Smith published his *Stratigraphical System of Organised Fossils* which contained a 'Geological Table of British Organised Fossils Which Identify the Course and Continuity of the Strata

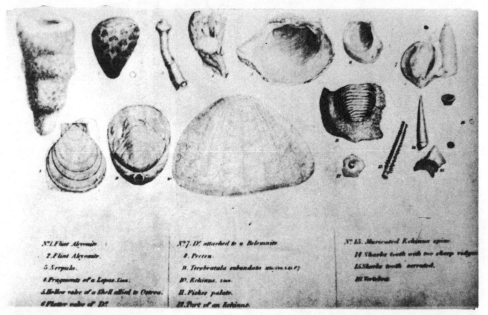

Fig. 4.6.3   Fossils from the Upper Chalk from William Smith's *Strata Identified by Organised Fossils*, 1816.

in Their Order of Superposition, as Originally Discovered by W. Smith, Civil Engineer; with Reference to His Geological Map of England and Wales'.

Smith's main interest was in the rocks and the uses to which they could be put. Alongside his stratigraphic column he lists the uses of the various strata— Cornbrash and Forest Marble Rock can be 'used for rough paving' and 'makes tolerable roads'; amongst the clays there is 'no building stone . . . but abundance of materials which makes the best bricks and tiles in the island'; and the granite and gneiss provide 'the finest building stone in the island for bridges and other heavy work'. For Smith, fossils were useful indicators for identifying strata but he never really considered what fossils were and how they might be related to one another.

Smith's approach was very different from that of the Frenchman Georges Cuvier (1769–1832)—a man who was Smith's contemporary but whose work in the Paris Basin was not known to Smith. Cuvier came to stratigraphy and palaeontology from comparative anatomy rather than from an interest in rocks. He was a professor at the best endowed institution of scientific research of the period—the Museum of Natural History in Paris. He reorganised and considerably extended the museum's large collection of skeletons and other specimens and developed his own ideas on comparative anatomy. These ideas stressed the interdependence of all the parts of an animal and the relationship between form and function. Cuvier claimed to be able to deduce the rest of the anatomy of an animal from the shape and size of just a few of its bones.

This passage clearly shows his style of thinking:

> Just as the equation of a curve entails all its properties so the form of the tooth entails the form of the knuckle, and the shape of the shoulder-blade that of the claws. And reciprocally, just as any element of a curve may be taken as the basis of the general equation, so too the claw, the shoulder-blade, the knuckle, the femur, and all the other bones taken separately, give the tooth as well as each other—so that whoever has a rational command over the laws of the organic economy may reconstruct the whole animal from any one of them. . . .[5]

These principles were just what is required when fragments of fossil skeletons are to be studied in the hope of establishing the form of the complete animal. When he was confronted by a very large number of fossil bones in the gypsum quarries of the Paris Basin, Cuvier describes his reaction like this:

> I was in the position of one who has been presented pell-mell with the incomplete and mutilated debris of hundreds of skeletons belonging to twenty kinds of animals. It was required that every bone should go find the beast to which it belonged. A miniature resurrection was called for, and I did not hold at my disposal the omnipotent trumpet. But the immutable laws prescribed for living beings filled that office, and at the voice of comparative anatomy, every bone and every fragment leaped to take its place.[6]

In 1811, Cuvier, with Brogniart, published an *Essay on the Mineralogical Geography of the Environs of Paris* which is one of the earliest systematic treatises

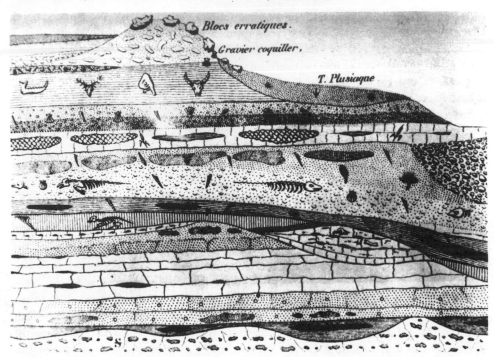

Fig. 4.6.4 Cuvier's diagrammatic section of strata with crude drawings of fossils associated with each stratum.

of stratigraphy. In this book, Cuvier used fossil evidence to distinguish successive formations of chalk, limestone, clay and sand lying in layers on top of one another. Fig. 4.6.4 shows a stratigraphic diagram from one of Cuvier's books. This essay of 1811 also formed part of a later work by Cuvier, the title of which emphasises one of his main conclusions. It was called *Research on Fossil Bones, Reestablishing the Character of Many Animals, of which the Species have been destroyed in the Revolution of the Earth*. It is interesting that, later, fossil bones of extinct species were to be one of the kinds of evidence which strongly influenced Charles Darwin's thinking during the voyage of the *Beagle* and later (see Chapter 4.8).

LYELL'S PRINCIPLES OF GEOLOGY

Another major influence on Darwin was the work of Charles Lyell (1797–1875), whose three-volume *Principles of Geology* is probably the most famous and influential book in the history of geology. The first edition was completed in 1833 and in the next 40 years it went through no less than twelve editions. Lyell's *Principles of Geology* contains so much that it is difficult to summarise. But perhaps the book's major contribution was to present a wealth of evidence in favour of Hutton's principle of uniformitarianism, which states that all the rocks we now see on the earth were formed by the same slow chemical and physical processes which we see forming them today. For Lyell the present was the key to the past and what he needed to establish was just how much time there had been for these slow processes to work. He did this in a variety of ways. For example, Lyell used historical records of volcanic activity on Mount Etna in Sicily to calculate the average rate of deposition of volcanic rocks, and hence he estimated how long it had taken for the volcanic strata and other strata to form. Or he estimated the present rate of sedimentation in the Mississippi delta, and hence tried to establish how long the whole delta had taken to form and how long other sedimentary rocks had been in the making. In ways like these Lyell was able to amass sufficient evidence to extend the geological time-scale to the extent required to make uniformitarianism plausible.

His approach has been summarised like this:

> Lyell surveyed the full range of processes which at present are altering the earth's surface—the eroding effects of running water in streams and rivers and of waves along the seacoast, the accumulation of sediments in deltas and on the sea bottom, and the cumulative effects of earthquakes and volcanoes in elevating the land. He showed that even the largest volcanoes, such as Vesuvius and Etna, were the product of a long series of eruptions distributed through immense periods of time, and the eruptions were never greater nor more frequent than in historic times. Lyell emphasized repeatedly that the magnitude of the geological changes which had occurred during the past was not a reason to postulate extraordinary convulsions or catastrophes. The greatest changes could be accomplished by ordinary geological processes acting gradually, if they were given sufficient time.[7]

One major novelty in Lyell's work was his classification of the Tertiary rocks into Pliocene, Miocene and Eocene. This classification was first put forward in 1833 and was based on precise counts of the number of still existing and extinct species which these strata contained. For example, the Pliocene (from the Greek *Pleios* or 'more recent') contained as many as 30–50 per cent of modern species, whereas the Eocene (from *Eos*, 'dawn' or 'recent') had only about 3 per cent. The Miocene (from *Meios*—'less recent') came somewhere between. Later, in 1839, Lyell added the Pleistocene (from *Pleiston* or 'most recent'). It is interesting to see the use of fossils as stratigraphic indicators, which had been pioneered by Smith and Cuvier, here being made quantitative by Lyell.

And Lyell's arguments, by their very nature, assumed that species became extinct and that new ones appeared. The philosopher William Whewell—the man who coined the word scientist—argued that 'Mr. Lyell should supply us with some mode by which we may pass from a world filled with one kind of animal forms, to another, in which they are equally abundant, without perhaps one species in common'. Whewell thought that it was 'undeniable . . . that we see in the transition from an earth peopled by one set of animals, to the same earth swarming with entirely new forms of organic life, a distinct manifestation of creative power, transcending the known powers of nature. . . .'[8]

That was in 1831, just as Charles Darwin was starting on his voyage in the *Beagle* and beginning the work which totally revised our ideas on the origins and transformation of species. And in that work the principles of the new geology, as set out by Lyell, were to play a major role.

REFERENCES

1   Ospovat, A., *Abraham Werner* in Gillispie, C. C. (ed.), *Dictionary of Scientific Biography*, (New York: Charles Scribner, 1976), Vol. XIV, pp. 259–60.
2   Ibid., p. 260.
3   Gillispie, C. C., *The Edge of Objectivity*, (New Haven: Princeton University Press, 1960), p. 293.
4   Hutton, J., *Abstract of a Dissertation ... Concerning the System of the Earth, Its Duration and Stability*, (Edinburgh: 1785), pp. 27–8.
5   Quoted in Gillispie, C. C., *The Edge of Objectivity*, (New Haven: Princeton University Press, 1960), p. 284.
6   Ibid., p. 282.
7   Wilson, L. G., *Charles Lyell* in Gillispie, C. C. (ed.), *Dictionary of Scientific Biography*, (New York: Charles Scribner, 1976), Vol. VIII, p. 568.
8   Whewell, W., 'Review of Lyell's *Principles of Geology*' in *The British Critic, Quarterly Theological Review and Ecclesiastical Record*, 1831.

SUMMARY

1   Between 1770 and 1840 the main outlines of modern geology began to emerge.
2   In the late 18th century, Abraham Werner thought that rocks were primarily shaped by the action of Water, while James Hutton thought that Fire was more important. Hutton also introduced the principle of uniformitarianism, which assumed that the same processes which had shaped the rocks in the past are still shaping them today.
3   William Smith used fossils to identify geological strata and published the first geological map of England and Wales in 1815.
4   George Cuvier brought his knowledge of comparative anatomy to the study of fossils and published one of the earliest systematic treatises on stratigraphy.
5   In his *Principles of Geology* (1733), Charles Lyell amassed sufficient evidence to extend the geological time-scale far enough to make uniformitarianism plausible.

4.7 Physics in the early 19th century

We saw in Chapters 4.5 and 4.6 that, by about 1830, the main outlines of modern chemistry and of modern geology had begun to emerge. At just about the same time, perhaps a little earlier, the same thing happened in two of the main areas of classical physics—the study of heat and of electricity and magnetism.

HEAT IN THE EARLY 19TH CENTURY

We have seen how one of the traditional Aristotelian elements, Fire, played an important part in the origins of chemistry and geology. Fire is also, of course, connected with the study of heat. Many different ideas concerning the nature of heat had been held at various times. One was that heat was related to motion—the motion of the particles of matter of which everything is made. Francis Bacon and John Locke both put forward this view of heat. Another common idea was that heat was a kind of fluid, often called *caloric*, which when added to a body made it hotter. Caloric, or 'matter of fire' as it was sometimes called,[1] was supposed to be a weightless fluid only detectable by its heating effects, and Lavoisier and Laplace both took this view of the nature of heat.

As with so much of early science, the early study of heat was descriptive and qualitative. It didn't start to become quantitative and so to involve careful measurements until about the middle of the 18th century. The man who mainly pioneered a quantitative approach to heat was Joseph Black, the same man who did so much to make chemistry quantitative by his use of the balance (see Chapter 4.5). Black developed thermometers with which he could accurately measure small temperature changes, and he used these thermometers and his balances to make the first measurements of both specific heats and latent heats. After this, many other measurements of this kind were made, their accuracy was increased, and a solid quantitative basis for the study of heat was gradually built up.

But these developments, important though they were, had little influence on another important aspect of the study of heat—the rapid development of heat-engines or steam-engines in the 18th century, in particular by James Watt and others in England (see Chapter 4.4 and Fig. 4.4.13, page 140). These developments were made by practical men who, by trial and error, made these engines more efficient and more versatile. Fig. 4.4.13 shows that the efficiency of Watt's engines increased considerably between 1770 and 1796 and that the later so-called Cornish engines were even more efficient.

After several decades of such increases in efficiency, people began to ask whether there were any limits to these improvements, and, if so, what these limits were. And would it be better to use some other working fluid, rather than steam?

Sadi Carnot
The first significant answers to these questions were given in the 1820s. But they

didn't come from England, where most of the development of steam engines took place. They came from France, and the man who gave them was Sadi Carnot (1796–1832) who studied at the École Polytechnique. His father, Lazare Carnot, himself a distinguished scientist, had been a member of the Convention during the French Revolution and was also one of Napoleon's chief supporters. These were unpopular connections in the 1820s when the French monarchy had been restored, and Sadi Carnot was very much an outsider in French society, which may explain why his work was neglected for so long.

In 1824, when Carnot was 28, he published a book of only 118 pages—called *Réflexions sur La Puissance Motrice du Feu* or *Reflections on the Motive Power of Fire*. This short book set out a completely new and more general way of approaching the study of heat-engines. Carnot's practical motivation is clear. In his Introduction he says:

> Already the steam-engine works our mines, impels our ships, excavates our ports and rivers, forges iron, fashions wood, grinds grain, spins and weaves our cloths, transports the heaviest burdens, etc. It appears that it must some day serve as a universal motor, and be substituted for animal power, waterfalls and air currents . . .
> . . . To take away today from England her steam-engines would be to take away at the same time her coal and iron. It would be to dry up all her sources of wealth, to ruin all on which her prosperity depends, in short to annihilate that colossal power. The destruction of her navy, which she considers her strongest defence, would perhaps be less fatal.[2]

Carnot's approach to the steam engine was totally new. As he put it, 'Notwithstanding the work of all kinds done by steam-engines, notwithstanding the satisfactory condition to which they have been brought today, their theory is very little understood, and the attempts to improve them are still directed almost by chance.'[3] Carnot would be different—he aimed to study the properties of 'all imaginable heat engines'.

How did he do it? Perhaps inspired by the basic elements of Watt's engine, the hot cylinder and the cold condenser, Carnot did a thought experiment. He imagined an idealised version of Watt's engine—a single cylinder containing gas which could absorb heat, or caloric, as Carnot called it, from a hot body, the boiler, and give up heat to a cold body, the condenser.

Carnot imagined his ideal engine working in four stages (see Fig. 4.7.1). First caloric was absorbed from the hot body and the gas expanded at constant temperature. Then the engine was isolated and the gas expanded further until its temperature fell to that of the cold body. Then caloric flowed from the engine to the cold body and the gas contracted, but with its temperature constant. Finally the engine was isolated again and the gas compressed until it reached its initial temperature and volume once more. All these changes were imagined to take place very slowly so that no unnecessary or wasteful turbulence occurred, and so Carnot's ideal engine was both *cyclic* and *reversible*. When the cycle was finished, the engine was in the same state as when it started. All that had happened was that

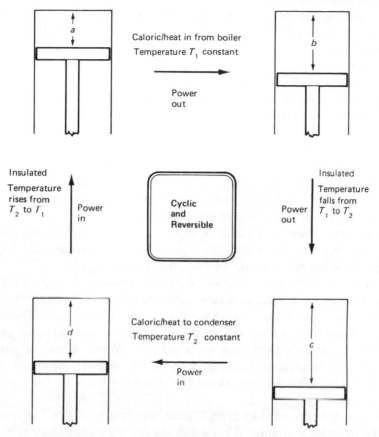

Fig. 4.7.1   The cyclic operation of Carnot's ideal reversible heat engine; see text for detailed explanation.

a certain amount of caloric had been taken from a hot body, the *same* amount of caloric had been given to a cold body and some useful mechanical power had been obtained. It seems likely that Carnot had in mind an analogy with the operation of a water wheel[4] in which a certain amount of water is taken from a reservoir and the *same* amount of water emerges from the wheel having done some useful mechanical work. Most water wheels too are cyclic and reversible.

Having imagined his ideal, cyclic and reversible heat engine, Carnot's next step was even more abstract and general. He did another thought experiment. He imagined two such reversible engines, each using the same amount of caloric but one being more efficient than the other or, in other words, one engine producing more mechanical power than the other while using the same amount of caloric. Carnot then asked what would happen if we now connect these two engines together so that the more efficient engine drives the less efficient backwards round its cycle which is perfectly possible if the engines are reversible. At the end of the cycle nothing will have changed. Each engine will be back in its initial condition and the caloric which has been removed from the hot body by one engine will have been put back by the other. *But* the more efficient engine will have produced some useful mechanical power. In other words we will have created a perpetual motion machine—for we could go on repeating the linked cycle as many times as we liked. But all experience shows that perpetual motion is impossible. So Carnot concluded that something was wrong with his argument. One ideal, cyclic and reversible engine couldn't be more efficient than another. They must ALL have the same efficiency. And this must be true whatever the working substance in the cylinder or the details of their construction. In other words Carnot had arrived at a very general result.

And Carnot didn't stop there. He repeated his thought-experiment, this time with an irreversible engine driving a reversible one backwards, and showed, by a similar argument about the impossibility of perpetual motion, that all irreversible engines must be less efficient than reversible ones.

What this meant was that Carnot's ideal, cyclic and reversible engine was the ultimate in efficiency. Its performance provided a limit which any real engine could approach but not exceed. Another very general result.

Carnot's remarkable general results were not appreciated in the 1820s and he did very little further work as he died in a cholera epidemic in 1832 at the age of 36 and most of his papers were burnt as was the practice at the time in epidemics. His compatriot Clapeyron developed Carnot's ideas in the 1830s and many years later, in the 1850s, Carnot's work and his arguments played a crucial part in the development of thermodynamics, especially the notoriously difficult second law of thermodynamics. General arguments like Carnot's are still used in modern thermodynamics.

In one crucial respect, however, Carnot was wrong. Caloric or heat is not conserved in an ideal engine. What actually happens is that some of the heat is turned into mechanical work. This became clear primarily through the experiments of James Joule (1818–89) after whom the modern unit of energy is named.

James Joule
Joule came from Manchester, and one of his teachers was John Dalton, then in his seventies. In a long series of careful experiments which he carried out during the 1840s, Joule established that mechanical work could be transformed into heat at a definite measurable rate of conversion. These experiments were extremely difficult—many of them involving precise measurements of small temperature changes to an accuracy of about one-hundredth of a degree. His most famous experiments involved measuring the heat produced by rotating paddle wheels in water and other liquids (see Fig. 4.7.2). Joule's paddle wheel experiments are mentioned in all the textbooks but they were only one of a whole series of careful experiments which he did in order to establish what came to be called 'the mechanical equivalent of heat'.

In 1849 Joule summed up all his experiments in a famous short paper in which he concluded:

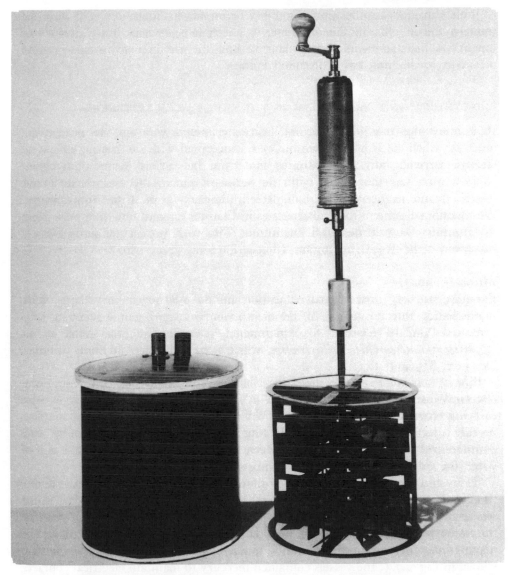

Fig. 4.7.2 Paddle wheels and calorimeter used in Joule's experiments on the mechanical equivalent of heat.

1st. That the quantity of heat produced by the friction of bodies, whether solid or liquid, is always proportional to the quantity of force expended.

and

2nd. That the quantity of heat capable of increasing the temperature of a pound of water (weighed in vacuo, and taken at between 55° and 60°) by 1° Fahrenheit requires for its evolution the expenditure of a mechanical force represented by the fall of 772 lb through the space of one foot.

Joule's accuracy can be appreciated if we compare his figure of 772 lb with the modern one of 778.2 lb. The difference is less than 1 per cent. Joule's work was one of the first statements of what later became the principle of the conservation of energy or the first law of thermodynamics.

ELECTRICITY AND MAGNETISM IN THE EARLY 19TH CENTURY

It is interesting that Joule's earliest heat experiments were not the mechanical ones for which he is best known but were concerned with the heating effects of electric currents both from batteries and from the earliest forms of dynamo. Joule's work was thus linked with the series of remarkable discoveries about electricity and magnetism that took place in the early years of the 19th century. We mentioned some of these discoveries in Chapter 4.5 and how they were used by Humphry Davy at the Royal Institution. This work was carried on by Davy's successor at the Royal Institution, Michael Faraday (1791–1867).

Michael Faraday

Faraday had very little formal education and he was never very happy with mathematics. But he was one of the most inventive experimental scientists who ever lived and he recorded his experimental researches in great detail in his *Experimental Researches in Electricity*, which were published in three volumes between 1839 and 1855.

But it was much earlier than this that Faraday made his first important electrical discovery. In 1819 Oersted in Copenhagen had shown that a wire carrying current from one of the new voltaic batteries could deflect a compass needle. Electricity in the wire was affecting the magnetic compass. Then in 1820 Ampère in Paris showed that, if an electric current was passed through a coil of wire, the coil acted just as if it was a magnet.

Davy and Faraday at once repeated these experiments and tried to investigate the connections between electricity and magnetism. As early as 1821 these researches led Faraday to the principle of the electric motor. He observed that the forces between a magnet and wire carrying an electric current tended to move the magnet in a circle round the wire. So Faraday constructed an apparatus like that shown in Fig. 4.7.3; the vessels contained mercury to maintain electrical contact. When the current was switched on, the magnet on the left moved continuously in a circle round the fixed wire and the wire on the right moved similarly in a circle around the fixed magnet. For the first time electrical energy had been converted into a continuous source of mechanical energy. This simple apparatus of Faraday's is the forerunner of all the electrical motors in the world and of all the moving coil instruments for measuring electric currents or voltages.

Ten years later, in 1831, Faraday did some more simple experiments which were to have far-reaching consequences. In one series of experiments he wound two separate coils onto the same soft iron ring; Fig. 4.7.4 shows his original apparatus of 1831. When Faraday changed the current in one of the coils, he found that an electric current was induced in the other coil. This was Faraday's well-known principle of electromagnetic induction and this simple apparatus is the prototype for all the transformers that are common features of nearly all modern electrical equipment.

Faraday went on to show that electromagnetic induction also occurs when a

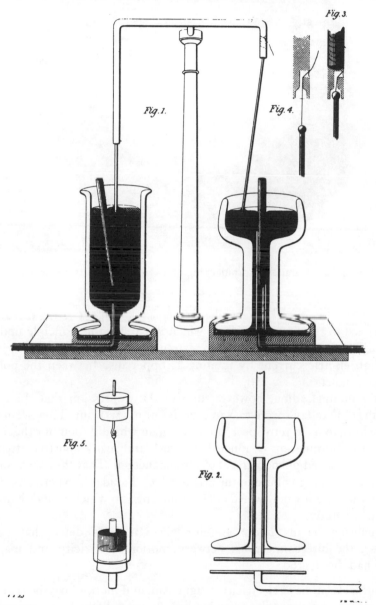

Fig. 4.7.3 Faraday's electromagnetic rotation apparatus. When the current is switched on the magnet on the left moves continuously around the fixed wire and the wire on the right moves continuously round the fixed magnet. The vessels contain mercury to maintain electrical contact. *From Royal Institution London, Michael Faraday and the Royal Institution, 1973.*

Fig. 4.7.4 Faraday's toroidal ring of 1831—the first transformer.

magnet moves near a coil or a coil moves near a magnet. And these further discoveries led to the first experiment which illustrates the principle underlying the operation of the dynamo. In an extremely simple experiment, Faraday produced an electric current by rotating a copper disc between the poles of a permanent magnet.

It would be misleading, however, to give the impression that these simple experiments of Faraday's were easy to conceive or to carry out. They seem simple, almost obvious, to us in retrospect, but to the man who did them for the first time they were the culmination of much hard work and many abortive attempts to understand the varied phenomena of electromagnetism. That they seem so simple to us is an outstanding tribute to the clarity which Faraday's experimental genius brought to the complex new world of electromagnetism which Volta's battery had opened up for study.

In conclusion let us quote an entry in Faraday's diary that concisely summarises the interconnections between motion, electricity and magnetism which he had discovered:

The mutual relation of electricity, magnetism, and motion may be represented by three lines at right angles to each other, any one of which may represent any one of these points and the other two lines the other points. Then if electricity be determined in one line and motion in another, magnetism will be developed in the third; or if electricity be determined in one line and magnetism in another, motion will occur in a third. Or if magnetism be determined first then motion will produce electricity or electricity motion. Or if motion be the first point determined, magnetism will evolve electricity or electricity magnetism.[5]

These principles have, ever since, been at the heart of the theory and practice of electromagnetism, and in time played their part in the discovery of electromagnetic waves (see Chapter 6.1).

REFERENCES

1 See Fox, R., *The Caloric Theory of Gases: From Lavoisier to Reynault*, (Oxford: Clarendon Press, 1971), pp. 9–10; see also Fig. 4.5.5.
2 Carnot, S., *Reflections on the Motive Power of Fire*, (Paris: 1824); quoted in Gillispie, C. C., *The Edge of Objectivity*, (New Haven: Princeton University Press, 1959), pp. 357–8.
3 Ibid., p. 358.
4 Water wheels, by Carnot's time, were not quaint rustic machines, but advanced, well designed machines which were widely used to drive the early mills and factories of the Industrial Revolution; see Cardwell, D. S. L., *From Watt to Clausius*, (London: Heinemann, 1971), ch. 3.
5 Gillispie, C. C., op. cit., p. 446.

SUMMARY

1 In the early years of the 19th century knowledge concerning heat and electricity and magnetism increased rapidly.
2 Carnot, in France, developed new general methods for studying the efficiency of 'all imaginable heat-engines'. He showed that his ideal, cyclic and reversible engine was more efficient than any other engine, and his methods later led to the formulation of the second law of thermodynamics.
3 Joule, in England, showed by a large number of very careful experiments that mechanical 'force' could be converted into heat at a definite rate. This work was the basis of the principle of the conservation of energy or the first law of thermodynamics.
4 Oersted and Ampère, around 1820, did experiments which showed that wires carrying electric currents could exert forces on magnets and also on other current-carrying wires, and soon after Faraday discovered the principle of the electric motor.
5 In the 1830s, Faraday did extensive experiments on the connections between electricity and magnetism in the course of which he discovered the principle of electromagnetic induction, which underlies the transformer and the dynamo.

4.8 Biology and Darwin's theory of evolution

Biology is a creation of the 19th century—the word itself dates from the year 1800. The great achievement of 18th-century natural history had been the classification system of Linnaeus in his *Systema Naturae*—a book which grew from the 14 pages of the first edition of 1735 to 2300 pages in the twelfth edition of 1766. Linnaeus' *Systema Naturae* or the *System of Nature* was an ambitious attempt to classify the whole of the natural world—the animal and mineral kingdoms as well as the world of plants and vegetables. Even while he was a student, Linnaeus had been a dedicated collector of specimens. One of his student friends described his study (see Fig. 4.8.1) like this:

> He had decorated the ceiling with spoils taken from birds, one wall with Lapp clothing and curiosities and another with large plants and a collection of bivalves; the two remaining walls were handsomely furnished, partly with medical and scientific books and partly with scientific instruments and stones. In the corners of the room were tall branches of trees, in which he had trained

Fig. 4.8.1   Artist's impression of Linnaeus' study.

tame birds of almost 30 species to nest. Finally, the inside window-ledges were occupied by large earthenware pots, filled with soil for the sustenance of rare plants.[1]

Linnaeus' classification system was based on the sexual characteristics of plants and animals and he was the first to introduce the binary system—genus and species—of naming species which is still in use today. He was also the first to use the name *Homo sapiens* for man in the tenth edition of his *Systema Naturae* published in 1758. Linnaeus' major contribution to the development of natural history was widely recognised and when the first British society for the study of natural history was formed in 1788, it was called the Linnaean Society—the name it still bears today.

Linnaeus' work was founded on the extensive *similarities* between different specimens which enabled them to be classified as members of the same species. By contrast, Charles Darwin's work depended crucially on *variations* between members of the same species.

CHARLES DARWIN

However, Linnaeus and Darwin had one essential feature in common. They both had a deep and encyclopedic knowledge of living things, based on extensive field work. Darwin (1809–82; see Fig. 4.8.2) had acquired most of his practical experience during the five years he spent on the *Beagle*. He had been an unsuccessful student—failing to qualify in medicine at Edinburgh and then going to Cambridge intending to become a clergyman. While at Cambridge he became interested in geology and went on a geological expedition to North Wales in 1831. Partly as a result of this, he was offered the post of naturalist on board a naval ship, the *Beagle* (see Fig. 4.8.3), which was about to sail round the world on a voyage of exploration. The voyage lasted nearly five years, from December 1831 to October 1836. It provided Darwin with so much new evidence—on geology, palaeontology, botany and zoology—that it took him many years to order and interpret it all. His chronological account *The Voyage of the 'Beagle'* did not appear until 1845 and *The Origin of Species* was first published in 1859.

Darwin took with him on the *Beagle* the first volume of Lyell's *Principles of Geology* and received the second volume when the ship called at Montevideo. He was able to confirm many of Lyell's ideas by direct observation of the geology of South America, but he gradually became more interested in the flora and fauna of the continent. He later wrote in his autobiography that:

> During the voyage of the *Beagle* I had been deeply impressed by discovering in the Pampean formation great fossil animals covered with armour like that on the existing armadillos; secondly, by the manner in which closely allied animals replace one another in proceeding southwards over the Continent; and thirdly, by the South American character of most of the productions of the Galapagos Archipelago, and more especially by the manner in which they differ slightly on each island of the group . . .[2]

Fig. 4.8.2   Charles Darwin (1809–82).

Fig. 4.8.3   The *Beagle* in the Straits of Magellan.

The fauna of the Galapagos Islands particularly interested him. These black volcanic islands in the middle of the Pacific were geologically and climatically virtually identical. They were inhabited by species which were related to some of those of South America. Yet, from island to island, the species exhibited slight but significant variations. The most famous examples were the giant tortoises—the natives could tell him unfailingly from which island any particular specimen came. And the finches too showed slight differences which were specific to each island. These facts puzzled Darwin:

> ... such facts as these, as well as many others, could only be explained on the supposition that species gradually became modified; and the subject haunted me. But it was equally evident that neither the action of the surrounding conditions (for these did not differ materially from island to island), nor the will of the organisms (especially in the case of plants), could account for the innumerable cases in which organisms of every kind are beautifully adapted to their habitats of life—for instance, a woodpecker or a tree-frog to climb trees, or a seed for dispersal by hooks or plumes. I had always been much struck by such adaptations, and until these could be explained it seemed to me almost useless to endeavour to prove by indirect evidence that species have been modified.[3]

When Darwin returned to England he set to work to order his data and to integrate it with that of other workers. Initially, in his own words, he operated 'on true Baconian principles, and, without any theory, collected facts on a wholesale scale'.[4] But in 1838 he read *An Essay on the Principle of Population* by Thomas Malthus—the book which emphasised that populations, since they tend to increase geometrically, will always outstrip food supplies, which can only be increased at a smaller rate (see Chapter 4.3). Thus competition for scarce food supplies is universal. Darwin summed up his reaction like this:

> In October, 1838, that is, fifteen months after I had began my systematic enquiry, I happened to read for amusement Malthus on *Population*, and being well prepared to appreciate the struggle for existence which everywhere goes on from long-continued observation of the habits of animals and plants, it at once struck me that under these circumstances favourable variations would tend to be preserved, and unfavourable ones to be destroyed. The result of this would be the formation of new species. Here, then, I had at last got a theory by which to work; but I was so anxious to avoid prejudice that I determined not for some time to write even the briefest sketch of it.[5]

So it was not until four years later, in 1842, that Darwin wrote a brief outline of his ideas, written in pencil and only 35 pages long. Two years later, in 1844, he grew a little bolder. He wrote a longer version of his theory, 235 pages this time, and deposited it with his wife with instructions for her to publish it if he died.

It was not until 15 years later, in 1859, that Darwin published his famous book *The Origin of Species*. Even then he was only induced to publish what he called an 'abstract' of his ideas because another man, Alfred Wallace, had arrived at virtually identical conclusions completely independently of Darwin.

Wallace (1823–1913) was a younger man than Darwin and in his early years he had worked as a surveyor during the railway boom of the 1840s (see Chapter 4.9). Later Wallace became a professional naturalist and, in search of specimens, he went on major expeditions to the Amazon basin in South America and to the Malay archipelago in the Far East. In February 1858, while in the Moluccas islands, Wallace fell ill. While he was feverish he recalled Malthus' *Essay on Population* and suddenly he hit on the same idea as Darwin had done 20 years earlier:

> It occurred to me to ask the question, Why do some die and some live? And the answer was clearly, that on the whole the best fitted lived. From the effects of disease the most healthy escaped; from enemies, the strongest, the swiftest, or the most cunning; from famine, the best hunters or those with the best digestion; and so on.
>
> Then I at once saw, that the ever present variability of all living things would furnish the material from which, by the mere weeding out of those less adapted to the actual conditions, the fittest alone would continue the race.
>
> There suddenly flashed upon me the *idea* of the survival of the fittest.
>
> The more I thought over it, the more I became convinced that I had at length found the long-sought-for law of nature that solved the problem of the Origin of Species . . . I waited anxiously for the termination of my fit so that I might at once make notes for a paper on the subject. The same evening I did this pretty fully, and on the two succeeding evenings wrote it out carefully in order to send it to Darwin by the next post, which would leave in a day or two.[6]

Four months later, in June 1858, Darwin received Wallace's paper. He was amazed. For 20 years he had worked on his theory and now it seemed that he would be forestalled. At once he wrote to Lyell:

> I never saw a more striking coincidence. If Wallace had my MS. sketch written out in 1842, he could not have made a better short abstract! Even his terms now stand as heads to my chapters.[7]

But Darwin's priority was preserved. It was agreed that Wallace's paper and some extracts from Darwin's work should be presented at the next meeting of the Linnaean Society on 1 July 1858.

Darwin immediately set to work to write out his ideas in full. The result was *The Origin of Species* which was published on 12 November 1859 and sold out all 1250 copies in a single day.

What then did this famous book contain? The full title of the first edition was *On the Origin of Species by Means of Natural Selection; or The Preservation of Favoured Races in the Struggle for Life.* In later editions the subtitle became *Or, The Survival of the Fittest in the Struggle for Life.* The phrase 'survival of the fittest' was not Darwin's own—he borrowed it from Herbert Spencer (see Chapter 4.10) and from Wallace.

The book—which Darwin called 'This Abstract which . . . must necessarily be imperfect'—was, again in Darwin's words, 'one long argument'. In the first five chapters Darwin sets out the core of his argument. As he puts it in his Introduction:

> I shall devote the first chapter of this Abstract to Variation under Domestication. We shall thus see that a large amount of hereditary modification is at least possible, and, what is equally or more important, we shall see how great is the power of man in accumulating by his Selection successive slight variations. I will then pass on to the variability of species in a state of nature; but I shall, unfortunately, be compelled to treat this subject far too briefly, as it can be treated properly only by giving long catalogues of facts. We shall, however, be enabled to discuss what circumstances are most favourable to variation. In the next chapter the Struggle for Existence amongst all organic beings throughout the world, which inevitably follows from the high geometrical ratio of their increase, will be considered. This is the doctrine of Malthus, applied to the whole animal and vegetable kingdoms. As many more individuals of each species are born than can possibly survive, and as, consequently, there is a frequently recurring struggle for existence, it follows that any being, if it vary however slightly in any manner profitable to itself, under the complex and sometimes varying conditions of life, will have a better chance of surviving, and thus be *naturally selected*. From the strong principle of inheritance, any selected variety will tend to propagate its new and modified form.
>
> This fundamental subject of Natural Selection will be treated at some length in the fourth chapter; and we shall see how Natural Selection almost inevitably causes much Extinction of the less improved forms of life, and leads to what I have called Divergence of Character.[8]

In his third chapter, 'Struggle for Existence', Darwin expands these points:

> Nothing is easier than to admit in words the truth of the universal struggle for life, or more difficult—at least I have found it so—than constantly to bear this conclusion in mind. Yet unless it be thoroughly engrained in the mind, the whole economy of nature, with every fact on distribution, rarity, abundance, extinction, and variation, will be dimly seen or quite misunderstood. . . . There is no exception to the rule that every organic being naturally increases at so high a rate, that, if not destroyed, the earth would soon be covered by the progeny of a single pair. Even slow-breeding man has doubled in twenty-five years, and at this rate, in less than a thousand years, there would literally not be standing-room for his progeny. Linnaeus has calculated that if an annual plant produced only two seeds—and there is no plant so unproductive as this—and their seedlings next year produced two, and so on, then in twenty years there should be a million plants. The elephant is reckoned the slowest breeder of all known animals, and I have taken some pains to estimate its probable minimum rate of natural increase; it will be safest to assume that it begins breeding when thirty

years old, and goes on breeding till ninety years old, bringing forth six young in the interval, and surviving till one hundred years old; if this be so, after a period of from 740 to 750 years there would be nearly nineteen million elephants alive, descended from the first pair.

But we have better evidence on this subject than mere theoretical calculations, namely, the numerous recorded cases of the astonishingly rapid increase of various animals in a state of nature, when circumstances have been favourable to them during two or three following seasons. Still more striking is the evidence from our domestic animals of many kinds which have run wild in several parts of the world; if the statements of the rate of increase of slow-breeding cattle and horses in South America, and latterly in Australia had not been well authenticated, they would have been incredible. So it is with plants; cases could be given of introduced plants which have become common throughout whole islands in a period of less than ten years. . . .[9]

In chapters six to ten of *The Origin of Species* Darwin sets out a vast number of difficulties and possible objections to his theory: 'the difficulties of transitions, or how a simple being or organ can be changed and perfected into a highly developed being or into an elaborately constructed organ . . . the subject of Instinct, or the mental powers of animals . . . Hybridism, or the infertility of species and the fertility of varieties when intercrossed'[10] and last but not least the imperfection of the geological record and the absence of fossil evidence of many of the stages in the evolution of separate species. In these chapters Darwin demonstrates the value of what he called his golden rule: 'that whenever a published fact, a new observation or thought came across me, which was opposed to my general results, to make a memorandum of it without fail and at once: for I had found by experience that such facts and thoughts were far more apt to escape from the memory than favourable ones. Owing to this habit, very few objections were raised against my views which I had not at least noticed and attempted to answer.'[11]

Only after trying to meet the main objections to his theory did Darwin point to a large number of facts which, while not a direct consequence of his theory, are much more plausibly explained by Darwin's ideas than by any existing alternative. These facts include the overall features and many details of the distribution of species and the geological succession of organic beings as revealed by the fossil record which Darwin schematically illustrated by a branching 'tree of life' (see Fig. 4.8.4).

The whole book is written in a clear, measured and almost pedestrian style which is totally in keeping with what we know of Darwin's character. There is no trace of polemic. The overriding impression is of a man, well aware of the importance of what he was writing and of how controversial it might be, striving not to overstate his case and wanting the totality of the evidence to speak for itself.

Yet, almost despite himself, Darwin's amazement at the dynamic nature of what he had conceived is apparent. In his own words again:

In looking at Nature, it is most necessary to keep the foregoing considerations

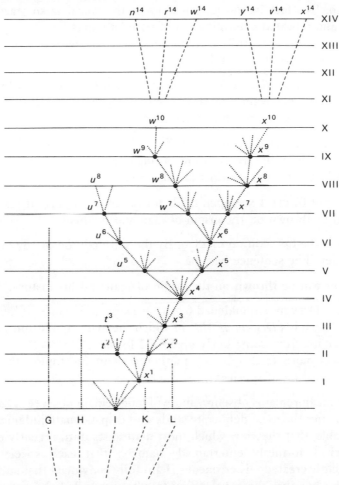

Fig. 4.8.4 Darwin's hypothetical tree of life showing how a single species might evolve over 14,000 generations. The diagram indicates that, given enough time, six distinct current species could evolve from a single ancestor and illustrates another extremely common event—the extinction of species as G, t^3, u^8 etc. *Redrawn from Coleman, W., Biology in the Nineteenth Century, Cambridge: Cambridge University Press, 1971.*

always in mind—never to forget that every single organic being may be said to be striving to the utmost to increase in numbers; that each lives by a struggle at some period of its life; that heavy destruction inevitably falls either on the young or old, during each generation or at recurrent intervals. Lighten any check, mitigate the destruction ever so little, and the number of the species will almost instantaneously increase to any amount.[12]

For Darwin it was the scale of the extinction—both of individuals and of species—that was so astounding. As he wrote of the very many species which had become extinct, more than eight million species compared with the existing million or so, 'No fact in the long history of the world is so startling as the widespread and repeated extermination of its inhabitants.'

To understand his ideas fully Darwin thought it was necessary to be staggered by them. As he wrote to a friend:

> If you are *in even so slight a degree* staggered, then I am convinced with further reflection you will become more and more staggered, for this has been the process through which my mind has gone.[13]

But the most staggering of the implications of Darwin's ideas—their bearing on the origin of man himself—receives only a single sentence in the *Origin of Species*. After nearly 500 pages Darwin had this to say:

> In the distant future I see open fields for far more important researches. . . . Light will be thrown on the origin of man and his history.[14]

That was in the first edition of 1859. By the sixth edition of 1872 Darwin had become bolder. The sentence became:

> Much light will be thrown on the origin of man and his history.[15]

But by then Darwin, emboldened by the growing acceptance of his ideas, had, in 1871, published his *Descent of Man* of which the main conclusion was '. . . that man is descended from some lowly organised form. . . .'[16]

Darwin was more forthright on another contentious topic—the fixity and independent creation of species:

> Although much remains obscure, and will long remain obscure, I can entertain no doubt, after the most deliberate study and dispassionate judgment of which I am capable, that the view which most naturalists until recently entertained, and which I formerly entertained—namely, that each species has been independently created—is erroneous. I am fully convinced that species are not immutable; but that those belonging to what are called the same genera are lineal descendants of some other and generally extinct species, in the same manner as the acknowledged varieties of any one species are the descendants of that species. Furthermore, I am convinced that Natural Selection has been the most important, but not the exclusive, means of modification.[17]

Cautious but firm.

Darwin also knew that he had overturned the traditional theological argument from Design for the existence of God. No longer was it possible to argue that the marvellous adaptations of living things to specific environments or functions was due to their having been explicitly designed by a Creator for that environment or function. After Darwin it was clear that if living things were not well adapted to their environment, they would not be there at all. But Darwin was also aware that this demolition of the detailed argument from Design did not dispose of the question of a Designer or Creator. In the last paragraph of the book—the only paragraph in which Darwin allows a hint of lyricism into his style—he had this to say:

It is interesting to contemplate a tangled bank, clothed with many plants of many kinds, with birds singing on the bushes, with various insects flitting about, and with worms crawling through the damp earth, and to reflect that these elaborately constructed forms, so different from each other, and dependent upon each other in so complex a manner, have all been produced by laws acting around us. These laws, taken in the largest sense, being Growth with Reproduction; Inheritance which is almost implied by reproduction; Variability from the indirect and direct action of the conditions of life, and from use and disuse: a Ratio of Increase so high as to lead to a Struggle for Life, and as a consequence to Natural Selection, entailing Divergence of Character and the Extinction of less improved forms. Thus, from the war of nature, from famine and death, the most exalted object which we are capable of conceiving, namely, the production of the higher animals, directly follows. There is grandeur in this view of life, with its several powers, having been originally breathed by the Creator into a few forms or into one; and that, whilst this planet has gone cycling on according to the fixed law of *gravity*, from so simple a beginning endless forms most beautiful and most wonderful have been, and are being evolved.[18]

Darwin always knew that his views would be controversial. A few days before *The Origin of Species* appeared, Darwin wrote, in a letter to Wallace, 'God knows what the Public will think'. And what did the public think? The public debate is usually thought to be summarised by the confrontation at the Oxford meeting of the British Association in 1860 between Bishop Samuel Wilberforce and Darwin's friend, the biologist Thomas Henry Huxley. Wilberforce is reputed to have asked Huxley to say whether he was descended from the apes on his father's or his mother's side—and Huxley is supposed to have replied that he would rather have an ape for an ancestor than a man who would abuse his intellect by using aimless rhetoric to cover his ignorance. Huxley is generally supposed to have had the better of the argument and his triumph at Oxford has been acclaimed as a victory for science over religion.

But, as in the conflict between Galileo and the Church, the issues were much too complex to be accurately described as a simple conflict between science and religion.

Even amongst biologists, there were many who were not convinced by Darwin—although by 1871 Darwin was able to write that 'it is manifest that at least a large number of naturalists must admit that species are the modified descendants of other species; and this especially holds good with the younger and rising naturalists. The greater number accept the agency of natural selection. . . .' But '. . . of the older and honoured chiefs in natural science, many unfortunately are still opposed to evolution in every form.'[19] And many physicists and chemists long remained hostile to evolution, with Lord Kelvin in particular giving convincing thermodynamic arguments which seemed to show that the earth could not possibly be old enough for evolution to be plausible (see Chapter 6.8).

The religious arguments were not straightforward either. To the layman,

Darwin's views on evolution were in direct contradiction to the biblical account of creation. But, if he had known about it, the layman would have been even more shocked by the increasing tendency amongst 19th-century theologians to subject the Bible to the same kind of critical analysis which the growing band of professional historians were learning to apply to any historical document. By the year 1900 both the book of scripture and the book of nature were not quite what they had been a century earlier.

Finally Darwin's ideas were also beginning to influence the social sciences. Here they were probably even more controversial than they had been in the natural sciences or in theology. In Chapter 4.10 we will see how Darwin's ideas influenced the social thinking of Herbert Spencer. Spencer's writings were an important factor in the development of a movement called Social Darwinism which is usually thought of as being on the political right. But Darwin's writings were also well regarded by Karl Marx, the prophet of the left, who hoped that his own works would be as important in politics and economics as Darwin's were in Biology.

Darwin's work resembles Newton's in the range of human activities which it influenced. And, like Newton's work in physics, Darwin's ideas are still of central importance in modern biology.

REFERENCES

1   Quoted in Broberg, G., *Carl Linnaeus 1707–1778*, (Uppsala: Uppsala University Press, 1978), p. 10.
2   Darwin, C., *The Autobiography of Charles Darwin*, (New York: Dover Publications, 1958), pp. 41–2.
3   Ibid., p. 42.
4   Ibid., p. 42.
5   Ibid., pp. 42–3.
6   Quoted in Bronowski, J., *The Ascent of Man*, (London: BBC Publications, 1973), pp. 306–8.
7   Letter from Darwin to Lyell in Darwin F. (ed.), *The Life and Letters of Charles Darwin* (London: 1887), Vol. II, pp. 116–17.
8   Darwin, C., *The Origin of Species*, (New York: Mentor, 1958), Reprint of Sixth Edition of 1872, p. 29.
9   Ibid., pp. 74–6.
10   Ibid., p. 29.
11   Darwin, C., *The Autobiography of Charles Darwin*, op. cit., p. 45.
12   Darwin, C., *The Origin of Species*, op. cit., p. 77.
13   Letter from Darwin to Henslow in Darwin F. (ed.), op. cit., Vol. II, p. 218.
14   Darwin, C., *Origin of Species*, (London: 1859), First edition, p. 488.
15   Ibid., Sixth edition, p. 449.
16   Darwin, C., *The Descent of Man*, (London, 1871); partly reprinted in *Darwin*, (New York: Norton, 1970), p. 275.
17   Darwin, C., *Origin of Species*, op. cit., Sixth edition, p. 30.
18   Ibid., p. 450.
19   Darwin, C., *The Descent of Man*, op. cit., p. 200.

SUMMARY

1   Darwin's work was based on an encyclopedic knowledge of living things mainly acquired from the round-the-world voyage of the *Beagle* from 1831–6.

2   In 1838 Darwin 'happened to read for amusement Malthus on *Population*' and this gave him 'a theory by which to work'; however it was not until more than 20 years later that he published his ideas.

3   In 1859 Darwin published his famous book which contains 'one long argument' which is summed up in its title: *On the Origin of Species by Means of Natural Selection; or the Preservation of Favoured Races in the Struggle for Life*.

4   *The Origin of Species* made almost no mention of man, but in 1871 Darwin published *The Descent of Man* which concluded that '. . . man is descended from some lowly organised form. . . .'

5   Darwin's theory of evolution aroused much controversy but within a few years its main premises were accepted by 'the younger and rising naturalists' and, more than a century later, it is still of central importance in modern biology.

In the last four chapters we considered some major areas of growth in scientific knowledge during the early and middle years of the 19th century. We described the emergence of chemistry, geology and much of physics as separate sciences in something like their modern form and we outlined Darwin's ideas on evolution which transformed the study of living things and are still at the heart of modern biology.

In the next three chapters we are going to look at some other changes which were taking place at roughly the same time. We will describe the further development of the Industrial Revolution in Britain, first discussing technical aspects of that revolution, and then considering its social consequences, including further developments in the social sciences which were partly stimulated by the spread of industrialisation. Finally we will describe the way in which the Industrial Revolution spread to most of Europe and thus contributed to the world-wide expansion of European influence.

4.9   Science and the Industrial Revolution in the 19th century

In Chapter 4.4 we described various developments which took place in the 18th century in those three pillars of the early Industrial Revolution: cotton, iron and steam. One question we asked then was: To what extent were these developments due to previously discovered scientific knowledge? The answer we gave was—not very much. It was much more the empirical, trial-and-error experimental attitude which drove the early pioneers of the Industrial Revolution rather than the direct application of scientific knowledge.

In this chapter we are going to look at the Industrial Revolution at a later stage—in the early and mid-19th century—and to ask the same question: What role did science play in its development? Once again, before trying to give an answer, we need to describe some of the major industrial developments of the period—the further development of the steam engine and the rise of railways; the various changes in the coal and iron industries; and the practical application of the new discoveries in electricity and magnetism (see Chapter 4.7).

THE RISE OF THE RAILWAYS

First, the steam engine. Fig. 4.4.13 shows that another big increase in efficiency took place in the early 19th century with the development of the Cornish pumping engine. These advances in efficiency were largely the work of the Cornishman, Richard Trevithick, who made two major changes in the design of steam engines. First, he used steam at high pressures, ten atmospheres or more. The technology of iron casting had advanced enough to make this feasible—a great change since Savery's high pressure engine a century earlier. Then Trevithick realised that he could now do away with James Watt's condenser. The difference between, say, ten atmospheres and one, and ten atmospheres and a partial vacuum was not very great, and it was more than counterbalanced by the more compact engine which Trevithick's changes made possible. His engine had only one cylinder and no condenser or pumps, and he also did away with the large beam which had been a major feature of steam engines since Newcomen's first engine at the beginning of the 18th century (see Fig. 4.4.10 and 4.4.11, pages 137–8). This was how the steam engine became portable.

The earliest railway locomotives used compact and efficient steam engines of this kind. Fig. 4.9.1 shows an early railway at Leeds in 1811 which was capable of moving 20 tons of coal at 10 miles per hour. In the early 1820s Robert Stephenson considerably improved the design of railway engines, and the first passenger line in the world, the Manchester and Liverpool Railway, opened in 1830. Fig. 4.9.2 shows various forms of railway travel which were becoming available in the 1830s—ranging from first-, second- and third-class passenger travel to trains for transporting goods and animals.

Fig. 4.9.1   Early goods railway in Leeds, 1811; this train could move 20 tons of coal at 10 miles per hour. *From 'The Industrial Revolution', Filmstrip 3, Common Ground, Longman Group Limited.*

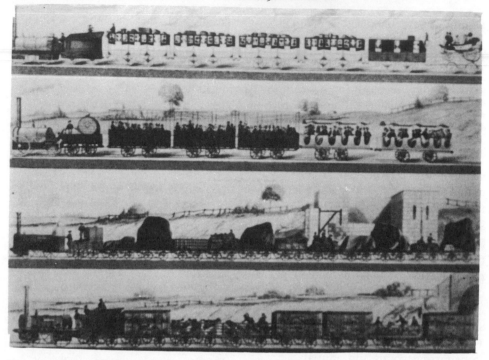

Fig. 4.9.2   Railway trains in the 1830s showing three classes of passenger travel (third-class passengers in open carriages), and trains for goods and animals. *From 'The Industrial Revolution', Filmstrip 3, Common Ground, Longman Group Limited.*

The early railways were not without their hazards—one man, Huskisson, the Home Secretary, was even killed at the official opening of the Manchester and Liverpool Railway and many more died in accidents during their construction. But, despite many problems, railways were built very rapidly all over Britain in the next two decades. In 1835 there were only 500 miles of track, but by 1845 this figure had risen to nearly 4000 miles, and by 1850 to 10,000 miles. This rapid growth is illustrated even more graphically in Fig. 4.9.3, which shows the railway network in 1845 and the further development by 1851. The 1840s have been described as a period of railway mania, so rapid was the rate at which they were both proposed and built. One of the most famous of the engineers who supervised these vast engineering projects was Isambard Kingdom Brunel (see Fig. 4.9.4)—a supremely practical and confident man. Brunel was both a superb engineer and a romantic. He built a tunnel under the Thames, the great suspension bridge at Clifton near Bristol and he pushed through and completed the Great Western Railway—an extraordinary feat of determination and engineering skill. It is said that when it was finished Brunel wanted to carry on across the Atlantic and so he built his 24,000 ton steamship, the *Great Eastern*. Fig. 4.9.4 shows Brunel in front of the huge chains used at the launching of the *Great Eastern*.

But in the 1840s even Brunel was upset by the pace of developments. As he wrote to a friend in 1845: 'Here the whole world is railway mad. I am really sick of

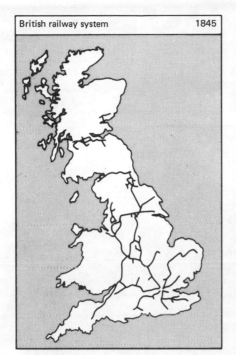
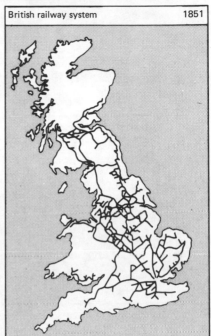

Fig. 4.9.3 The 'railway mania' of the 1840s; the growth of the railway network between 1845 (left) and 1851. *Redrawn from Open University Arts Foundation Course A100 (replaced in 1978). Units 29–30, The Industrialisation Process, 1830–1914, copyright © The Open University Press.*

hearing proposals made. I wish it were at an end. . . . The dreadful scramble in which I am obliged to get through my business is by no means a good sample of the way in which work ought to be done . . . I wish I could suggest a plan that would greatly diminish the number of projects; . . . it would suit my interests and those of my clients perfectly if all railways were stopped for several years to come'.[1]

Nevertheless, the boom went on and railway building continued during the rest of the century until in 1920 the network was over 30,000 miles long.

COAL AND IRON—THE MATERIAL BASIS OF THE INDUSTRIAL REVOLUTION

Coal was the basis of much of 19th-century technology, and its increasing importance is closely connected with the railway boom. The annual coal output of British mines rose from about 10 million tons per year in 1750 to 280 million tons in 1913. Even more striking is the rapid rise from about 1835 onwards (see Fig. 4.11.2, page 214). This coincided with the railway boom, which made coal easily available all over the country and not just in mining areas. What was all this coal used for? Much of it was used for domestic heating. Coal replaced timber which was in increasingly short supply, and by 1800 over 80 per cent of the

Fig. 4.9.4   Isambard Kingdom Brunel in 1857 in front of the chains used in the launching of his vast steamship—the *Great Eastern*.

world's coal was mined in Britain. Over half the output was used for domestic heating in 1800 (about 7 million tons) and so was about a third of the much larger output in 1900 (about 70 million tons). But remember that the population rose from 10 million in 1800 to nearly 40 million a century later in 1900. So they weren't really keeping themselves ten times as warm!

Large as this increase in domestic coal-burning was, it doesn't account for all the increase in coal consumption. What else was it used for? Some, of course, was used by the railway engines themselves. But various other large-scale uses also developed. For example, coal was used as a source of coke, of gas and of various chemicals. Let us look briefly at these different uses.

A PEEP AT THE GAS LIGHTS IN PALL-MALL.

Fig. 4.9.5 Gas lighting in Pall Mall, London; cartoon by Rowlandson, 1809.

The rising demand for coke was a result of the changes in iron smelting techniques which took place in the 18th century (see Chapter 4.4). In the 18th century coke had replaced charcoal as the fuel in most blast furnaces and this led to a considerable increase in total iron production. In the 19th century iron production increased even more dramatically (see Fig. 4.11.3, page 215) and of course this meant that coke was needed in similarly increasing quantities.

Initially coke was produced from coal in much the same way as charcoal was from wood—by heating it in a confined space from which air was partially excluded. The smoke and gases which were given off were allowed to disperse into the atmosphere and must have caused appalling pollution, particularly as more and larger coke ovens were built. But it was soon realised that these 'waste products' need not be wasted. As a result, two new coal-based industries developed—one concerned with coal-gas and the other with coal-tar. The gas industry developed early in the century and much of the early experimental work was done at Boulton and Watt's factory in Birmingham (see Chapter 4.4). As early as 1807 a demonstration of street lighting by gas took place in Pall Mall in London; Fig. 4.9.5 shows a contemporary cartoon of this event. In 1814 the parish of St Margaret's, Westminster, was lit by gas and two years later there were 26 miles of gas mains in London. And by 1823 streets, homes and factories in 52 English towns were lit by gas. One reason given for lighting the streets in this way was 'as a means of Police' for London, and one London gas company appears to have been formed in an attempt to get rid of highwaymen from the Whitechapel Road! It is interesting that these large-scale applications of coal-gas, involving mass production, storage and distribution, were taking place only 30 years or so after Priestley's pioneering laboratory studies on the handling and properties of gases (see Chapter 4.5).

The coal-tar industry developed much later. Up to about 1850 the large amounts of coal-tar produced in coking and gas plants were an unpleasant and not particularly useful by-product—and one which was not easily disposed of. But from the mid-1850s this situation changed. Why? In the main, it was because of the growth of chemical knowledge, and in particular of what we now call organic chemistry, over the previous 30 years. One of the starting points was the discovery of benzene by Michael Faraday at the Royal Institution in 1825—the same Faraday whose work on electromagnetism we discussed in Chapter 4.7. Faraday isolated benzene from whale-oil by the long and tedious process of fractional recrystallisation. It was this technique, combined with fractional distillation, that led to the discovery of benzene and various other organic compounds, such as aniline and toluene, in coal-tar during the 1840s. This work was carried out at the Royal College of Chemistry, which had been founded in 1845 under the direction of the German chemist, A. W. Hofmann, who had been brought to London by Queen Victoria's husband, Prince Albert. It was this work that underlay the famous discovery by William Perkin in 1856 of a mauve dye which he made from aniline. This mauve, or aniline purple, was the first of a series of synthetic dyestuffs to be manufactured on a large scale in the second half of the 19th century. And the basic raw material for this new industry was the formerly useless waste-product—black, evil-smelling coal-tar.

However, much of the development of the coal-tar industry, and its later diversification into other organic chemicals such as pharmaceutical drugs, saccharin and explosives, took place not in England, where many of the key discoveries had been made, but in Germany. The new German chemical industry was much quicker to develop new products and to employ large numbers of trained chemists. But, ironically, it still imported most of its coal-tar from England.

APPLICATIONS OF ELECTROMAGNETISM

Finally, let us consider some of the applications of the electromagnetic discoveries made by Faraday in the 1830s (see Chapter 4.7).

One such application was in communications. The electric telegraph was first introduced in 1837, primarily for railway signalling purposes, and submarine cables to carry telegraph messages were successfully laid across the English Channel in 1850 and, much more difficult, across the Atlantic in 1866.

But probably the most important engineering uses of electromagnetism were the development of the electric motor and the dynamo. The principles underlying these machines had been established by Faraday early in the century—1821 for the motor and 1831 for the dynamo. But it was not until well into the second half of the 19th century that they were used in large numbers.

The early motors were technically successful but they were hopelessly uneconomic compared with steam, primarily because they had to be driven by bulky and expensive batteries. A cheaper source of electricity was needed and Faraday's discovery of the principle of the dynamo pointed to one obvious possibility. The first magneto–electric machines or magnetos were hand-driven and used bulky permanent magnets (see Fig. 4.9.6). Later machines were driven by steam engines and used electromagnets. They were christened dynamos and their efficiency was considerably increased by improvements in design. By about 1870, dynamos were being used for the electric arc lighting of public places.

However, the demand for electricity did not really rise rapidly until the invention of the incandescent filament light bulb by Edison and Swan in the 1870s. After this it was not long before the first power stations were built—in New York and London in 1882, and in Berlin in 1888. Fig. 4.9.7 shows the dynamo room at Edison's power station in New York which opened in 1882; the dynamos were driven by steam. In the growing cities of the late 19th century, it was economic to lay out electrical mains circuits for lighting, and electricity rapidly displaced gas lighting.

Once supplies of mains electricity were available, it also became economic to develop powerful electric motors which were used in the first electric trams and electric railways. These operated first in the cities of the United States during the late 1880s, while the first London tube line opened in 1890 and is still in use—it now forms part of the Inner Circle.

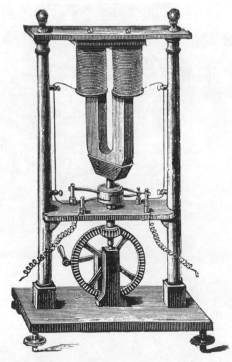

Fig. 4.9.6 Hand-driven magneto–electric machine using a permanent magnet, 1833—a forerunner of the dynamo.

Fig. 4.9.7 Steam-driven dynamos in Edison's first power station for electric lighting, New York, 1882.

Fig. 4.9.8 Electrically-powered weaving factory, about 1880.

Another major development was the harnessing of the energy of the Niagara Falls and the transmission of the electric power along power lines to the city of New York. Electricity was first generated from Niagara in 1896.

Finally, electric motors were now being widely used in factories to drive the machines and this had a dramatic effect on the organisation and efficiency of these factories. Compare the factory shown in Fig. 4.4.6 (page 133), in which the machines are driven by belts all running from a single steam engine, with the electrically driven weaving factory of 1880 shown in Fig. 4.9.8. The machines now have their own electric motors and can be situated anywhere on the factory floor, while the whole factory is lit by electricity.

Now let us try to answer our original question—how much were all these developments caused by the increase in scientific knowledge? In the development of the steam engine and railways, probably not very much. Carnot's work (see Chapter 4.7) was little read until the 1850s. But in the coal-gas industry knowledge about gases and their properties was important, and later developments in the iron and steel industry involved some application of chemical knowledge. However, it was the coal-tar industry which was probably one of the earliest clear examples of a science-based industry. And the extensive developments in the new electrical industry were firmly based on the electrical and electromagnetic discoveries of Faraday, which had been made many years earlier.

In all these industrial developments the empirical, experimental scientific attitude was still important. But many industries depended to an increasing extent on previously discovered scientific knowledge. And this is a trend which has continued right up to the present day.

REFERENCE

1 Letter from I. K. Brunel, 1845; quoted in Rolt, L. T. C., *Victorian Engineering*, (Harmondsworth: Penguin, 1974), p. 23.

SUMMARY

1 In 19th-century Britain the Industrial Revolution continued and many new large-scale industries developed.
2 More efficient steam engines were made and these led to the dramatic expansion of the railways—from 100 to 10,000 miles of track in about 20 years.
3 Coal became the basis of much of the new 19th-century technology, both as a fuel and as a source of coke for the iron industry, of coal-gas for the gas industry and of coal-tar for the chemical industry.
4 Faraday's electromagnetic discoveries led to major applications in electric telegraphy, in the generation of electric power for lighting and in the development of electric motors for factories, railways and trams.
5 In all these developments the empirical, experimental scientific attitude was important, but many of them also depended to an increasing extent on previously discovered scientific knowledge.

4.10 The rise of the social sciences

The Industrial Revolution had led to enormous changes in the way people lived and worked. So it is not surprising that the development of the social sciences in the 19th century was greatly influenced by the changes brought about by industrialisation. In this chapter we will first describe some of these social changes and then see how these changes influenced the work of some of the founding fathers of the social sciences—Comte, Spencer, Marx, Weber and Durkheim. Finally we will describe early attempts to identify social factors in the causes and spread of disease.

SOCIAL CHANGE IN THE 19TH CENTURY: URBANISATION AND INDUSTRIALISATION

The Industrial Revolution depended on power-driven machinery. Water was the first source of power for some of the large enterprises like mills. In the early days mills had to be built in river valleys because they depended on the fall of water. This meant that they could not be too close together. However, with the use of the rotary steam engine, factories could be built close together in towns, where coal supplies and abundant labour were readily available. During the 19th century people from the countryside poured into the nearest industrial town and urbanisation proceeded at a reckless pace, storing up social problems, many of which still await solution. These figures tell some of the story:

Table 4.10.1   Population growth in the 19th century

	1801	1851	1910
England and Wales	10·5 m	20·8 m	40·8 m
Birmingham	71 000	233 000	526 000
Manchester–Salford	95 000	401 000	714 000
Glasgow	84 000	329 000	784 000
London	1 117 000	2 689 000	7 256 000

Thus the numbers living in towns were increasing much faster than the overall population, which itself doubled between 1800 and 1850 and doubled again by 1910.

Among the tremendous social problems which were created by this move to the towns were those of bad housing, overcrowded work conditions, diseases and bad sanitation.

Fig. 4.10.1   Back-to-back housing built in Lancashire in the early 19th century. *From 'The Industrial Revolution', Filmstrip 4, Common Ground, Longman Group Limited.*

Housing

The problems of housing arc still seen in some remnants of the appalling slum conditions of the notorious 'back-to-back' houses in many large cities. Small, dreary and with totally inadequate sanitation, they provided cramped and unhealthy homes for vast numbers of the urban working classes (see Fig. 4.10.1). Also, in many cities, the environment was literally blackened by industrial processes such as those of the iron industry, which was growing very fast—for example, the output of pig-iron rose from 678,000 tons in 1830 to 3,218,000 tons by 1855 (see Fig. 4.11.3, page 215). All day long, smoke belched from blast furnaces, and towns in regions such as the Black Country, became blacker and blacker (see Fig. 4.10.2). This is Dickens' description of the Black Country in about 1840:

> On every side, and as far as the eye could see into the heavy distance, tall chimneys crowding on each other, poured out their plague of smoke, obscured the light and made foul the melancholy air. . . . Men, women and children, wan in their looks and ragged in their attire, tended the engines, . . . begged upon the road, or scowled half-naked from doorless houses. . . . But night-time in this dreadful spot! Night, when the smoke was changed to fire, when every chimney spurted up its flame, and places that had been dark vaults all day, now shone red hot, with figures moving to and fro within their blazing jaws, and calling to one another with hoarse cries—night when the noise of every strange machine was aggravated by the darkness; when the people near them looked wilder and more savage. . . .[1]

Fig. 4.10.2   The John Wilkinson steelworks at Bilston in 1836. *From 'The Industrial Revolution', Filmstrip 2, Common Ground, Longman Group Limited.*

Conditions of work

The factory system and mass production made rapid headway, employing increasing numbers of people, often in horrifying conditions. In the cotton mills the worst features of the new industrialism prevailed: long hours, insanitary conditions, child labour. Large numbers of poor children were handed over to employers from the age of seven, to work for over 12 hours a day, Saturdays included, under the control of overseers who often used the whip on them. Sometimes children worked for 14 or 15 hours a day for six days a week, with meal times being given up to clean machinery. Attempts to control this situation included the Health and Morals of Apprentices Act of 1802 and the Factory Act of 1819, but conditions in many factories remained deplorable. Listen to this evidence given to the government Committee on Factory Children's Labour in 1831–2:

'At what time in the morning, in the brisk time, did those girls go to the mills?'
'In the brisk time, for about 6 weeks, they have gone at 3 o'clock in the morning, and ended at 10, or nearly half-past, at night'.

'What intervals were allowed for rest or refreshment during those 19 hours of labour?'
'Breakfast $\frac{1}{4}$ hour, and dinner $\frac{1}{2}$ hour, and drinking $\frac{1}{4}$ hour.'

'Was any of that time taken up in cleaning the machinery?'
'They generally had to do what they call dry down; sometimes this took the whole of the time at breakfast or drinking, and they were to get their dinner or breakfast as best they could; if not, it was brought home.'

'Had you not great difficulty in awaking your children for this excessive labour?'
'Yes, in the early time we had to take them up asleep and shake them when we got them on the floor to dress them, before we could get them off to their work . . .'

Such was the story of factory life for many workers; but for the outworkers, who still worked in their own homes, conditions could be equally harsh. It was about these workers, in jobs such as domestic handicrafts, that Hood sang his 'Song of the Shirt':

> With fingers weary and worn,
> With eyelids heavy and red,
> A woman sat in unwomanly rags,
> Plying her needle and thread—
> Stitch—Stitch—Stitch!
> In poverty, hunger and dirt,
> And still with a voice of dolorous pitch,
> She sang the Song of the Shirt.

Equally appalling conditions of work were common in other areas too, such as the coal mines. A Royal Commission of 1842 revealed a shocking state of affairs.

. . . . instances occur in which children are taken into these mines as early as 4 years of age . . . . while from 8 to 9 is the ordinary age at which employment in these mines commences.

Such children were often employed to open and shut ventilation traps, which meant they had to sit for hours in small niches cut out in the coal, where, in the words of the Commissioners, their work was 'solitary confinement of the worst order'.

The setting up of this Royal Commission is one example of many government-sponsored enquiries which led to legislation designed to reduce some of the worst horrors of the time.

Disease and death
Given these glimpses of some of the unhappy aspects of life in early 19th-century Britain, it is not surprising that there were also great problems of disease and very high death rates. Apart from the obvious dangers to life associated with such appalling conditions—such as overtired children dropping asleep at work and falling into moving machinery—there were many health hazards in the bad sanitary arrangements in the towns and in the country. Infectious diseases like tuberculosis took great toll in families that lived crowded together, inadequately

fed and grossly overworked. Other epidemic diseases spread like wild-fire through insanitary drainage systems or contaminated water supplies. Thus, outbreaks of cholera or typhoid fever were common. They could ravage a town or village and suddenly kill off many of the population. One such epidemic hit a little village in Essex as late as 1867 and is recorded for us in letters written by the local people at the time. The main writers lived at the manor house, called Terling Manor, and they describe the way in which typhoid fever suddenly afflicted their village. These letters also show the state of medical knowledge at that time.

Lord Rayleigh, *Dec. 21:*
'We are in a great "muddle"—more than 100 people in the Parish attacked with a low fever. . . . Mr. Gimson (the local doctor) is very attentive and purifies the whole place with "disinfecting" liquid—chlorine of lime or zinc. . . .'

Dec. 26: 120 cases yesterday we heard and Dr. Gimson does not feel sure we have seen the worst of it . . . the scullery maid after struggling bravely against it for some days has at last succumbed . . . that makes the 8th servant and 6 of them are in the house . . . (Clara Strutt, Lord Rayleigh's daughter).

Dec. 26: (Lord R.): 'The bell was tolling for another death—the 4th . . . Mr. Gimson tells me we may expect a death every day for the next week . . .'

Dec. 30: 'the Thurgoods have lost their 3 daughters . . . they have only 1 child left—a boy—poor things are in a dreadful way. . . .'

Jan. 4: 'Mary Ann our head housemaid is dead . . . our kitchen maid also is very bad and the 2 Avis are dead, making the numbers of deaths 15 in all. . . . Dr. G. was over here . . . and said there were 170 cases exclusive of the 15 who died and there are more since. . . . A geologist is coming tomorrow to examine the soil. . . .'

Jan. 17: (C.S.): 'Poor young Thurgoods—the only remaining child has got it severely, 2 men . . . died this week; many more men have got it. Mrs. Smith at the mill . . . has got it very badly and her boy died. . . .'

Feb. 1: (C.S.): 'There have been some more deaths . . . this makes 28 deaths, 6 this week. . . .'

Feb. 10: (C.S.): 'Alfred Turner died last week, and also a Mrs. Webb which makes 33 deaths. . . .'

It was not until April 4th that Dr. Gimson could reply:

April 4: (Dr. Gimson): 'All the cases are improving . . . how glad I shall be when all is over and quiet reigns once more—I am weary. . . .'

May 6: (Lord R.): 'Our thanksgiving day and sermon went off well. . . . We had a very full church—we had 5 baptisms and 3 fatherless—very affecting.'[2]

These illustrations of certain aspects of 19th-century life in Britain highlight some of the social problems of the age and provide some of the background against which the men of ideas were developing their thinking. We will now turn to see what was happening in the world of ideas.

THE WORLD OF IDEAS

In Chapter 4.2 we described the optimism which reigned at the end of the 18th century, with the hopes that people would be able to use their new knowledge of both natural and social sciences to create a Heavenly City on Earth. But, as we have just seen, the cities which grew so fast in the 19th century were not very heavenly. Inevitably the thinkers of the 19th century were deeply affected by the events occurring around them and we will see many strands in their work which reflect concern over the situations we have been describing.

The social scientists were also influenced by the natural sciences, but in ways rather different from their 18th-century predecessors. For example, some of them, such as Durkheim, wanted to apply the methods of the natural sciences directly to the social sciences. Others, such as Max Weber, felt that a science of society would need to be rather different, as it must also take account of different kinds of phenomena—people's feelings, beliefs and will—which are more difficult to see and measure than many phenomena in the physical world.

One of the interesting features which we will see is that, as the social sciences developed, a great deal of disagreement emerged. We will note how many social scientists argue with each other about fundamental matters, and we shall see how deep fissures developed which are still present today. Subjects such as sociology or psychology do not have a generally agreed core of knowledge such as you find in geography or biology or physics. They consist of a number of different approaches or perspectives which may shed light on various aspects of human behaviour and experience, but as yet they have not been brought together into a generally agreed body of knowledge. So, if you feel a little confused by the different ways in which social scientists approach their subjects, don't be worried: the confusion is in the social sciences themselves and not just in your mind. Perhaps this is not surprising—given the relative newness of the subjects and the enormous complexity of the areas with which they deal.

Auguste Comte (1798–1857)

Comte was the son of a French government official. In 1814 he entered the École Polytechnique in Paris which was famous for the study of subjects such as science, engineering and politics (see Chapter 4.3). Here he gained insights into the Industrial and French revolutions and developed an enthusiasm for one of his main missions: the attempt to apply the methods of the physical sciences to society. Comte was greatly influenced by the 18th-century thinkers Adam Smith and Condorcet—remember Condorcet's claim: 'What we can do for bees and beavers, we ought to do for men' (see Chapter 4.2). And Comte was deeply concerned about the social changes occurring around him, believing that one kind of society was dying and a new one being born. He described the one that was dying as being 'theological' and 'military' while the newly emerging one was 'scientific' and 'industrial'. He saw industrialists replacing warriors, and scientists replacing priests with, in due course, the social scientists gaining pre-eminence and becoming the new high priests.

Comte wanted to reform society in ways which would reconcile progress with order. He inherited a belief in human progress from the ideas of the Enlightenment, but he wanted to prevent the destructive aspects of the various revolutions which were taking place around him from breaking up those features of society which he cherished: community, family, authority and the preservation of social order. Comte thought that sociology could establish a set of laws and principles which would be just as successful in ordering society as the laws of the physical sciences had been in ordering the world of machines. He believed that the social sciences, and especially sociology—it was Comte who christened the subject—represented the high peak of intellectual and scientific achievement.

As the French sociologist Raymond Aron puts it:

> One might say that all the sciences converge on sociology, because the whole hierarchy of being culminates in the human species, which represents the highest level of complexity, nobility and fragility among being. Thus, when Comte established the synthesis of the sciences, to culminate in sociology, he was merely following, as it were, the natural direction of the sciences, which tended towards the science of society as their end, in the double sense of conclusion and goal.[3]

So it was Comte who founded sociology and described it as the new religion of humanity. He had such an important picture of himself that, in his later years, he actually signed himself: 'The Founder of Universal Religion, Great Priest of Humanity.' However, Comte was not a prophet of violence, like Karl Marx, and his subsequent influence has hardly lived up to his own estimate of his importance.

Herbert Spencer (1820–1903)

Spencer was another 19th-century thinker who was also very concerned by the changes occurring in the society around him, and deeply influenced by writers of the 18th century such as Adam Smith and Thomas Malthus. But the conclusions of his thinking were very different, almost diametrically opposite, to those of Comte.

Spencer was English; his father was a schoolteacher in Derby. From the age of 17 to 21, he worked as an engineer for a railway company and so, like Comte, he developed an approach to learning which was influenced by science and engineering. Spencer's first book *Social Statics* was published in 1851, the year of the Great Exhibition. He became increasingly impressed by recent discoveries in biology and geology and was especially influenced by the publication of Darwin's *Origin of Species* in 1859 (see Chapter 4.8).

In his key ideas and themes Spencer differed fundamentally from Comte. Whereas Comte wanted to see sociology influencing government and shaping society in accordance with its scientific principles and laws, Spencer wanted to see as little government interference as possible. Here his ideas resemble those of the 18th-century economist Adam Smith, who argued for a free market with a minimum of government interference. This was because Spencer was greatly influenced by contemporary biology and especially by the theory of evolution; he took some of the key elements of this theory and applied them to the analysis of human society.

For example, Spencer saw society as an organism which becomes increasingly complex as it develops. This complexity is associated with structural differentiation and specialisation of functions. He also suggested that these developments make societies more adaptable and able to adjust to changing conditions. Finally, he adopted the idea of natural selection, or the survival of the fittest, suggesting that a process of natural elimination would 'weed out' less healthy or adaptable peoples.

Spencer was a prolific writer and his works were widely read in the Victorian era. For example, his *The Study of Sociology* (1873) was published in both England and America in book and serial form. And his ideas pleased the Victorians for they supported many of the political and economic policies of the day such as 'laisser-faire' policies of minimum government intervention in industry and in health and welfare.

Spencer's ideas were thus very different from those of Comte. But as we turn to our next thinker, Karl Marx, we find yet another very different interpretation of the way in which society works, which this time is linked with specific proposals for change.

Karl Marx (1818–83)

Karl Marx was born in Trier in Germany. He was the son of a lawyer and received his main education at the University of Berlin. He failed to obtain a university teaching post and turned to journalism, becoming editor of a radical newspaper in the Rhineland. After marrying in 1843, he left Germany to spend the rest of his life in exile. He went first to Paris where he stayed until 1845, meeting many socialist thinkers including Friedrich Engels who collaborated with Marx in many publications and gave him financial support. Marx then moved on to Brussels where, in 1848, he and Engels published *The Communist Manifesto*. The 1840s were a time of considerable social unrest which led to political uprisings in many parts of Europe. Marx's work was intended to promote revolution and was used for this purpose during the revolutions of 1848. Marx then came to London, where he lived for the rest of his life. He spent a great deal of time studying in the British Museum and published many books and articles. In 1863 he helped to found the International Working Men's Association, which was designed to provide the basis for a revolutionary workers' movement throughout Europe. Marx also spent much of the next ten years of his life working on the affairs of the Communist International. It was in an attempt to prevent the revolutionary movement from splitting into factions that Marx wrote the first volume of *Das Kapital* (1867), which was meant to provide a political programme for the movement. He died in 1883 and is buried in Highgate Cemetery in London.

It is clear that Marx was not just an academic writer but was actively involved in trying to bring about revolution and the seizure of power which his ideas

Fig. 4.10.3   Title-page of the *Communist Manifesto* published by Marx and Engels in 1848.

predicted. As with all the writers discussed in this chapter, it is impossible to do justice to all aspects of Marx's ideas. All we can do is to highlight a few key themes.

Marx's ideas concerning both progress and social change differed fundamentally from those of Comte and Spencer. These writers felt that the revolutions which had already occurred had contributed to human progress and that some further changes would follow. Yet they were also concerned to try to preserve some features of the old order and to keep a balance between order and change. But for Marx, the revolutions had not gone far enough. He felt that, if anything, the mass of the population were worse off than previously and it was in this context that he wrote about alienation and class conflict.

Alienation   For Marx, alienation was concerned with the extent to which a person becomes estranged from the results of his or her own work. If we think of the changes which had occurred as a result of the Industrial Revolution we can get some feeling for what Marx was worried about. Adam Smith, too, had to a certain extent anticipated Marx when he described the possibly dehumanising effects of the division of labour. In place of the creative process associated with a craftsman's work, or the responsibility involved with caring for one's own land, however humble and poor, work in the new industries was often dominated by the division of labour. A person might just play a minute part on a production line, endlessly repeating a tedious task, with no feeling of responsibility for the finished product and no satisfaction in using skill or initiative. Also, at the time when Marx was writing, physical conditions in many factories were appalling, hours of work very long and the pace of work dominated by the demands of the new machines.

So Marx referred to a person's alienation from the products of his or her own labour. But he also used the term to describe the harmful effects which he believed the whole system of production under capitalism had on people's relationships with each other. The system whereby a person would sell his or her labour to a capitalist employer, who would extract profits from the surplus value of that person's work, would, for Marx, necessarily induce a condition of estrangement and conflict.

Class conflict This essential part of Marx's thought was based on his belief that the most important characteristic of any society is the way in which its economy is organised. For Marx, what really matters is who owns what he called the means of production, such as land, factories and wealth of various other kinds. Marx argued that if these means of production are owned by private individuals then there will inevitably be conflict between the owners and the non-owners—the 'bourgeoisie' and the 'proletariat'. Marx went on to claim that the only way of resolving this conflict would be by a revolution which would take all private property away from individuals and make it publicly owned. He saw the whole of human history in terms of such conflict between classes. He believed it was inevitable that capitalist society would be overthrown in due course by the working class, who would replace it by a socialist society. Then, because the means of production would be publicly owned, there would be no more conflict and people could find their true selves in a 'heavenly city on earth'. This was a new way of realising the utopian aspirations of the 18th century and a new interpretation of 'perfectibilism'—the idea that people can have such a complete vision of both natural and social reality that they can create a perfect world for themselves. It will be interesting when later we study the 20th century, to see to what extent those societies which have put Marx's ideas into practice, following their socialist revolutions, have or have not achieved the happy state which Marx prophesied.

Max Weber (1864–1920)
As a young man, most of Weber's university studies were as a law student at the University of Berlin, where he also practised as a barrister and then became a teacher. In Berlin, Weber developed a vast knowledge of law, economics and history, and published many articles. Later, he was appointed Professor of Economics, first in Freiburg and then in the prestigious University of Heidelberg.

Weber suffered a nervous breakdown in 1897 which prevented him from writing for five years. However, just after the turn of the century he was able to work again and it was then that he published many of his most famous books. For example, in 1904 he published two very famous works—*The Protestant Ethic and the Spirit of Capitalism* and *The Methodology of the Social Sciences*. Later he wrote two profound studies of the sociology of religion—*The Religion of China* in 1916 and *Ancient Judaism* in 1917.

Towards the end of his life Weber became very active in politics, serving on the committee of experts which drafted the Weimar Constitution for the new German state after World War 1. Now let us look at one or two key aspects of Weber's work.

Weber took issue with Marx in a number of ways. Perhaps the most significant was that he challenged the Marxist view that the economic characteristics of societies are the most important factors affecting human life and social change. While he agreed that economic factors are important—and he had studied and taught economics for many years—Weber argued that people are also influenced by other considerations and especially by ideas, such as religious beliefs. For Marx, religion was largely a result of the class structure, a tool of the ruling class, the 'opiate of the people'. For Weber, religion could be an independent reality which could motivate people very profoundly and could actually bring about changes in the way they live and organise their society. In summary, Marx was basically an *economic determinist* who believed that people's lives are largely 'determined' by economic conditions and forces while Weber was more of an *idealist* who argued that people's ideas may play a greater part in shaping their own destinies. This is reflected in Weber's research methods. He argued that social scientists should consider people's subjective experiences, feelings and beliefs as well as the more objective features of society such as the distribution of wealth and poverty. Weber believed, therefore, that social scientists should both observe, like natural scientists, but that they must also *understand* the meaning of what they observe. For example, a stranger going into a bank may see people handing over pieces of paper and metal; but unless he *understands* what this is all about, this observation will not be very helpful.

Like other social scientists of his time, Weber was deeply affected by the changes occurring around him. In particular, he was concerned about the effects of the growth of large organisations and of bureaucracies. Weber was worried that the pervasive influence of a bureaucratic way of organising society and of organising people's jobs would be very dehumanising. Although he appreciated that bureaucracies may be efficient and may help to increase the standard of living, Weber was afraid that many jobs would be routine. Individuals could lose their liberty by not being able to use their own creative initiative, and they could get so caught up in the proverbial 'red tape' that they might lose any wider vision and freedom of spirit.

Emile Durkheim (1858–1917)

Durkheim was born in France. He was the son of a Jewish rabbi and attended the high status École Normale Supérieure in Paris. His first book *The Division of Labour in Society* was published in 1893 and shows that Durkheim was influenced by the ideas of Herbert Spencer. After this he taught in schools for five years, before being appointed to the staff of the University of Bordeaux in 1887. There he taught the first social science course in France and was, nine years later, to become the first ever full Professor of Sociology. During the 1890s Durkheim published two other famous books, *The Rules of Sociological Method* in 1895 and *Suicide* in 1897.

In his studies Durkheim was passionately keen to demonstrate that sociology is necessary for our understanding of human behaviour and that sociology can be a science like the natural sciences, using similar methods of observation and analysis.

It is in this context that we can look at Durkheim's famous study of suicide, for here he is trying to do both of these things. He is taking a human act which one might think is intensely individual, intimate and personal and showing that suicide may have social causes, and also that sociological data can be studied in a scientific way. Durkheim compiled statistics for suicide rates in different societies and succeeded in showing that suicide rates varied significantly for different social groups. He found, for example, that members of some religious groups are more prone to commit suicide than others and that suicide rates are related to other social factors such as loneliness and economic well-being. Table 4.10.2 shows some recent data on suicide rates compiled by the World Health Organisation.

Table 4.10.2   Suicide death rates per 100 000 population

Country	Rate	Year
Australia	11.1	1977
Austria	24.8	1978
East Germany	36.2	1974
France	15.8	1976
Hungary	40.3	1977
Italy	5.6	1975
Japan	17.6	1978
New Zealand	9.3	1976
Poland	12.1	1976
Republic of Ireland	4.6	1977
Sweden	19.0	1978
UK England/Wales	8.0	1977
Scotland	8.1	1977
N. Ireland	4.6	1977
USA	12.5	1976
West Germany	21.7	1976

(*Source: World Health Statistical Report, WHO, Geneva, 1980*)

As you can see, the variations do not always occur in ways which might have been expected. Sometimes suicide rates are high for societies with a high standard of living and lower among those which are poorer. Some wealthy countries have relatively high rates as do some of the socialist societies. There seems to be a general tendency for Roman Catholic countries to have low suicide rates—but why should Austria's rate be more than twice that for Australia and why should East Germany's rate be significantly higher than that for West Germany? Some clues to these puzzles may be found in Durkheim's book, and as he himself said many years ago 'at each moment of history, therefore, each society has a definite aptitude for suicide'.[4]

In conclusion it is clear that Durkheim's work was path-breaking in its demonstration that sociology was a subject deserving serious study.

SOCIAL MEDICINE AND EPIDEMIOLOGY

As a final example of social science in the 19th century, let us turn from the famous writers to the development of social medicine and epidemiology—the study of the distribution of disease in a population.

This approach was graphically described by a 19th-century medical researcher: 'Medical statistics will be our standard of measurement: we will weigh life for life and see where the dead lie thicker, among the workers or among the privileged.'[5]

We saw earlier how the 19th century had experienced a massive growth of towns with people living in very squalid, cramped and overcrowded conditions and often working in equally unhealthy environments. These conditions were conducive to disease, and particularly to those infectious diseases which spread in the form of epidemics, like smallpox, typhoid or cholera. Remember Carnot had died in a cholera epidemic in 1832 (see Chapter 4.7). Here is an interesting account of how an intelligent analysis of the social factors associated with one such epidemic of cholera led to a greater understanding of what caused it, and to a practical policy helping to prevent it.

> In 1855, there occurred in London a terrible outbreak of cholera which had caused over 500 deaths in less than ten days. At the time Sir John Snow attempted to identify the source of the epidemic in an effort to stop it. In poring over the lists of deceased and afflicted persons while looking for common factors, he noted that the epidemic was most severe in a particular part of London and although cases of cholera had been reported in several areas, the majority of them had occurred in the neighbourhood of Broad Street [see Fig. 4.10.4]. Upon interviewing members of the families of the deceased, Snow was able to isolate a single common factor, namely the Broad Street pump, from which victims had drunk in every case. Corroborating evidence was made from the observation that in the local workhouse, also in the Broad Street area, only a few inmates had contracted cholera and that in every case they had contracted it before being admitted to the workhouse. Snow hypothesised (and found) that the workhouse drew water from a separate well. Similar findings were made in other establishments. The pay off for Snow's careful investigation occurred when, finally convinced that impure water from the Broad Street pump was the cause of the cholera, Snow appealed to the authorities to have the pump closed. What makes this investigation unusual is the fact that the cholera bacillus was not discovered, by Robert Koch, until some twenty-eight years after Snow's investigation. Thus, even though Snow did not know the precise cause of cholera, by following a logical sequence of steps he was able to isolate and identify the source of the infection and to prescribe ways of combating it.[6]

Fig. 4.10.4 Part of Sir John Snow's map of Soho showing cholera cases (each small black line is one death) and the position of the Broad Street pump.

Fig. 4.10.5 shows a cartoon of the time which succinctly illustrates the results of Sir John Snow's pioneering work.

In conclusion, we can see that the situation at the end of the 19th century was very different from that at the end of the 18th. The glorious Utopia so happily anticipated by many 18th-century writers had not materialised. Instead, the industrial societies of Europe were experiencing the very mixed blessings of industrialisation and urbanisation. What a poet described as England's 'dark, satanic mills' had brought the horrors of slums and appalling working conditions to many people, especially in the earlier decades. But they had also brought new wealth and higher standards of living for large numbers of the population by the end of the century. There was great and widespread perplexity about the future. The old order had been breaking up and many people felt very apprehensive about what would take its place. The social scientists reflected this situation. Their own studies led them to very different diagnoses of the social problems and so to very different prescriptions for remedying them. As we move into the 20th

MISTAKING CAUSE FOR EFFECT.

Boy. "I SAY, TOMMY, I'M BLOW'D IF THERE ISN'T A MAN A TURNING ON
THE CHOLERA."

Fig. 4.10.5   Punch cartoon of 1849.

century, we will have a chance to see to what extent their predictions came true
and what were the results of their prescriptions for change. What is certain is that
by the end of the 19th century, the social sciences were forces to be reckoned
with—intellectually, politically and in practical terms. Think of Durkheim's
achievement in obtaining official recognition of sociology as an academic subject
in France and the lasting influence of his studies of suicide; or of Marx's thought
being translated into action and eventually changing the political map of the
world; or the advances of social medicine with the first steps being taken towards
the eradication of many epidemic killer diseases such as cholera, tuberculosis or
smallpox. All these facts show that social factors now had to be taken seriously in
any attempts to understand human behaviour and that the social sciences had an
important and distinctive contribution to make in the understanding of society.

REFERENCES

1 Dickens, C., *The Old Curiosity Shop*, (Geneva: Edito-Service S.A., Centennial Edition, 1950), Vol. II, Ch. XLV, pp. 73-5.
2 From 'The Terling Fever, 1867, and an Early Community Hospital'; quoted by permission of Dr R. Pilsworth.
3 Aron, R., *Main Currents in Sociological Thought*, (Harmondsworth: Pelican, 1970), Vol. 1., p. 106.
4 Durkheim, E., *Suicide: A Study in Sociology*, (London: Routledge and Kegan Paul, 1952), p. 48.
5 Virchow, R., *Scientific Methods and Therapeutic Standpoints (1849)*; quoted in Cox, R. M. *Sociology of Medicine*, (New York: McGraw Hill, 1970), p. 31.
6 Cox, R. M., op. cit., p. 41.

SUMMARY

1 In the 19th century social scientists responded in many different ways to the drastic changes in the ways people lived and worked, which had been caused by industrialisation and the growth of towns.
2 Both Comte and Spencer derived inspiration from the natural sciences in developing their social theories. Comte saw sociology becoming a master science, rather like Newton's mathematical physics, while Spencer drew analogies from Darwin's theory of evolution and saw society as evolving with a minimum of government interference.
3 Marx saw the structure of society as being dominated by economic factors such as the ownership of the means of production, and advocated a revolution which would take these means into public ownership. By contrast, Weber laid much more emphasis on the power of ideas in shaping human society and was concerned that the growth of bureaucracies might stifle the human spirit.
4 Durkheim believed that sociology was necessary for an understanding of human society and that it could use objective data like the natural sciences. He illustrated these principles in his work on suicide and became the first ever professor of sociology in 1896.
5 Social factors were starting to be used in interpreting patterns of disease as in the work of Sir John Snow on the spread of cholera in London. Social medicine and epidemiology were thus beginning to be able to solve practical social problems.

4.11 Science, industrialisation and the world-wide expansion of European influence

During the 19th century, the Industrial Revolution spread across Europe and played a major part in the world-wide expansion of European power and influence. In this chapter we will first consider the industrialisation of Europe and then step back from the detailed sequence of events to review some of the changes which had taken place in Europe since the scientific revolution in the 17th century.

THE INDUSTRIAL REVOLUTION IN EUROPE

We have already looked at the origins and development of the Industrial Revolution in Britain (see Chapters 4.4 and 4.9). We saw that it was a complex process involving a rapidly rising population, many more people living in towns, improvements in agriculture and, above all, the rise of factories and industrialisation. This complex of changes took place first in Britain from about 1750 onwards. Then, during the 19th century, similar changes took place all over Europe. We can see this very clearly if we make a comparison between four major European countries—Britain, France, Germany and Russia—in the three major areas of population growth, the move from the countryside to the towns, and the growth of industrialisation.

Population

The pessimistic predictions of Thomas Malthus—that the increase of population would outstrip the food supply—did not come true in the 19th century. Fig. 4.11.1 shows the population changes in Britain, France, Germany and Russia between 1750 and 1910. The population of France doubled during the 19th century, that of Germany almost tripled, while in both Britain and Russia the population quadrupled. These changes reflect a significant decrease in death from harvest failures, and subsequent famines, and from epidemic disease. The kind of rapid increase in death-rate due to epidemic disease or a famine year, shown in Fig. 4.2.1 (page 104), became much less common in the 19th century. In England and Wales between 1840 and 1900 age-specific death rates practically halved for most of the population—the exceptions were those in their first year of life and those over 50. Similar changes took place over most of Europe, which meant that the average expectation of life also rose, as Table 4.11.1 shows.

Fast though the overall rate of population growth was—on average about a threefold increase over the century—it was much less than the rate of growth of towns. As we saw in Chapter 4.10, towns and cities grew in size by about a factor

Table 4.11.1   Expectation of life at birth in western countries in the 19th century (in years)[1]

Year	Male	Female	Year	Male	Female
1840	39.6	42.5	1880	43.9	46.5
1850	40.3	42.8	1890	45.8	48.5
1860	41.1	43.4	1900	48.9	52.1
1870	42.3	44.7	1910	52.7	56.0

Note: average of Denmark, England and Wales, France, Massachusetts, Netherlands, Norway and Sweden.

of 10 during the century—or about three times as fast as the overall population increase. This means, of course, that the proportion of people working on the land fell (see Table 4.11.2) and also that agriculture was becoming more efficient and productive.

Table 4.11.2   Percentage of population working in agriculture 1850–1950

	1850	*1900*	*1950*
Britain	22	9	5
France	52	42	30
Germany		35	24
Russia	90	85	56

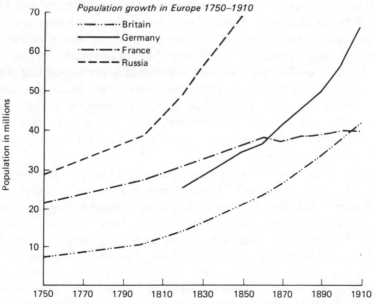

Fig. 4.11.1   Population growth (millions) in Britain, France, Germany and Russia, 1750–1910.

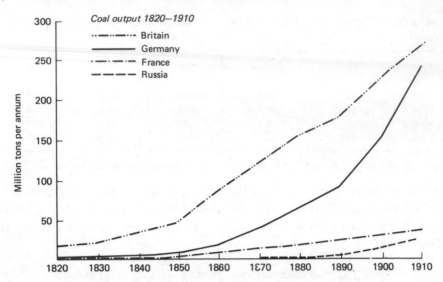

Fig. 4.11.2   Annual coal output (million tons) in Britain, France, Germany and Russia, 1820–1910.

Industrialisation

Yet it was the pace of industrialisation that was the most characteristic feature of the Industrial Revolution. Here we are going to use as indicators the outputs of coal, iron and steel—what we might call the material basis of the Industrial Revolution—and the growth of railway networks. Fig. 4.11.2 shows coal output from 1820 to 1910. British output was dominant throughout and had increased five times between 1800 and 1850 and by another five times between 1850 and 1900. France and Germany started from very low levels, which increased as the century progressed—gradually in France but dramatically in the case of Germany. German coal output increased twentyfold in the second half of the 19th century and was almost equal to Britain's output by 1910. Finally Russian output, while starting at very low levels, had begun to increase rapidly by 1900. Throughout the 19th century coal provided well over 95 per cent of energy supplies, and even as late as 1960 more than half the world's energy still came from coal (see Fig. 8.2.1, page 441).

The output of pig-iron between 1820 and 1910 (Fig. 4.11.3) showed similar trends to coal output. Once again Britain clearly led the way, particularly before 1870, but German output again rose rapidly in the second half of the century and had overtaken Britain by 1910. France and Russia showed steady rises throughout. For steel, the picture is a little different, since the steel industry developed later in the century as the underlying chemistry was better understood and applied. Fig. 4.11.4 shows steel output only between 1875 and 1910. In the early years Britain is in the lead but is overtaken by Germany in the 1890s and by 1910 the German steel output was more than twice that of Britain. Again France and Russia showed steady increases throughout.

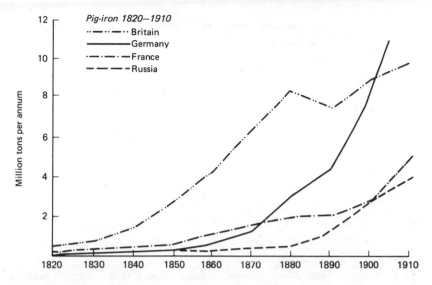

Fig. 4.11.3   Annual output of pig-iron (million tons) in Britain, France, Germany and Russia, 1820–1910.

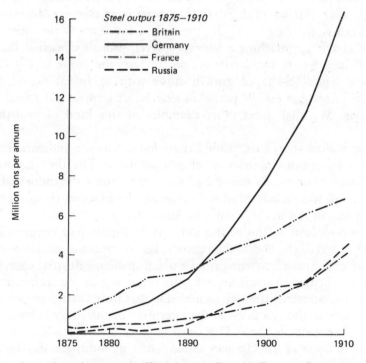

Fig. 4.11.4   Annual steel output (million tons) in Britain, France, Germany and Russia, 1875–1910.

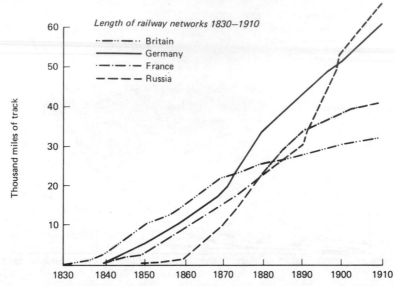

Fig. 4.11.5 Length of railway networks (thousands of miles) in Britain, France, Germany and Russia, 1830–1910.

Railway networks also expanded rapidly as Fig. 4.11.5 shows. Again Britain led the way—remember the railway 'mania' of the 1840s (see Chapter 4.9 and Fig. 4.9.3, page 189). But by 1875 Britain had been overtaken by Germany and by 1890 by France and Russia too. By the end of the century the British railway network is clearly approaching a saturation level, probably because there wasn't much of Britain left to build railways on. In fact, the curve in Fig. 4.11.5 for Britain is a typical S-shaped growth curve with an initial period of almost exponential increase, a middle period of near-linear growth and a final approach to saturation. We shall meet other examples of this kind of growth in later chapters.

All these indices show that, while Britain led the way in industrialisation, the other major European countries developed similarly. The time lag was only a couple of decades or so. By the end of the 19th century the industrialisation of much of Europe was far advanced and this made the European countries the most powerful group in the world. France no longer had the dominant position it had held under Napoleon in the early 1800s, while Russia had developed rapidly, particularly towards the end of the century. But the most dramatic development was the rise of Germany. Starting from a much smaller industrial base, Germany had overtaken Britain in many areas by the beginning of the 20th century. This rise of German industrial strength stemmed in part from two developments which originated early in the century—in the aftermath of the overwhelming defeat of Prussia by Napoleon in 1806. The German government made great efforts to rebuild the strength of the German nation—first by industrial development and then by establishing new universities. Industrial development was fostered by setting up many technical institutes, such as the Berlin Technical Institute, founded in 1821. These institutions took British industrial developments as their models and spread technical information about them throughout the country. New universities were also set up, the first being the University of Berlin in 1808, and these universities combined teaching and research in a way that was unique at the time. Although not specifically intended as scientific institutions, it was the scientific departments in these universities which grew most rapidly. Many research laboratories were established and some became world-famous—such as that of the physiologist Müller at Berlin and the organic chemists Liebig at Giessen and Bunsen, of Bunsen burner fame, at Heidelberg. Both the technical institutes, in the early part of the 19th century, and the university research laboratories, in the second half of the century, made great contributions to the growth of German industry, illustrated in Figs. 4.11.2 to 4.11.5.

THE WORLD-WIDE EXPANSION OF EUROPEAN POWER AND INFLUENCE

Now let us consider another aspect of the rise of population and the industrial development shown in Figs. 4.11.1 to 4.11.5. Taken together, these changes represent an enormous increase in the power and wealth of Europe as a whole. European populations rose steeply and, by the end of the 19th century, they were considerably healthier and more prosperous than they had been in 1800. This rise of European wealth and power led to a rapid increase in the spread of European influence across the world. This process had started in a small way with the great voyages of discovery of the Portuguese and the Spanish in the 15th and 16th centuries, but it now grew enormously in scale and scope. Many more people were involved and they went to many more parts of the world. Instead of setting up relatively isolated trading posts, the tendency was now for large areas to come under direct European government. We can't describe this process in any detail, but we can briefly understand its scale just by looking at the map of the world. In 1770, just as the Industrial Revolution was starting, the main European colonial powers were Spain in South, Central and North America, and Portugal in South America and parts of India. Britain had its North American colonies and a few coastal settlements in West Africa and India, Holland had some colonies in the East Indies, while France had a few scattered around the world. By 1914, the map had completely changed. The old Spanish and Portuguese empires in the Americas had disintegrated while the USA had been independent for over a century. But the power of newly industrialised Europe had extended so much that the world was dominated by Europeans. Britain held vast territories in Canada, India, Australia and New Zealand, and had divided South-East Asia with France, Holland and Germany. The most rapid expansion took place in Africa. In 1880 colonisation by Europeans—British, French, Dutch, Portuguese, Spanish—was well under way but was mainly confined to coastal regions. By 1914 the whole of the African continent, except for Abyssinia and Liberia, was completely divided up between the European powers. Britain and France had most territory but Germany, Italy, Spain, Portugal and Belgium all had major African colonies.

Another way of realising the scale of European expansion across the world in the 19th century is to look at the emigration figures. By far the largest emigration was from Britain. Between 1850 and 1920 more than 15 million people emigrated from Britain—an extraordinary figure when we remember that the total population was only 10 million in 1800, had risen to 20 million in 1850 and 40 million by 1920. Somewhere between a third and a half of the population emigrated. Emigration from other European countries was not so large, but even so many millions did emigrate in those decades—about 4 million from Germany, 8 million from Italy and 3 million from Spain.

The result was that in 1900 Europe dominated the world—and continued to do so until well into the 20th century. Amongst the European powers Britain was still dominant, despite the increasing strength of Germany. This pre-eminence was due, in large part, to the industrial developments we have described, and these were increasingly seen at the time to be dependent on the growth of science and of scientific knowledge. In 1897, the *Illustrated London News* published a special edition to mark the sixtieth Jubilee of Queen Victoria. This edition contained an article 'Science of the Reign' which is worth quoting:

SCIENCE OF THE REIGN

When the Queen ascended the throne a considerable capital of scientific knowledge had been collected. In her time it has grown with the pleasant and ever-increasing rapidity of a sum at compound interest. Science in England has become less insular and more and more cosmopolitan. The Victorian Era has been a time of giants in science; but that is not the peculiarity of the period. Darwins, Daltons, Faradays, Stephensons, Kelvins and Listers there have been in every century, but never before has the world seen the growth and organisation of a vast army of scientific workers that add their quota to the general sum and disappear nameless into the unknown. They are as necessary to the giants as workers to the queen bee. Scientific literature is now so vast, and expanding so astoundingly that it threatens, like an avalanche, to descend and bury its votaries.

The sum and substance of all the efforts of Victorian science has been to put the body and mind of man more at ease. Thanks to the Stephensons, the Brunels, and their Disciples, to Faraday, Siemens, Kelvin, Rowland Hill, and their followers, space has been almost annihilated, and the ends of the earth brought together. Simpson, Lister, the Continental descendants of the great Jenner, Parks and the pioneer band of English sanitarians have alleviated the sting of disease, and put our bodies more at ease. More than any nationality, Englishmen have been busy to put our minds at rest by teaching us the nature of the world we live in, and making us less the creatures of chance. The seas have been charted, the depths of space fathomed, the heavens mapped, the earth and its plants and animals examined. Darwin, Lyell, Spencer and Huxley have replaced empirical theories by a history written in rocks, bones and living tissue.[2]

That passage gives some idea of how science was regarded in Britain at the end of the 19th century.

EUROPE AND THE SCIENTIFIC REVOLUTION

Now let us look back from the Europe of the end of the 19th century—industrialised, urbanised and increasingly ruled by powerful nation states—and compare it with the Europe of Copernicus and Galileo—agricultural, rural, largely feudal, and dominated by the Church. The Europe of the 16th century was just beginning to explore the rest of the world. By the 19th century a much more powerful Europe was settling overseas a large fraction of its greatly increased population—and dominating the world.

In all these dramatic changes science had played a significant part—not necessarily directly, definitely not in a planned and predictable way, but in many subtle and often indirect ways. The movement started by Galileo had come of age. Now science, rather than the Christian religion, could be said to represent the spirit of the age. A large claim, but let us remember some of the changes which had taken place.

First there was the enormous growth in the body of scientific knowledge. The world of the natural philosophers of the 16th century was turned upside down, almost literally, by Copernicus and Galileo and completely recast by Isaac Newton's *Principia*—a mathematical masterpiece which unified astronomical and terrestrial motion in one harmonious whole. It was Newton, too, in his *Opticks*, who inspired much of the fruitful experimental work of the 18th century; examples include the use of quantitative methods in the study of heat and in the many changes which gave rise to modern chemistry. The 18th century also saw the collection, organisation and classification of a large amount of observational data concerning rocks, plants and animals of all kinds. This work led in the 19th century to the emergence of separate specialised sciences—first geology and biology and then subspecialities like zoology and botany. A similar specialisation took place in physics, with the study of electricity, magnetism and thermodynamics, and in chemistry, which divided into organic and inorganic specialities in the 19th century. It was the 19th century, too, which saw the coining of the word 'scientist' by William Whewell in 1837.[3]

Secondly, in parallel with this vast increase in scientific knowledge, there had been a steady development of what today we would call the organised scientific community—that mixture of scientific societies, scientific journals and institutions of higher education and research which make up the professional world of today's practising scientist. In the 16th century no organised scientific community existed. It began to arise in the 17th century with the foundation of scientific journals and scientific societies like the Royal Society, and it developed significantly in the 18th century, as we saw in Chapter 3.5.

These early societies and journals were nearly all concerned with any aspect of science, but later the rise of scientific specialities was reflected in the development of specialist journals and societies with the Linnaean Society for natural history being the first in 1788. Darwin and Wallace read their famous papers on evolution to a meeting of the Linnaean Society. This society was followed by the Zoological, Geological, Astronomical, Chemical and Physical Societies in the 19th century.

Many of these societies also published specialist journals of their own. By the end of the 19th century these and other developments, like the rise of scientific departments in universities and of research institutes, had made it possible for numbers of people to earn their livings as professional scientists—not yet in anything like the numbers who do so today but a significant change compared with earlier periods. This change partly reflects the growing importance of science in many of the practical activities of the wider society, such as the development of the science-based industries discussed in Chapter 4.9. And it led to that growing involvement of governments with science which is such a feature of the 20th century.

So by the end of the 19th century science had come to represent the spirit of the age. This is, in part, connected with the growth of scientific knowledge and with the growing importance of science and its application for all areas of society. It is also related to the attempts to build social sciences on similar principles (see Chapters 4.3 and 4.10) to those of the natural sciences. No one would have even thought of doing that in the 17th century.

But, more fundamentally, it is connected with the growth of the sceptical and experimental scientific spirit. The scientific revolution really began with Galileo and his overriding urge to publish and publicly defend his views. Galileo established that there was a definite area of knowledge in which publicly defended scientific authority had, by its very nature, to be respected. That revolution reached another climax with the work of Darwin. Darwin was very different in temperament from Galileo. Who could imagine Galileo voluntarily withholding publication of his major ideas for nearly 20 years as Darwin did? Yet Darwin's work was just as important as Galileo's. After Darwin the area of scientific study necessarily included man, his nature and his origins. The methods of science— careful experiment and observation, public defence and criticism—which had been first established by Galileo in mechanics, had now spread to the whole of the natural world, including man.

So far in this book, we have traced that development from the 17th to the 19th century. We will follow a similar course in the 20th century. Once again we will start in physics, with some of the work of Albert Einstein, the man who dominated science in the first half of this century, and then move into biology, describing, amongst other things, some recent developments in molecular biology such as the elucidation of the molecular structure of enormous biological molecules like haemoglobin and DNA. These and other major developments in science in the 20th century will receive more detailed discussion in later chapters.

REFERENCES

1 United Nations, *The Determinants and Consequences of Population Trends*, A Summary of the Findings of Studies on the Relationships between Population Changes and Economic and Social Conditions (United Nations publication, Sales No. 53.XIII.3), (New York: UN Population Division, 1953), Report No. 17, Table 5, p. 54.
2 *Illustrated London News*, Jubilee edition, 1897, p. 33.
3 See Ch. 3.5, Note 1 (page 85).

SUMMARY

1   During the 19th century the Industrial Revolution spread from Britain to most of the other European countries.
2   The spread of the Industrial Revolution was reflected by increasing populations, the growth of towns, steeply rising outputs of coal, iron and steel, and the expansion of the railways.
3   The industrialisation of Germany was particularly rapid and was stimulated by direct government support for new technical institutes and universities.
4   Industrialisation enormously increased the power and wealth of Europe as a whole, which in turn led to the spread of European influence across the world.
5   The scientific revolution was a major factor in the rise of European power and influence; by the end of the 19th century, science rather than the Christian religion could be said to represent the spirit of the age.

PART FIVE

Science and technology in traditional China

In traditional China science and technology were highly developed and for many centuries were more advanced than in Europe. However, science in China developed in almost total isolation from science in Europe, and Chinese culture and language are totally different from those of Europe. In this chapter, therefore, we will first outline the main structure of traditional Chinese society. Then we will trace some of the most important scientific developments in that society from the earliest times until the 19th century when traditional Chinese society began to break up as a result of the world-wide spread of European power and influence (see Chapters 4.11 and 7.6). The discussion of these topics will necessarily be more simplified and compressed than our account of the development of science in Europe. But it is preferable to devote some attention to Chinese science, however brief, rather than to miss out the whole fascinating subject.

TRADITIONAL CHINA

There are records of civilisation in China as early as 1500 BC and its greatest philosopher Confucius (551–479 BC) lived at roughly the same time as Pythagoras and Socrates, two of the greatest philosophers of ancient Greece. And as early as about 200 BC, at the beginning of the Han dynasty of emperors, Chinese society took on its traditional structure. This structure was remarkably stable, lasting without major changes for more than 2000 years. It was only in the 19th century that it began to break up and it was not until 1911 that the emperor and the imperial system were replaced by a republic. This stability is in striking contrast with the changes which took place in Europe over the same period—the rise and fall of the Roman Empire, the Dark Ages, the rise of Christianity, the Renaissance and the Reformation, the rise of separate nation states and the scientific and industrial revolutions.

Throughout this period, Chinese civilisation developed to a very high level in almost complete isolation from that of Europe. The Chinese regarded their country as the centre of the world and looked on other people as barbarians. The very name for China in Chinese is *Júng Gwó* which means the Middle Country or Middle Kingdom. During all this time China has had a significantly greater population than Europe. Fig. 5.1 shows the population of China between 400 BC and the present.

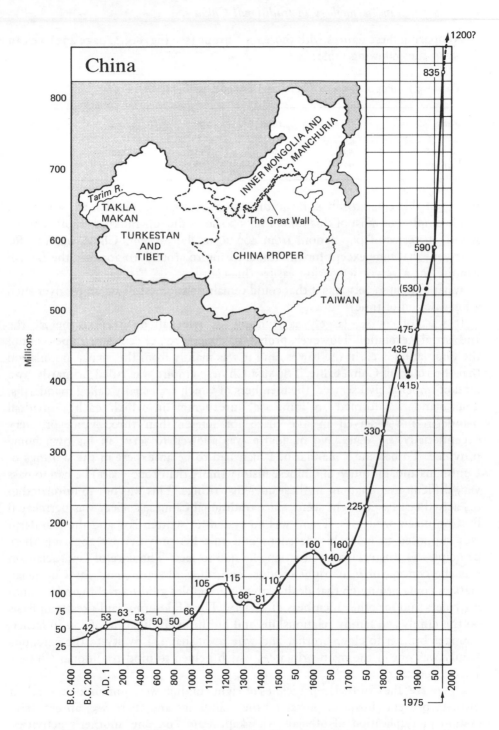

Fig. 5.1 Population of China from 400 BC to 1975. *Redrawn from McEvedy, C. and Jones, R., Atlas of World Population History, page 167, copyright © Colin McEvedy and Richard Jones, 1978, reprinted by permission of Penguin Books Ltd.*

Comparing these figures with those for Europe (see Fig. 8.5.1, page 472) we can draw up the following table:

Table 5.1   Populations of China and Europe, millions, AD 1–1975

Year, AD	I	1000	1500	1800	1975
China	53	66	110	330	835
Europe	31	36	81	180	635

But even this comparison is misleading since Europe, unlike China, was not a single empire for most of this time. Fig. 5.2 shows the top-ranking empires of the world, in terms of population, from 400 BC to the present. China was top for nearly all that time except for the Roman Empire around AD 200 and the British Empire for a very brief period earlier this century.

What kind of society was it that could remain so successful and stable over such a long period of time?

It was, of course, largely agricultural, as were all civilisations before the Industrial Revolution. However, probably between 10 per cent and 20 per cent of the population lived in the towns, and it was mainly from this urban population that the rulers of China came. The vast Chinese empire was ruled primarily by a kind of imperial civil service, the members of which are usually called mandarins. The mandarins formed an immense bureaucracy in which each individual mandarin never stayed in one place for longer than three years or, very exceptionally, six years. No mandarin was allowed to serve in his own home province. To become a mandarin, a man had to be an expert in the writings of Confucius and, for much of Chinese history (more than 1000 years) he had to pass very difficult examinations in these ancient writings. This was not so unrelated to his actual tasks as it might seem, since traditional Chinese society was permeated by Confucian principles. There was no system of written law and the mandarin was supposed to make his decisions and rule his province by applying these principles to whatever new situations he had to face. The core of Confucianism was the concept of *li*, or right conduct according to status, and the Confucian classics contained much that dealt with the principles guiding right conduct both for rulers and for other members of society. The philosophy laid great emphasis on the family and family relationships and on the subordination of son to father, younger brother to elder brother and wife to husband. Loyalty to the extended family was one of the main principles which helped to bind traditional Chinese society together.

Of course, the Confucian principles of right conduct were not enough, on their own, to prevent abuses of power by the mandarins and there was an extensive system of duplication of officials who kept watch on one another's activities. There was even a body of men called the censors who didn't actually have to do anything—except to report on what the other mandarins were doing!

Above all was the emperor who appointed all the senior mandarins and lived in

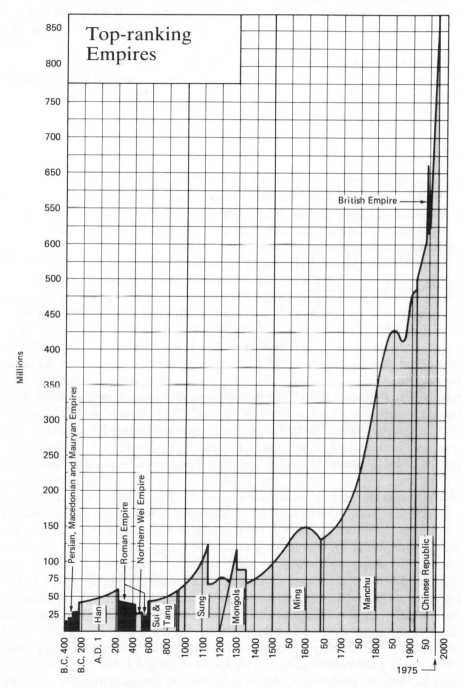

Fig. 5.2 Top-ranking empires of the world, 400 BC to 1975. *Redrawn from McEvedy, C. and Jones, R., Atlas of World Population History, page 127, copyright © Colin McEvedy and Richard Jones, 1978, reprinted by permission of Penguin Books Ltd.*

isolated splendour in his court—a sort of symbolic father to the whole empire. It was held that the emperor derived his mandate to rule from Heaven and that this 'mandate of heaven' could be withdrawn if he ruled badly. The sign that the 'mandate of heaven' had been withdrawn was simply that a state of misrule existed and this principle could be, and was, used to justify rebellion and a change of emperor. In a strongly hierarchical and centralised society like traditional China such upheavals didn't happen very often, but the possibility did enable society to change without destroying its essential structure.

THE CHINESE LANGUAGE

The Chinese language played a very important part in traditional China, not least because the mandarins had to be expert in it before they could pass their examinations in Confucianism. Since it is very different from most European languages, we will outline its main features and discuss its importance in the government of the Chinese empire.

Written Chinese has no alphabet. Like the hieroglyphics of ancient Egypt, its symbols or characters were originally pictograms or outline sketches which were meant to look like the thing they represented. This is obvious in some characters. For example, the characters in the left-hand column of Fig. 5.3 are the numbers one to five. The first three are clearly pictographic but the others are not. However, even the simplest characters have long since lost any obvious pictographic clues to their meanings. For example, the first character in the right-hand column of Fig. 5.3 is supposed to represent a crescent moon and the second, a man. There are more than 40,000 characters in the largest Chinese dictionaries, while the printing of a modern newspaper needs about 7000 characters. Anyone who wants to be reasonably literate needs a working knowledge of about 3000 characters. And, as well as learning to recognise the character, it is also vital to learn the order in which the strokes are made if recognisable characters are to be written.

Since the language has no alphabet, it is not easy to construct a Chinese dictionary. However, some of the same elements do recur in many characters. For example, the last three characters in the right-hand column of Fig. 5.3 all have the same element on the left—a modified version of the character for man. Common elements like these—or *radicals* as they are called—are used to classify the characters. Even this is not easy as there are 214 radicals in modern Chinese, and more than one radical may appear in the same character. As with the man radical, many radicals can also be characters in their own right. Examples include the characters for one, two and month shown in Fig. 5.3. For purposes of classification, the radicals are numbered in order of the number of strokes they contain so that, for example, the one character is radical 1, the two character is radical 7, the man character is radical 9, and so on for more complex radicals. Characters in a Chinese dictionary are listed under their radicals, again in order of the number of strokes they contain.

The written language is common to all parts of China but there are a number of

Character	Explanation	Character	Explanation
一	yī NU: one, a (pronounced yī in counting only; yi' before a falling tone; yì before the other three tones)	月	ywè N: month
二	èr NU: two	人	rén N: person, people, human being
三	sān NU: three	你	nǐ N: you (singular)
四	sż NU: four	他	tā N: he, she, him, her
五	wǔ NU: five	們	men BF: (pluralizing suffix for pronouns and nouns denoting persons when no definite number is mentioned)

Fig. 5.3   Some simple Chinese characters with meanings, approximate pronunciations and tones (see text). *From Lee, P., Read Chinese (New Haven: Yale University, 1961).*

different spoken languages in the different regions. These are not just dialects for they often show much greater differences than European languages like English and, say, German or French. So, though all Chinese write their language similarly, they may speak it in very different ways. The nearest equivalent for us is the way in which the symbol '5' is read as 'five' in England, *cinq* in France, *fünf* in Germany and *cinque* in Italy. Of course, this means that the way a Chinese character is written is no guide to the way it is pronounced. And spoken Chinese has another feature not found in European languages. The tone in which a word is said can *completely* alter its meaning. For example, in the Mandarin language which is spoken in Peking, there are four tones. The syllable 'ma' can be pronounced $\overline{ma}$ (a monotone), *má* (a rising tone), *mǎ* (falling then rising) or *mà* (falling). One of these does mean 'mother' but another means 'horse', another means 'hemp', while the fourth is a kind of curse. Cantonese, the language of much of southern China, including Hong Kong, has even more tones than Mandarin.

It should now be clearer why mastery of the written Chinese language was so important for the mandarins who ruled traditional China. It provided a common and unifying language which was recognised throughout the Chinese empire, enabling all regions to communicate with the capital, Peking. And Mandarin, the spoken language of Peking, became an essential second spoken language for the mandarins of all regions.

Although the written Chinese language is complex, requiring perhaps ten years of study for full mastery, it was used to produce a greater volume of recorded literature than any other language before modern times. It is estimated that before 1750 there had been more books published in Chinese than in all the other languages in the world put together.

SCIENCE AND TECHNOLOGY IN TRADITIONAL CHINA

Until relatively recently the general belief in the West was that the science and technology which had existed in traditional China was of relatively little importance compared with that of Europe. Now we know that this is not true. Traditional China had developed a substantial body of knowledge about many scientific and technological topics. Much of this knowledge predates that of Europe, in some cases by several centuries, and was acquired in a society which knew very little, if anything, of what was taking place in Europe. The knowledge we now have of science and technology in traditional China is largely the result of the work of the biochemist and embryologist Joseph Needham, who since the 1940s has published many volumes of his pioneering work, *Science and Civilisation in China*. Much of this chapter is necessarily based on Needham's work, since there is no other published source which is in any way comparable to it.[1] This, of course, is very different from the case of European science and technology, where the problem is rather that there are too many different sources rather than too few.

Printing, gunpowder and magnetism

Needham in one of his articles prints this quotation from Francis Bacon—the 17th-century English writer whom we mentioned in Chapter 3.5:

> It is well to observe (said Lord Verulam) the force and virtue and consequences of discoveries. These are to be seen nowhere more conspicuously than in those three which were unknown to the ancients, and of which the origin, though recent, is obscure and inglorious; namely, printing, gunpowder, and the magnet. For these three have changed the whole face and state of things throughout the world, the first in literature, the second in warfare, the third in navigation; whence have followed innumerable changes; insomuch that no empire, no sect, no star, seems to have exerted greater power and influence in human affairs than these mechanical discoveries.[2]

These three discoveries—printing, gunpowder and magnetism—were all made much earlier in China than in Europe but, unlike in Europe, these discoveries were not followed by major changes in the structure of Chinese society.

Block printing was invented in China in the 9th century AD and printed books

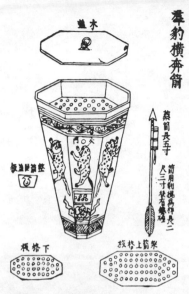

Fig. 5.4 Chinese rocket arrow dating from 11th century AD. *From Needham, J., The Grand Titration, George Allen and Unwin, 1969.*

began to appear in the later years of that century. The oldest surviving printed book is a Buddhist text, which dates from AD 868, and a complete printed edition of the Classical Books of Confucius was commissioned in AD 932 and completed in AD 953. Moveable type was developed in the 11th century, even though a separate piece of type was needed for each of the thousands of characters. Moveable type was not introduced in Europe until 400 years later, when Gutenberg printed his Latin Bible in 1456.

The origins of gunpowder in China also date from the 9th century AD. The earliest written formula for a form of gunpowder, a mixture of charcoal, saltpetre and sulphur, appeared in a Chinese book published in AD 1044. It was not until the early 14th century that any similar reference can be found in Europe. And the new invention was soon applied to weapons such as the rocket launcher and the barrel gun. Fig. 5.4 shows a rocket arrow, together with a launching box for 75 rockets. Such devices date from the 11th century, as does a type of flame-thrower based on naphtha as a fuel.

The first mention of magnetism and the equivalent of a magnetic compass are even earlier. There is a reference to a 'south-controlling spoon' in a text dating from AD 83, while Fig. 5.5 shows a reproduction of such a spoon. The spoon itself was carved from lodestone and, when placed on a highly polished bronze plate, always rotated until it pointed south. Chinese compasses always point south! There are many references to a 'south-pointer' in the following centuries—well before the first European mention of magnetic polarity in 1180. Fig. 5.6 shows another south-pointing compass, this time based on remanent magnetism, which dates from AD 1044. The floating iron fish was magnetised by being heated to red heat and allowed to cool in the earth's magnetic field.

Fig. 5.5 Modern reproduction of a Chinese 'south-pointing spoon' dating from 1st century AD. *From Needham, J., The Grand Titration, George Allen and Unwin, 1969.*

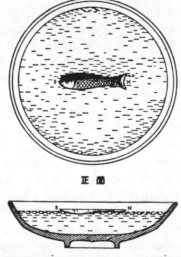

Fig. 5.6 Floating iron compass using remanent magnetism; modern reproduction from a description of AD 1044. *From Needham, J., The Grand Titration, George Allen and Unwin, 1969.*

It seems likely that these magnetic compasses were used in navigation as early as the 10th century, and there is some evidence that the Chinese knew of magnetic declination—the fact that compasses do not point exactly North–South and that the difference varies with time—before Europeans knew of magnetic polarity.

Other scientific and technological discoveries

The Chinese also made important discoveries in many other branches of science and technology, but we only have space to mention a few of them.

In mathematics the Chinese never developed anything comparable to the geometry of the ancient Greeks, and their work lacked the idea of rigorous proof which is inextricably bound up with formal Euclidean geometry. But the Chinese were always strong in arithmetical techniques and calculations. By the 5th century AD they had calculated that the value of π lay between $3 \cdot 1415926$ and $3 \cdot 1415927$, an accuracy surpassing that of the Greeks and one not achieved in Europe until the 17th century. And by the 14th century Chinese mathematicians had made significant advances in algebra, in the numerical solution of equations and in the rudiments of the binomial theorem. Fig. 5.7 shows a diagrammatic representation of the binomial coefficients up to the eighth power, which appeared in a book published in AD 1303. It is identical with Pascal's Triangle, which was known in 16th-century Europe and fully worked out by Pascal in 1665.

The Chinese were also very thorough astronomical observers, but their astronomy concentrated on the celestial pole star and the stars near it rather than the stars of the zodiac around the ecliptic the path of the sun through the heavens. This meant that observations of the planets, which are always near the ecliptic, did not occupy the central part in Chinese astronomy that they did in the astronomy of ancient Greece and medieval Europe (see Chapters 2.1 and 2.2). However, the Chinese did map the stars with great thoroughness and kept detailed records of eclipses, comets, novae and meteors which are still useful to modern astronomers. For example, every reappearance of Halley's comet (see Chapter 3.4) in its 76-year cycle is recorded in the Chinese texts right back to 87 BC and possibly even further back than that. Their emphasis on the celestial pole, and the stars near it, meant that they traditionally used the same celestial co-ordinates, based on the pole star and the celestial equator, as are used by modern astronomers, rather than co-ordinates based on the ecliptic as used by the Greeks (see Fig. 5.8).

The Chinese also developed astronomical instruments which incorporated the equatorial mounting and clock drive used in all modern large telescopes. An example of such a mounting is shown in Fig. 5.9 of an astronomical clock built between AD 1088 and 1092 by Su Sung. This clock is driven by a water-wheel and linkwork escapement. It contains an elaborate clockwork mechanism which strikes the hours, turns a celestial globe and drives the armillary sphere on the roof which was used for astronomical observations. The whole complex mechanism resembles that of the Dondi clock (see Fig. 2.2.1, page 20) which was built in Italy nearly 300 years later, around 1350. Moreover, the principle of the water-

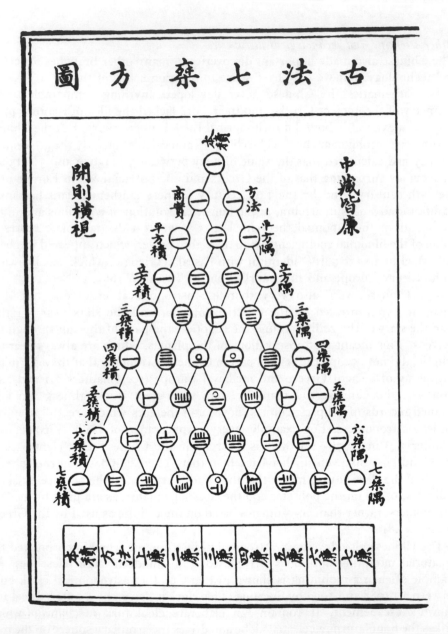

Fig. 5.7   A diagram of the Pascal Triangle from a Chinese book published in 1303; it shows the binomial coefficients up to the eighth power. *From Ronan, C. A., The Shorter Science and Civilisation in China, vol. 2, Cambridge University Press, 1981.*

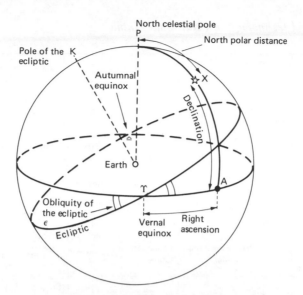

North celestial pole
P
North polar distance
Pole of the K
ecliptic
Autumnal
equinox
X
Declination
Earth O
A
Obliquity of
the ecliptic
ε
Ecliptic
Vernal
equinox
Right
ascension

Fig. 5.8 Celestial co-ordinates: the Chinese and modern astronomers use co-ordinates referred to the celestial pole and celestial equator (e.g. right ascension and declination): the Greeks based their celestial co-ordinates on the ecliptic and the pole of the ecliptic. *Redrawn from Ronan, C. A., The Shorter Science and Civilisation in China, vol. 2, Cambridge University Press, 1981.*

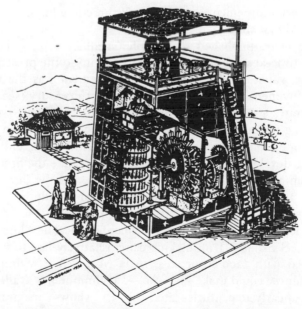

Fig. 5.9 Reconstruction of an astronomical clock built in China between AD 1088 and 1092. The water-wheel rotated a celestial globe, an armillary sphere which was used for observations and an elaborate series of figures which announced the time. This type of hydro-mechanical clockwork had been invented in AD 725, six centuries before the first European mechanical clocks. *From Needham, J., The Grand Titration, George Allen and Unwin, 1969.*

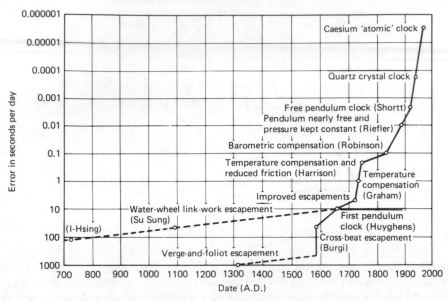

Fig. 5.10   Chart showing the rise in accuracy of mechanical clocks from AD 700 to the present. *Redrawn from Needham, J., The Grand Titration, George Allen and Unwin, 1969.*

wheel and linkwork escapement, which appears here in a highly developed form, was known as early as AD 725.

Fig. 5.10 is a chart published by Joseph Needham which shows the most accurate known time-keeping devices from AD 700 up to the present. As you can see, the Chinese water-wheel and linkwork escapement was the most accurate right up until Huyghens' pendulum clock in the 17th century (see Chapter 3.2). The chart illustrates a common pattern identified by Needham—a high level of steadily improving technical precision in China which is overtaken by very rapid progress in Europe after the scientific revolution of the 17th century.

Other striking developments in Chinese science include the first seismograph, which dates from as early as AD 132 (see Fig. 5.11), and the development of an early form of vaccination or immunisation against smallpox as early as the beginning of the 16th century, and possibly even earlier.

Now let us briefly look at some large-scale technological developments, starting with iron and steel making. Iron making came relatively late to China—around 600 BC, compared with 1200 BC or so in Asia Minor. Yet within two or three centuries the Chinese could make cast iron—a technique not available in Europe until about 1380, nearly 17 centuries later. Fig. 5.12 shows a magnificent cast-iron pagoda, built in AD 1061 and still standing. Iron was also used very early in bridge construction. A segmental arch bridge, which was built in AD 610 and still stands, made extensive use of iron clamps between the staves of the arches. The earliest European segmental arch bridge was built at Florence in 1345. Suspension bridges using iron chains were also built from about AD 600 onwards—centuries before the first iron-chain suspension bridges in Europe, which date from around

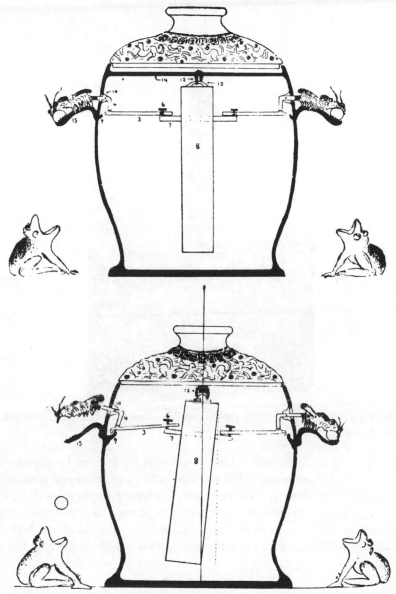

Fig. 5.11   Attempted reconstruction of Chang Heng's seismograph of AD 132. *From Ronan, C. A., The Shorter Science and Civilisation in China, vol. 2, Cambridge University Press, 1981.*

1740. Steel, too, was well known in ancient China, and steel bits were used in deep drilling for brine and natural gas to depths of around 2000 feet from around the first century AD onwards.

Two other technologies which must be mentioned, if only briefly, are hydraulic engineering and porcelain and ceramics. Many massive hydraulic projects were carried out in China from very early in its history. The purposes of these projects were both irrigation and the building of canals for transport. As early as the 13th

Fig. 5.12   Cast iron pagoda from AD 1061. *From Needham, J., The Grand Titration, George Allen and Unwin, 1969.*

century a Grand Canal was built which stretches as far as from London to Athens. Finally, the Chinese porcelain industry—famous for centuries and from which we get our word 'china'—must have been very advanced technologically and must have developed very precise methods for controlling the conditions inside the firing furnaces. Fig. 5.13 shows one example of the kind of high-quality craftsmanship which the Chinese achieved—a vase which dates from the 12th or 13th century.

SCIENCE IN CHINA AND IN EUROPE

Although there is much more that could be said, it should now be clear that many areas of science and technology were highly developed in traditional China. There is also a wealth of evidence, some of which we have mentioned, to show that for many centuries Chinese society was much more effective in the practical application of natural knowledge than were contemporary societies in Europe or elsewhere.

But what didn't happen in China was anything which could be compared to the scientific revolution of 17th century Europe and to the subsequent industrial,

Fig. 5.13   Chinese vase from the 12th or 13th century.

social and political revolutions which we have been discussing in this book. Fig. 5.14 shows a schematic diagram drawn by Joseph Needham which attempts to show the level of scientific achievement in China and in Europe from 300 BC to the present. The diagram clearly shows the exponential growth of various sciences in Europe after 1500 and tries to identify fusion points after which the Chinese and European traditions merge into one universal science. Note that the fusion point in medicine is set at some time in the future!

We could ask the question, as Needham does: Why didn't anything like the 17th-century scientific revolution take place in China? Possible answers to this question will clearly depend on what we take to be the central features of the scientific revolution and what characteristics of European society we believe enabled that revolution to occur. Some have argued that mathematics was the key factor in the scientific revolution. They claim that Chinese science was hampered because Chinese mathematics was largely algebraic and lacked anything comparable to the Euclidean geometry of classical Greece. Others suggest that the complex Chinese language inhibited science since it tends to be imprecise and allusive and so is not well adapted to precise logical expression. Another suggestion is that the absence in traditional Chinese philosophy of anything

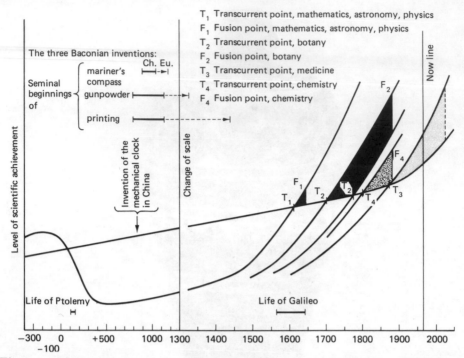

Fig. 5.14   Schematic diagram to indicate the levels of scientific achievement in China and Europe. *Redrawn from Needham, J., Clerks and Craftsmen in China and the West* (*Cambridge: Cambridge University Press, 1970*).

resembling a divine law-giver made it less likely for men to seek for universal laws of nature. Or it has been claimed that the centralised bureaucratic structure of traditional Chinese society restricted the development of independent centres of thought and of power, while at the same time giving stability to that society when it was faced with major technological changes.

All of these are fascinating questions which are very difficult to answer. Some people think that it is not fruitful to ask why an event like the scientific revolution did *not* happen unless there was reason to expect it. In a letter to a friend, Einstein put it like this:

> The development of Western Science has been based on two great achievements, the invention of the formal logical system (in Euclidean geometry) by the Greek philosophers, and the discovery of the possibility of finding out causal relationship by systematic experiment (at the Renaissance). In my opinion one has not to be astonished that the Chinese sages have not made these steps. The astonishing thing is that these discoveries were made at all.[3]

It may be more important for the future to consider a different kind of question. The civilisation of 17th-century Europe, the first to develop modern science, had only to adapt to the scientific revolution itself. All other civilisations which encounter modern science also have to adapt to the alien civilisation from which science reaches them—that is, to a civilisation which has much less in common with themselves or with any previous civilisation, including the Europe of the past.

As we shall see in Chapter 7.6, it was the rising power of industrialised Europe in the 19th century and its increasingly world-wide influence (see Chapter 4.11) that eventually led to the break-up of traditional Chinese society and to the establishment of a republic there in 1911.

So, rather than asking why the original scientific revolution took place when and where it did, it may be more fruitful to seek to understand the problems and strains of adapting to its effects. It is the attempt to understand this kind of problem that gives an additional importance and interest to the development of science and technology in China in the 20th century, both before and after the Communist revolution of 1949.

In the rest of the book we shall be trying to illuminate these problems by describing the growth of science and technology in the 20th century and by comparing the way in which different societies have responded to the challenges posed by that growth.

REFERENCES

1   A summary of part of Needham's multi-volume work *Science and Civilisation in China*, (Cambridge: Cambridge University Press, 1954 to present) is now available—Ronan, C. A., *The Shorter Science and Civilisation in China, 1 & 2*, (Cambridge: Cambridge University Press, 1978, 1981). It is worth emphasising that Needham's work is the source for most of this chapter. If his facts or interpretations are wrong, then my summaries of them are wrong too because there are virtually no sources against which Needham's work can be checked.

2   Bacon, F., *Novum Organum*, Book 1, aphorism 129; quoted in Needham, J., *The Grand Titration*, (London: Allen and Unwin, 1969), p. 62.

3   Ibid., p. 43.

SUMMARY

1   In traditional China, civilisation reached a very high level under a bureaucratic system of government which remained essentially unchanged for about 2000 years.

2   Due largely to the work of Joseph Needham, we now know that many branches of science and technology were highly developed in traditional China and that much of this knowledge predates that of Europe by many centuries.

3   Three major technological discoveries—printing, gunpowder and magnetism—were made much earlier in China than in Europe and, unlike in Europe, these discoveries were not followed by major changes in the structure of Chinese society.

4   The Chinese were also well advanced in some branches of mathematics, in the development of scientific instruments such as mechanical clocks and seismographs and in iron and steel technology, hydraulic engineering and the manufacture of ceramics and porcelain.

5   Science and technology in traditional China developed steadily over the centuries but China never experienced anything comparable to the 17th-century scientific revolution and the subsequent exponential growth of scientific knowledge.

PART SIX

The growth of science in the 20th century

6.1　Science and the 20th century

In earlier chapters we traced the development of science from its beginnings in ancient Greece and in the scientific revolution of the 16th and 17th centuries through to the end of the 19th century. We saw how the methods of science—careful experiment and observation, public criticism and defence—which had first been established by Galileo in mechanics, had now, after the work of Darwin, spread to the study of the whole of the natural world including the study of man. We saw how the applied science and technology of the Industrial Revolution raised European living standards, and populations, to higher levels than ever before and how the very success of the natural sciences led men to try to apply similar methods to the understanding and organisation of society. And we saw how the prestige of science—deriving from its increasing ability to understand, control and manipulate nature—had risen so much that, at least in Europe, science rather than religion had come to represent the spirit of the age.

All these trends have continued throughout the 20th century and the interactions between science and society have grown ever more pervasive and complex. So the task of describing and understanding the role of science and scientists in the 20th century is not easy.

In this introductory chapter we will review four main topics. The first two are the very rapid growth of science and the increasing interrelatedness of the different sciences. Here we will be primarily concerned with the development of science itself. Then we will consider two social and political topics—the increasing involvement of governments in science and the dangers involved in both overestimating and underestimating the importance of science.

THE GROWTH OF SCIENCE

At the end of the 19th century science had just experienced the most rapid period of growth in its history. At this time some scientists had even been heard to complain that all the major discoveries had been made and that only relatively trivial tasks were left—rather like Diderot predicting the end of mathematics in 1754 (see Chapter 4.1). Yet, rather than stagnating, scientific knowledge has continued to grow at an extremely rapid rate. Let us consider a few examples.

Fig. 6.1.1   The explosion of the first atomic bomb in the New Mexico desert, USA, in 1945.

In 1900 the electron had only just been discovered, by J. J. Thomson in 1897, soon after being predicted on theoretical grounds by the Irishman, G. J. Stoney. Now the enormous electronics industry pervades much of our lives—from TV to computers and microprocessors. All this depends on our ability to control and manipulate electrons. Also, in 1900 the discovery of the nucleus was ten years or more in the future. Rutherford first suggested its possible existence in 1911. Now we live with the twin consequences of that discovery—nuclear reactors and nuclear weapons. Fig. 6.3.3 (page 271) shows the world's first nuclear reactor, which was built in Chicago in 1942, while Fig. 6.1.1 shows the first explosion of an atomic bomb in the New Mexico desert in 1945.

These two discoveries—the electron and the nucleus—have completely transformed physics by enabling us to understand the forces both within and between atoms. And these discoveries have transformed chemistry too because our understanding of the electronic structure of atoms has explained the regularities of Mendeleev's Periodic Table in physical terms. It is fascinating to recall that in 1900 outstanding physicists and chemists like Ernst Mach and Wilhelm Ostwald did not believe in the existence of atoms.

In other fields too there have been dramatic changes in this century. In pure mathematics the *Principia Mathematica* of Bertrand Russell and A. N. Whitehead, which examines the logical foundations of the subject, dates from 1910 while much modern statistical theory developed in the 1920s and 1930s.

Turning to biology it was in 1900 that the work of Mendel on genetics, dating from 1866, was rediscovered after more than 30 years of neglect. Since then genetics has been probably the fastest growing area of biology. Major landmarks have been the identification of chromosomes (see Fig. 6.1.2) and the discovery, in 1953, of the structure of the key genetic material—deoxyribonucleic acid, or DNA as it is usually known. Biochemistry too has developed rapidly, and the detailed structure of many complex organic molecules is now known. A major example is the protein haemoglobin (see Fig. 6.1.3) which as oxy-haemoglobin gives our blood its colour and carries oxygen to all parts of our bodies.

Biochemistry too has made great contributions to medical knowledge—and here it is worth remembering that in 1900 there were only a handful of effective drugs which doctors could use. It is not until the 1940s that antibiotics like penicillin, became widely available. Nor were there in 1900 many sophisticated aids to diagnosis. X-rays had only just been discovered, in 1896, and instruments like electrocardiographs, a common sight in today's hospitals, were inconceivable.

In the other applied sciences we have only space to mention two things. One is the rapid development of communication using radio waves and the other is the growth of powered flight. Radio waves had first been detected in 1888 by Hertz in Germany, and it was in 1901 that the young Marconi succeeded in sending a radio message across the Atlantic for the first time. The first aeroplane flight, by the Wright brothers in 1903, lasted only 12 seconds and covered all of 40 metres. Today we take for granted space flights powered by enormous rockets like those which launched the Apollo spacecraft on their journeys to the moon (see Fig. 6.1.4). And these spacecraft send back by means of radio waves, pictures of the earth as seen from space (see Fig. 6.1.5) or even pictures of distant planets like Jupiter and Saturn.

Another way to illustrate the growth of scientific knowledge is to consider how the literature of science has grown. Remember Fig. 3.5.1 (page 78) which shows the increase in the number of scientific journals and journals of abstracts from about 1700 to the present. Over all that time the literature of science has grown exponentially with a doubling time of roughly 15 years. A similar growth has taken place in the number of scientists—again with a doubling period of about 15 years. This fact leads to the remarkable conclusion that more than 90 per cent of all the scientists who have ever lived are alive now and *that this has also been true at any time in the last 250 years or so*. If such exponential growth were to continue, it would not be long before we were all scientists. However, in the real world, exponential growth does not continue for ever. What happens is shown in Fig. 6.1.6. The growth rate slows and approaches a saturation limit just as, for example, the railway network in Britain did in the 19th century (see Fig. 4.11.5, page 216).

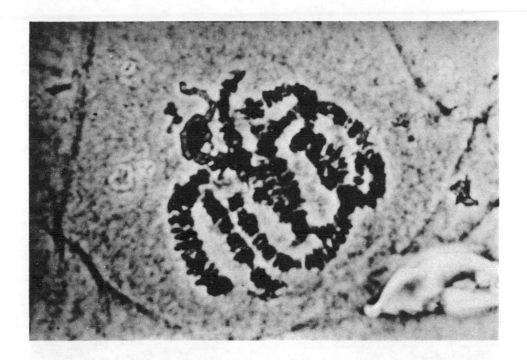

Fig. 6.1.2 Chromosomes.

Fig. 6.1.3 Model of the haemoglobin molecule.

Fig. 6.1.4 Launching of Apollo spacecraft to the moon.

Fig. 6.1.5 Earth seen from space.

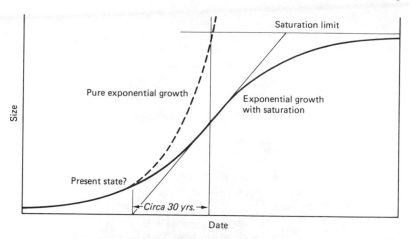

Fig. 6.1.6   Exponential growth with saturation. *Redrawn from de Solla Price, D. J., Little Science, Big Science (Columbia University Press, 1963).*

THE INTERRELATEDNESS OF THE DIFFERENT SCIENCES

In the 19th century we saw that separate specialised sciences started to emerge —
and in one sense it is true that this process has gone on ever since. Indeed, many
people complain that today there are too many separate scientific specialisms
which often fail to communicate with one another. But in another sense the
separate sciences have converged and become more interdependent as time has
passed, particularly if you take an overall view of their development as we are
trying to do in this book. Later, we shall be looking in more detail at some
examples of this interdependence both in pure science—as in the unification of
chemistry and physics or the plate tectonic revolution in the earth sciences—and
in applied science—as in the development of the semiconductor industry or the
growth of scientifically based medical knowledge. In this chapter we have space to
mention only one example—but a very significant one. This is the discovery of the
structure of DNA and of its importance in transmitting genetic information. This
key biological discovery was only possible because of work in a whole range of
different sciences. These included the development of physical techniques such as
electron microscopy, mass spectroscopy and the ultracentrifuge; great advances in
chemical analysis and in the chemistry of very large organic molecules—eventually
developing into the mushroom growth of biochemistry in the 1940s; and advances
in genetics and in the handling of material from the nuclei of living cells. However
the key technique used by Watson and Crick to find the DNA structure was X-ray
crystallography. Fig. 6.1.7 shows the X-ray diffraction pattern given by a
crystalline form of DNA. The details of the famous double helix structure for
DNA were deduced from photographs like this using methods which in essence
were the same as those used in 1912 by the Braggs to deduce the very much
simpler structure of common salt crystals.

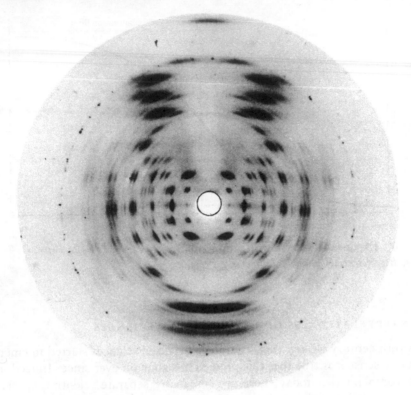

Fig. 6.1.7　X-ray diffraction pattern of the A form of DNA.

SCIENCE, GOVERNMENTS AND POLITICS

In earlier centuries governments were involved relatively little in the development of science and technology. The efforts of the German government in the 19th century (see Chapter 4.11) were the exception rather than the rule. In the 20th century that has all changed. Now government expenditure, direct or indirect, on science and technology forms an important part of the budgets of all the industrialised nations of the world.

Why has this happened? It wasn't because, in the early stages, governments wanted it and intended it to happen. Most governments in the early part of this century knew little of science and cared even less about it. The major reason for the rise in government science budgets has been the very success of science which has made it a major source of economic and political power. Science works, so it has largely forced itself on the attention of governments. This can be seen most clearly in the links between the major wars of the 20th century and government spending on science. National defence and the conduct of wars have for centuries been the prerogative of governments. All the important theories of politics, while they differ in nearly everything else, agree that national defence is an area which *must* involve governments. What is new in the 20th century is the increasing importance of scientific and technical knowledge in wars. World War 1 (1914–18) was the first major conflict in which a decisive factor was the mass production of weapons such as rifles and machine guns and of tanks and submarines. The chemical industry, too, was deeply involved in the manufacture of large quantities of explosives and of poisonous gases. This war has sometimes been called the chemists' war. By contrast, World War 2 (1939–45) was the physicists' war. The development of new devices like radar systems and nuclear weapons were based on the application of physical knowledge. The successful production of devices like these required the organisation of scientists and engineers on a scale which had never been seen before. The Cold War between the United States of America and the Soviet Union and their respective allies started in 1946 and has been going on with various degrees of coolness ever since. In this war electronics and rocketry have joined nuclear physics as major factors in the balance of military power. And expertise in electronics and rocketry are vital components in that extension of the Cold War—the Space Race of the last two decades.

The watershed in these developments was the successful application of science in the large-scale projects of World War 2. It was after this that science budgets rose so dramatically and began to approach their present levels.

After the question 'Why have governments invested heavily in science?' comes the question 'How?' How have governments chosen to involve themselves in science and how is such involvement best organised? The earliest 20th century government to give a wholehearted welcome to science was the Russian Communist revolutionary government established by Lenin in 1917. The new Soviet government aimed to sweep away the old order and to make a completely new start. And they intended science and technology to play a major part in this new start, as can be seen from Lenin's famous phrase 'Communism is nothing but Soviet rule plus the electrification of the entire country'—a statement made in November 1920! But, as we shall see, many of the later policies of the Soviet government towards science and scientists have been disastrous, and this is also true of some of the policies of the National Socialist government in Germany during the 1930s. Later we will look in more detail at various examples of government policies towards science. Yet the overall record seems to show that government sponsorship of science is most successful when it is arranged in ways which recognise the autonomy of the scientific community in *scientific* matters. In other words, 20th-century governments have had to learn the same lesson that the Catholic Church was taught by Galileo.

SCIENTISM AND THE REJECTION OF SCIENCE

Scientism is the tendency to extend the methods or results of science to areas in which their application or relevance is doubtful. In other words, to overestimate the power or scope of science. One commonly quoted example of scientism is the 19th-century tendency to extend Darwin's ideas on natural selection and the survival of the fittest to the organisation of human societies. Such arguments, often called Social Darwinism, were frequently used to justify extreme laisser faire social policies. Another example of scientism is the modern tendency to try to develop precise quantitative techniques for social forecasting as in computer simulations of the future direction of the world economy such as those used in the widely publicised book *The Limits to Growth*.[1] But perhaps the clearest example of scientism in the modern world lies in the claim of many Marxists to be able to organise society along scientific lines. In this context, it is interesting to remember that Karl Marx wanted to claim the prestige of science for his own work in politics and economics.

Scientism overestimates the power of science. At the opposite extreme is the view which, shocked by some disastrous consequences of applied science such as the thalidomide tragedy or some severe environmental pollution, ends by rejecting science and all its works. This view, which was often linked with a desire to return to a simpler kind of society, was particularly strongly held by many young people during the late 1960s and early 1970s, and it contributed then to a swing away from science as a subject for study by many students.

Such a wholesale rejection of science is as misguided as the scientistic idea that science is the answer to all the world's problems. It forgets the fact that the world population has risen so dramatically in the last 200 years (see Fig. 6.1.8) and that this rise has been closely connected with the rise of science and technology. Without science and technology, the gloomy predictions of Thomas Malthus of widespread famine and death would probably have come true in the 19th century. And they would certainly come true in the last quarter of the 20th century if science and technology were abandoned. Science has now made itself indispensable for any future society, however societies may evolve or be organised. The question we need to ask is not 'Are we for or against science?' Rather we should ask 'How can science best be used?' and 'How can the misuse of science be avoided?' or, better still, 'What kind of social and political system will best enable us to check the misuse of science *and* retain its benefits?' Questions such as these are particularly important for scientists to consider because their specialised knowledge and training should enable them to make a unique contribution to informed public discussion about scientific matters. This is a contribution that students of science ought to make since no-one else is as well placed as they are to make it.

Finally consider this quotation:

> The bourgeoisie, during its rule of scarce one hundred years, has created more massive and more colossal productive forces than have all preceding generations together. Subjection of nature's forces to man, machinery, application to chemistry, to industry and agriculture, steam navigation, railways, electric telegraphs, clearing of whole continents for cultivation, canalization of rivers, whole populations conjured out of the ground—what earlier century had even a presentiment that such productive forces slumbered in the lap of social labour.

That was Karl Marx in the *Communist Manifesto* of 1848. Since then the

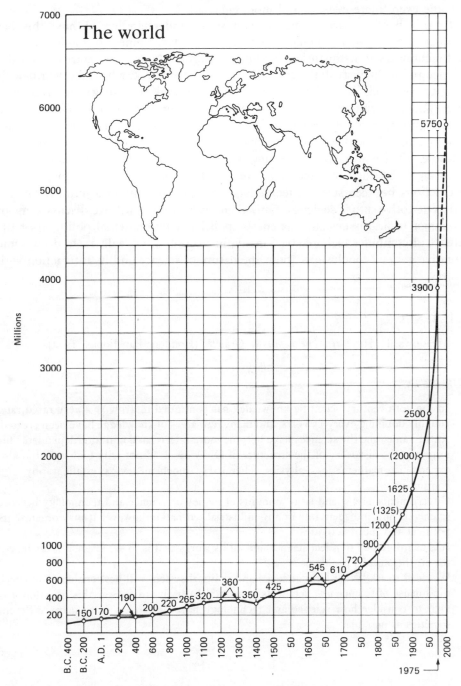

Fig. 6.1.8 World population, 400 BC to the present. *Redrawn from McEvedy, C. and Jones, R., Atlas of World Population History, page 342, copyright © Colin McEvedy and Richard Jones, 1978, reprinted by permission of Penguin Books Ltd.*

'productive forces' associated with the 'subjection of nature's forces to man' have become even 'more massive and more colossal'.

In this book we are trying to gain some understanding of how this has happened. So far we have mainly considered the development of modern science and technology in its historical birthplace, Europe. Now as we start to consider science and technology in the 20th century, we need to take a broader, world-wide perspective. This is not only because science and technology are the most international of cultural products. It is also because many of the problems involved in the application of science and technology can only be appreciated if we consider how they can be and have been tackled in as many different types of society as possible. That is why, after outlining some major developments in 20th-century scientific knowledge in Section 6, we will turn in Section 7 to consider the interactions between science, technology and governments in a number of very different 20th-century societies. Only then, in Section 8, will we discuss some of the major topical issues such as energy policies, environmental pollution or the impact of computer technology. Finally, in Section 9, we will try to draw some tentative general conclusions about the nature of science and its interaction with society.

REFERENCE

1   Meadows, D. H. *et al.*, *The Limits to Growth*, (London: Pan Books, 1972).

SUMMARY

1   In the 20th century scientific knowledge has continued to grow at a very rapid rate, roughly doubling every 15 years, and many new areas of knowledge have been created.
2   At the same time the sciences have become more interrelated and interdependent; for example, the discovery of the structure of DNA involved not only biologists but also experts in mathematics, physics, chemistry, biochemistry, crystallography and computing.
3   Governments now spend large amounts of money on science and technology because science is now indispensable in any industrialised nation however it may organise its political and social activities.
4   It is as misguided to reject science and all its works as it is to see science as a panacea for all the world's problems.
5   We need to ask 'What kind of social and political system will best enable us to check the misuse of science *and* retain its benefits?' In trying to answer this question it is vital to consider how science and technology have developed in as many different societies as possible.

6.2 Electromagnetic waves and relativity: Maxwell and Einstein

Newton and Einstein are such towering figures in the history of science that their names are familiar to many who know little of science itself. Both men, while in their early twenties, did work which changed the whole course of science. That is why Newton is often taken, rightly, as the symbol of the 17th-century scientific revolution and why Einstein has dominated science in the first half of the 20th century.

In this chapter we will try to gain some understanding of Einstein's major work—the special theory of relativity—while in the next chapter we will look at other aspects of Einstein's life and work and at some of the turbulent social and political events in which Einstein became involved in his later years.

Einstein's special theory of relativity of 1905 does not stand in isolation. It is part of the continuous development of physical science from the time of Galileo onwards. In particular special relativity derived from some inconsistencies between two of the major triumphs of classical physics—Newton's dynamics and Maxwell's theory of electromagnetism. So, in order to appreciate what Einstein did, we need first to look briefly at these two areas and then try to see how Einstein's work both superseded and synthesised that of Newton and Maxwell.

NEWTON'S DYNAMICS

We have discussed Newton's dynamics in some detail in Chapters 3.4 and 4.1. We discussed, in particular, his laws of motion and his inverse square law of gravitational force. We mentioned then how his ideas were generalised in the 18th century by such great French mathematicians as Lagrange and Laplace. In the 19th century Newton's laws became even more deeply entrenched in all branches of physics. The general application of the principles of the conservation of momentum and the conservation of kinetic and potential energy in many simple systems—both implicit in Newton's laws—contributed greatly to this process.

One striking example of the application of Newton's laws raised their prestige during the 19th century to even greater heights. This was the discovery of the planet Neptune in 1846. Up till then, the outermost known planet in the solar system was Uranus, discovered by William Herschel in 1781. When the orbit of Uranus was observed, it was found that it could be predicted on the basis of Newton's laws of motion and gravitation—just as the orbits of the other planets had been in the *Principia*. However, more accurate observations showed that a small discrepancy or wobble, of only 1·5 minutes of arc, existed between the predicted and observed orbits. Two theoretical astronomers, Leverrier in France and Adams in England, put forward the suggestion that this discrepancy was due to an undiscovered planet beyond Uranus. And they both used Newton's laws to calculate the position and size of this unknown planet. In 1846 a new planet, later named Neptune, was observed in just the place in the sky that had been predicted. Later, the much smaller planet, Pluto, was discovered by a similar method in 1930. So, at the end of the 19th century, Newton's laws were more widely accepted than ever before.

MAXWELL'S ELECTROMAGNETIC THEORY

James Clerk Maxwell (1831–79) was responsible for at least two major scientific developments in the 19th century—one practical and one theoretical. The practical one was his founding at Cambridge University of the Cavendish laboratory after his appointment as the first Professor of Experimental Physics at Cambridge. It was in the Cavendish laboratory that many crucial experiments in atomic and nuclear physics were first performed. But it is Maxwell's greatest theoretical achievement, his theory of electromagnetism which dates from the early 1860s, that we will discuss in this chapter.

The laws of electricity and magnetism had been slowly and laboriously worked out over the previous century—from about 1760 onwards. The process had involved many careful experiments and much inspired theoretical work. The laws of force between static electric charges and between permanent magnets came first. These laws were quite separate but had a very similar mathematical form. Both were inverse square laws like Newton's law of universal gravitation. These laws of force were established in the second half of the 18th century by both direct and indirect measurements. The electrostatic law is today called Coulomb's law but others, including Priestley and Cavendish, played a significant part in establishing it. The magnetostatic law was also established experimentally by Coulomb in 1785 and confirmed indirectly by Gauss in 1833.

At this stage electric and magnetic forces were regarded as separate phenomena. But after 1800 this all changed when the study of steady electric currents was made possible by the widespread use of Volta's early batteries (see Chapter 4.5). The result of many years of work was that electric and magnetic forces were shown to be closely linked. Oersted and Ampère in the 1820s showed that electric currents gave rise to magnetic forces, while a little later, in the 1830s, Faraday showed that changing magnetic fields gave rise to electric currents (see Chapter 4.7); in both cases simple equations were found which precisely described the observed effects. Now we come to Maxwell.

What Maxwell did was to take the four mathematical equations of Coulomb, Gauss, Ampère and Faraday,[1] put them together and rewrite them in a different but equivalent mathematical form. He also added a term of his own to one of them. When he had finished, the four equations in their simplest form—for free space—looked like this.

(1) $\operatorname{div} E = 0$ \qquad (Coulomb's Electrostatic Law)

(2) $\operatorname{curl} E = -\dfrac{1}{c}\dfrac{\partial H}{\partial t}$ \quad (Faraday's Law of Electromagnetic Induction)[1]

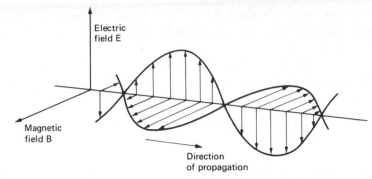

Fig. 6.2.1 Properties of electromagnetic waves as predicted by Maxwell's electromagnetic theory.

(3) $\operatorname{div} H = 0$ (Gauss' Magnetic Law)

(4) $\operatorname{curl} H = \dfrac{1}{c}\dfrac{\partial E}{\partial t}$ (Ampère's Current Law)

   E represents electric field
   H represents magnetic field

   In these equations t represents time and c is a constant which could be calculated from the experimentally measured constants in the original equations. For our purposes the details of the equations are not important. All we need to note is that they specify relationships between electric and magnetic *fields*.

   The constant c had the dimensions of velocity and, when Maxwell calculated its value from the already known experimental constants, he found that it was equal to $3 \cdot 1 \times 10^8$ metres per second. This was very close to the value for the velocity of light, $3 \cdot 15 \times 10^8$ metres per second, which had been obtained by Fizeau in 1849.

   That was extraordinary enough. But Maxwell's next step was even more extraordinary. He rearranged his equations—in effect solving them separately for the electric field E and the magnetic field H in turn. This, in its simplest form, was the result.

$$\frac{\partial^2 E}{\partial x^2} = \frac{1}{c^2}\frac{\partial^2 E}{\partial t^2}$$

and

$$\frac{\partial^2 H}{\partial x^2} = \frac{1}{c^2}\frac{\partial^2 H}{\partial t^2}$$

in which x represents position, t represents time and c is the same constant as before.

   What do these equations mean? Readers who are mathematicians may recognise them as examples of the standard wave equation—familiar, for instance, in the theory of sound or water waves. In other words, Maxwell's equations predicted the existence of a new kind of wave, involving both electric and magnetic fields, which later came to be called electromagnetic waves. Fig. 6.2.1 indicates some of the properties of these waves which Maxwell's equations predicted. As shown, the electric and magnetic fields are at right angles to each other and to the direction of propagation of the waves.

But it is even more interesting that the constant c in the wave equations gives the speed of the waves. Remember that c in Maxwell's equations was numerically equal to the velocity of light. As Maxwell himself put it:

> The velocity of transverse undulations in our hypothetical medium, calculated from the experiments of Kohlrausch and Weber, agrees so exactly with the velocity of light calculated from the optical experiments of M. Fizeau, that we can scarcely avoid the inference that *light consists in the transverse undulations of the same medium which is the cause of electric and magnetic phenomena.*[2]

At this time, in the early 1860s, these 'transverse undulations' of the electromagnetic field had never been observed. Maxwell never lived to see his predictions confirmed; he died from cancer in 1879 at the early age of 48. It was not until the mid-1880s that Heinrich Hertz, working in Karlsruhe, succeeded in both producing and detecting these invisible electromagnetic waves. Between 1886 and 1888, using very simple apparatus, just an electrical oscillator and a spark gap detector, Hertz produced waves with wavelengths ranging from about 30 centimetres to 10 metres. These waves travelled at the speed of light and had all the properties predicted by Maxwell's equations. Thus, after 25 years, Maxwell's theory was triumphantly confirmed. Hertz was to die even younger than Maxwell, in 1894 at the age of 36. But the waves he had discovered were soon applied in wireless telegraphy, or radio as it came to be called. Hertz was thus the direct founder of the enormous expansion in radio communication which has taken place in the 20th century and which we take so much for granted today. And, of course, the modern SI unit of frequency is named after him. Later it was realised that γ- and X-rays, and ultra-violet and infra-red radiation, were also electromagnetic waves of just the same kind as visible light and radio waves (see Fig. 6.2.2).

So Maxwell's electromagnetic theory was one of the great unifying achievements of classical physics. Fig. 6.2.3 shows a succinct summary of classical physics made by the Nobel prizewinner Richard Feynmann who gives Maxwell's equations pride of place over Newton's Laws of Motion.

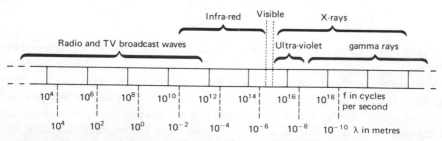

Fig. 6.2.2 Spectrum of electromagnetic waves from radio waves to γ-rays.

Maxwell's equations

I. $\nabla \cdot E = \dfrac{\rho}{\epsilon_0}$ (Flux of E through a closed surface) = (Charge inside)$/\epsilon_0$

II. $\nabla \times E = -\dfrac{\partial B}{\partial t}$ (Line integral of E around a loop) = $-\dfrac{d}{dt}$ (Flux of B through the loop)

III. $\nabla \cdot B = 0$ (Flux of B through a closed surface) = 0

IV. $c^2 \nabla \times B = \dfrac{j}{\epsilon_0} + \dfrac{\partial E}{\partial t}$ c^2 (Integral of B around a loop) = (Current through the loop)$/\epsilon_0 + \dfrac{\partial}{\partial t}$ (Flux of E through the loop)

$\left[\begin{array}{l} \textit{Conservation of charge} \\[4pt] \nabla \cdot j = -\dfrac{\partial \rho}{\partial t} \quad \text{(Flux of current through a closed surface)} = -\dfrac{\partial}{\partial t}\text{(Charge inside)} \end{array} \right]$

Force law

$F = q(E + v \times B)$

Law of motion

$\dfrac{d}{dt}(p) = F$, where $p = \dfrac{mv}{\sqrt{1 - v^2/c^2}}$ (Newton's law, with Einstein's modification)

Gravitation

$F = -G\dfrac{m_1 m_2}{r^2} e_r$

Fig. 6.2.3 Summary of classical physics given by Richard Feynman. From Feynmann/Leighton/Sands, The Feynmann Lectures on Physics, vol. II, © 1964. California Institute of Technology. Table 18.1. Published by Addison–Wesley Publishing Company, Inc., Reading, MA. Reprinted with the permission of the publisher.

NEWTON'S LAWS OF MOTION AND MAXWELL'S THEORY

Now at last we can start to discuss Einstein's work on relativity. To the scientist the term 'relativity' has a fairly precise meaning. It deals with how the world appears to observers who are moving with respect to one another and in particular with how this relative motion affects the measurements made by the observers and the physical laws they deduce from these measurements. For many years this subject was not considered particularly interesting or significant since the effects of relative motion were thought to be well understood on the basis of a few simple commonsense ideas. Most physical measurements, when analysed in detail, involve measurements of the positions of various objects. These positions, as seen by different observers at different times, were related to one another using the ideas on relative motion which we all learnt at school—that distance travelled is equal to speed times time, and that the relative speed of two objects is found by adding their speeds if they are moving in opposite directions and by subtracting them if they are moving in the same direction. Time was considered to be no problem. A universal absolute time-scale, the same for all observers, was assumed to exist. These were the assumptions of Galileo and Newton and, like the rest of their ideas, they were very successful when applied to the problems of dynamics, in the sense that predictions and observations were in close agreement.

However, these classical ideas on relative motion weren't so successful when they were applied to Maxwell's equations which, remember, show how electric and magnetic fields vary in space and time. Maxwell's equations changed their form for different observers, and various effects were predicted which were not observed. For example, the equations predicted that electromagnetic waves should travel at different speeds in different directions. So the two great triumphs of classical physics turned out to be inconsistent with one another. It was Newton or Maxwell. One or the other had to give, and it turned out to be Newton's laws of motion—the laws which had been at the heart of physics for so long—that needed to be changed. So physics was in fundamental trouble and it is not surprising that Einstein's work had to be of a very fundamental kind.

EINSTEIN'S THEORY OF SPECIAL RELATIVITY

In 1905 Einstein published his special theory of relativity which succeeded in reuniting the laws of Newton and Maxwell. Fig. 6.2.4 shows him at that time. He was just 26. How did he do it? We know that the inconsistencies between Newton's laws and Maxwell's theory were very much in his mind. Indeed, the title of Einstein's paper was 'On the electrodynamics of moving bodies'. What we don't know, in detail, is how he arrived at his new ideas of relative motion. All we can do, therefore, is to try to set out Einstein's ideas as clearly as possible.

Einstein's special theory deals with relative motion of a particular kind—that of two observers moving at a constant velocity with respect to one another. Such observers are called inertial observers. The theory starts from three general principles:

Fig. 6.2.4   Einstein in 1905, aged 26; 1905 was the year in which Einstein published his special theory of relativity and his work on light quanta and Brownian motion.

(1) The first principle is that physical laws are the same for all inertial observers. This principle is not new; it is carried over from the classical relativity theory of Galileo and Newton.

(2) The second principle is that all quantities used in a theory, even the simplest and apparently most obvious such as time, should be specified operationally; that is, a definite experimental procedure should be laid down for measuring the quantity.

(3) The third general principle is that the speed of light is the same in all directions, for all inertial observers and is independent of any motion of the body emitting the light.

Principles (2) and (3) were introduced by Einstein. Principle (2), the operational principle, seems an obvious commonsense one in an experimental subject like physics. But principle (3) is not obvious at all and it is only relatively recently that it has been confirmed by direct experiments. What Einstein did was

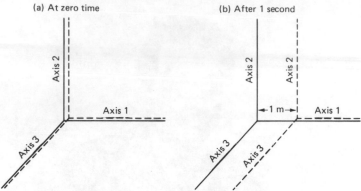

Fig. 6.2.5 Cartesian co-ordinate axes for two observers moving with respect to one another at a velocity of 1 metre per second along axis 1. *Redrawn from Marks, J., Relativity, by permission of Geoffrey Chapman, a division of Cassell Ltd.*

to use it as an hypothesis in order to see what its indirect consequences were. These could then, in principle, be tested by direct experiments.

What were these consequences? The most startling one was that absolute time vanished. This seemingly obvious commonsense assumption of Newton's was destroyed by the operational principle and the constancy of the speed of light. Detailed analysis showed that the assumption that time is the same for all observers turns out to be equivalent to the assumption that information can be transmitted from one observer to another instantaneously—or, in other words, that signals can travel at an infinitely great speed rather than at a finite speed like that of light.

To put it more precisely, Einstein showed that if two inertial observers, perhaps represented by the two co-ordinate frames in Fig. 6.2.5, measured the position and time co-ordinates of the same event, then these co-ordinates would be related by the following equations, which are usually called the Lorentz transformations:

$$x' = \frac{x - ut}{\sqrt{1 - \frac{u^2}{c^2}}} \qquad t' = \frac{t - \frac{ux}{c^2}}{\sqrt{1 - \frac{u^2}{c^2}}}$$

in which u is the relative velocity of the two observers, c is the speed of light and (x, t) and (x', t') are the position and time co-ordinates of the event as measured by the two inertial observers. Contrast these equations with those of Newton and Galileo, which are usually called the Galilean transformations:

$$x' = x - ut \qquad t' = t$$

As you can see, in the Lorentz transformations t' is not equal to t which means that time is no longer absolute and the same for all observers as it is in the Galilean transformations. And notice one other thing about these equations. If u is a lot smaller than c, Einstein's equations reduce to those of Newton and Galileo. In other words the older laws are a limiting case of the newer ones for speeds which are small compared with the speed of light. This turns out to be true of all Einstein's results.

Having destroyed absolute time, Einstein obviously couldn't stop there. Time was much too deeply entrenched in Newton's laws of motion and all that had been deduced from them. What he had to do was to construct a new dynamics with new laws of motion which nevertheless reduced to Newton's laws at low speeds. And this is just what Einstein succeeded in doing. However, when he had done this Einstein found that his theory predicted that mass, as well as time, ought to behave in a new and peculiar way. Instead of being constant, the mass of body ought to vary with its speed in the manner shown in line II of Fig. 6.2.6. Notice that line II only differs significantly from line I at speeds which are close to the speed of light. At ordinary earthly speeds there is no detectable change in mass. The prediction of the variation of mass was an extraordinary one for Einstein to make. And it was all the more extraordinary when we remember that the constancy of mass and the conservation of matter had been one of the great unifying principles of 19th-century science and had played such an important part in the development of quantitative chemistry.

But Einstein's next prediction seemed even more extraordinary. This was that mass and energy could be converted into one another according to the famous formula, $E = mc^2$, in which E represents energy, m, mass, and c the velocity of light.

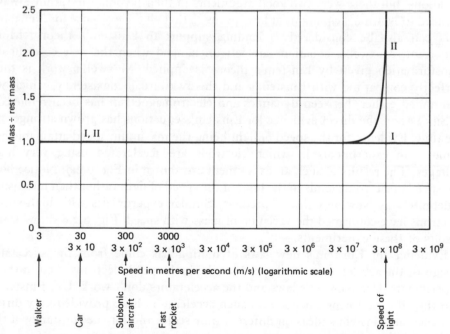

Fig. 6.2.6 Variation of mass with speed: I—according to Newton's laws; II—according to Einstein's special theory of relativity. *Redrawn from Marks, J., Relativity, by permission of Geoffrey Chapman, a division of Cassell Ltd.*

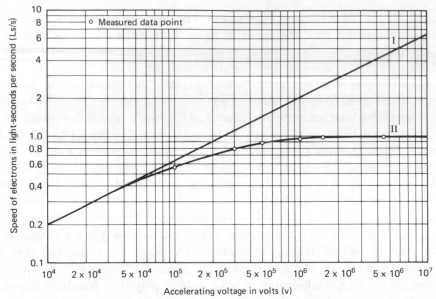

Fig. 6.2.7 Speed of electrons as a function of accelerating voltage: I—according to Newton's laws; II—according to Einstein's special theory of relativity. *Redrawn from Marks, J., Relativity, by permission of Geoffrey Chapman, a division of Cassell Ltd.*

In 1905 there was virtually no direct evidence for any of Einstein's new equations, but there were two good arguments in their favour. First, they always reduced to Newton's equations at low speeds so that all the evidence for Newton's laws can also be considered as lending support to Einstein's theory. More important, Maxwell's equations are left unchanged when the new co-ordinate transformation given by Einstein's theory is applied. Maxwell's work is thus perfectly compatible with Einstein's and the awkward inconsistency, which we mentioned earlier, between dynamics and electromagnetism has been removed.

Since 1905, the direct evidence for Einstein's equations has grown stronger all the time. Evidence for the speed of light being the maximum speed attainable has come from experiments in which electrons are accelerated using very high voltages. The results of such an experiment are shown in Fig. 6.2.7. Notice how the speed of electrons tends to the limit of the speed of light rather than increasing indefinitely as Newton's theory predicts. Similar experiments with high-speed electrons have confirmed the variation of mass with speed. Fig. 6.2.8 shows some results of these experiments.

Evidence for Einstein's new laws of motion has come from the successful design of the accelerators which produce the high-speed electrons used in these experiments. Use Newton's laws and the accelerators don't work. Use Einstein's and they do. And experiments with such accelerators have provided very direct evidence for Einstein's ideas on different time scales and for the constancy of the speed of light.

Evidence for the equivalence of mass and energy has come, on the small scale, from the study of nuclear reactions, and, on the large scale, from nuclear weapons

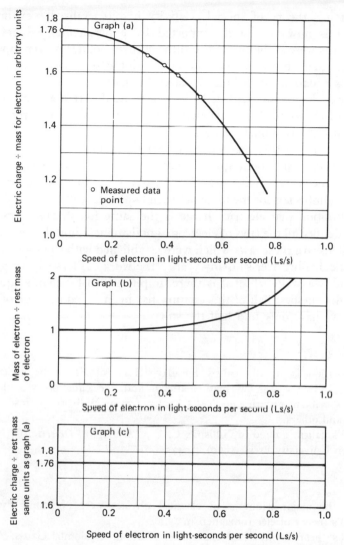

Fig. 6.2.8 Graphs showing (a) the ratio of charge to mass for an electron, (b) the ratio of mass to rest mass predicted by special relativity and (c) the ratio of charge to rest mass for an electron, all as a function of speed. Graph (c) is obtained by point-by-point multiplication of graphs (a) and (b); it confirms that the variation of mass predicted theoretically by Einstein is in agreement with the measured variation. *Redrawn from Marks, J., Relativity, by permission of Geoffrey Chapman, a division of Cassell Ltd.*

and nuclear reactors. On an even larger scale we now know that nuclear energy is the source of the energy of the stars; for example, a typical star like the sun is converting mass into energy at the rate of more than a thousand million (10^9) kilograms every second. However, the sun's mass is so large (about 10^{30} kg) that, even at that rate of loss of mass, it would last for another million million years or so.

So Einstein's theory of special relativity has generalised Newton's laws of motion and is now very well supported by experimental evidence—just as Newton's ideas were at the end of the 19th century. So, if it is ever superseded, its equations are likely to remain as a special limiting case of some wider theory just as Newton's equations are a limiting case of Einstein's.

In the next chapter we shall be looking at some different aspects of Einstein's work and at the way his life developed after his success as a theoretical physicist had made him internationally famous.

But before ending this chapter, let us briefly mention one more little known and extraordinary fact about special relativity. Suppose we start with the basic principles of special relativity, including the new co-ordinate transformation; add to them Coulomb's law for the force between two electric charges; and then throw in the assumption that electric charge is the same for all observers. It is now possible to deduce all the laws of electromagnetism including the very existence of magnetic fields. All these laws which were established with so much experimental and theoretical effort over so many years—the work of Oersted, Ampère, Gauss, Faraday and Maxwell—all of it is there, implicitly, in Einstein's equations.

No wonder the first half of this century has been called the Age of Einstein—and so far we have only told half the story.

REFERENCES

1 This equation is usually called Faraday's Law of Electromagnetic Induction. However, Faraday was no mathematician and the equation was derived from Faraday's brilliant experiments (see Chapter 4.7) by William Thomson (later Lord Kelvin) and others.

2 Maxwell's italics; quoted in Gillispie, C. C., *The Edge of Objectivity*, (New Haven: Princeton University Press, 1960), p. 473.

SUMMARY

1 Two of the major triumphs of classical physics were Newton's dynamics and Maxwell's theory of electromagnetism.

2 Maxwell's theory synthesised the earlier work of Coulomb, Gauss, Ampère and Faraday on electricity and magnetism and predicted the existence and properties of radio waves 25 years before they were observed; however, Maxwell's theory was inconsistent with Newton's ideas on relative motion.

3 Einstein's special theory of relativity superseded Newton's laws of dynamics and removed the inconsistency between dynamics and the laws of electromagnetism; however, Newton's laws of motion remain as a special case of Einstein's equations for velocities which are small compared with the velocity of light.

4 Einstein's theory predicted that time was not the same for all observers, that the mass of a body should vary with its speed and that mass and energy are interconvertible according to the equation $E = mc^2$.

5 All these predictions have since been confirmed in many experiments and we now know that the conversion of nuclear mass into energy is the source of the energy of the sun and other stars as well as of the energy produced in nuclear reactors and nuclear weapons.

6.3 Einstein's work and some of its social implications

In the last chapter we looked at Einstein's work on special relativity and its connections with electromagnetism. The emphasis was on the physics. In this chapter we are going to talk more about Einstein the man and about the influence of his work on society.

Einstein was born in 1879, just over a hundred years ago, in Ulm in Germany. His family moved to Munich when he was one and Einstein went to school there. He was not at all outstanding at school and he disliked it intensely—so much so that, when he was 15 (see Fig. 6.3.1) Einstein decided to leave school without taking his final examination. At the same time he renounced his German citizenship! He thus showed early signs of that independence of mind, verging at times on obstinacy, that was one of his most outstanding characteristics.

A few years later Einstein tried to enter the Polytechnic at Zurich in Switzerland to study electrical engineering but he failed the entrance examination. The next year he was successful but now he had decided to study physics

Fig. 6.3.1 The schoolboy Einstein in 1893—at the age of 14.

and mathematics. Once again, just as at school, he wasn't a particularly successful student. He did graduate but with no great distinction, and then he couldn't get a job. Eventually, on the recommendation of a family friend, he was taken on as a patent examiner in the Swiss Patent Office at Berne. Einstein worked there full-time for seven years, from 1902 to 1909, and during this time he had virtually no contact with other physicists or mathematicians. It was in his spare time, evenings and weekends, that Einstein did the work which has totally changed physics in the 20th century.

The year 1905, when he was still only 26 (see Fig. 6.2.4, page 257) was Einstein's golden year. In that single year he did three pieces of work, any one of which would have been enough to win him a Nobel prize. It was rather like Newton's burst of creative activity during the plague years of 1665–6. There was Einstein's work on special relativity, which we outlined in the last chapter. He published two important papers on special relativity in 1905. Then there was some revolutionary work on the nature of light, for part of which Einstein was awarded the Nobel prize in 1921. It is interesting that Einstein, like Newton, did fundamental work on both motion and light. Finally, there was a paper which enabled the existence of atoms to be directly demonstrated for the first time.

In his work on light, Einstein analysed the energy spectrum of the light emitted by any hot body. He showed that this spectrum could best be explained if light consists of independent particles of energy, later called quanta or photons. Einstein treated light in much the same way as Maxwell and Boltzmann had treated gases in their kinetic theory of gases—as a very large number of individual particles whose overall behaviour could only be interpreted statistically. His quanta of energy were also able to explain some other newly discovered phenomena such as the photoelectric effect. These quantum ideas of Einstein's were really an extension of those put forward by Max Planck (see Fig. 6.3.2) in 1900. But Planck had only supposed that quanta of energy were involved when light was being absorbed or emitted. Einstein went much further; he assumed that light energy was always in the form of quanta. In an image which Einstein himself used, it was not only that beer was bought and sold in pints from the keg, but that the beer in the keg consisted *only* of pints.

Einstein's third 1905 paper was called 'On the Movement of Small Particles Suspended in a Stationary Liquid Demanded by the Molecular-Kinetic Theory of Heat'. It gave a theoretical explanation for a well-known phenomenon called Brownian motion. This was the random dancing motion seen when a suspension of fine pollen grains is observed under a microscope. It had first been observed in the 1820s by a botanist called Brown and he thought that the particles were alive!

What Einstein did was to treat the suspended particles as if they were very large atoms or molecules in equilibrium with the invisible atoms or molecules of the liquid. He then deduced equations which described the average motion of the suspended particles. Einstein's formula was verified experimentally by Perrin in 1908 and this convinced even Ostwald and Mach of the existence of atoms.

All this was more than enough one might think for one lifetime, let alone a single year. But that wasn't all. In 1916 Einstein produced his General Theory of

Fig. 6.3.2 Max Planck (1858–1947)—the originator of the quantum theory.

Relativity which, in some ways, is a generalisation of the special theory. But in other ways, and particularly in its treatment of gravitation, the General Theory is something completely new. It now forms the basis of much of modern cosmology.

General relativity is even more difficult to test experimentally than the special theory. But it did suggest one dramatic experiment. The theory predicts that when the light from a distant star passes close to the sun, it will be deflected by a very small amount—a bit less than 2 seconds of arc (3600 seconds = 1 degree). Normally the sun is so bright that such measurements can't be made. But on 29th May 1919, there was a total eclipse of the sun during which two separate expeditions confirmed a bending of light by the sun of just about the predicted amount. This positive confirmation made the headlines and Einstein became a public figure, almost overnight. It also had a great effect on Karl Popper, then 17 years old, who has since become one of the best-known philosophers of science. In his recent intellectual autobiography Popper writes:

... at the same time I learned about Einstein; and this became a dominant influence on my thinking—in the long run perhaps the most important influence of all. In May, 1919, Einstein's eclipse predictions were successfully tested by two British expeditions. With these tests a new theory of gravitation and a new cosmology suddenly appeared, not just as a mere possibility, but as a real improvement on Newton—a better approximation to the truth.[1]

And Popper concludes that:

... what impressed me most was Einstein's own clear statement that he would regard his theory as untenable if it should fail in certain tests ...

Here was an attitude utterly different from the dogmatic attitude of Marx, Freud, Adler, and even more so that of their followers. Einstein was looking for crucial experiments whose agreement with his predictions would by no means establish his theory; while a disagreement, as he was the first to stress, would show his theory to be untenable.

This, I felt, was the true scientific attitude. It was utterly different from the dogmatic attitude which constantly claimed to find 'verifications' for its favourite theories. Thus I arrived, by the end of 1919, at the conclusion that the scientific attitude was the critical attitude, which did not look for verifications but for crucial tests; tests which could *refute* the theory tested, though they could never establish it.[2]

We shall be discussing Popper's philosophy of science again in Chapter 7.1.

EINSTEIN—THE PUBLIC FIGURE

The year 1919, when he was 40, was a turning point in Einstein's life. His best work was now behind him and, as we shall see, he grew increasingly out of sympathy with the way physics developed from the mid-1920s onwards. And his new status as a public figure led him into much deeper waters.

Einstein had been a professor at the University of Berlin since early in 1914 and he was one of the few German intellectuals who would not join in the widespread support given to the German cause in World War 1. This may have made him mildly unpopular but it was as nothing compared with what he had to endure from the early 1920s onwards. At this time Einstein was a pacifist and a keen supporter of the League of Nations—an international organisation which had been formed after World War 1 in much the same way as the United Nations after World War 2. Einstein was also a Jew and a supporter of Zionism.

All these facts made him the target for personal attacks, particularly now that his work had given him world-wide prestige. In 1920 an Anti-Einstein League was formed in Germany, which offered substantial sums of money to anyone who would refute Einstein's work. The Anti-Einstein League held public meetings at which Einstein and his work were attacked; and the League became associated with the anti-semitism of Adolf Hitler's National Socialist Party.

In the early 1920s Einstein and his work were also attacked by the physicist

Phillipp Lenard. Lenard had won the Nobel Prize in 1905 for his work on the photoelectric effect but he publicly protested after Einstein had been awarded the prize in 1921—also for work on the photoelectric effect. Later, in his autobiography Lenard stated that relativity 'was a Jewish fraud, which one could have suspected from the first with more racial knowledge than was then disseminated, since its originator Einstein was a Jew.'[3] When the National Socialists came to power in 1933, Einstein was in America. He refused to return to Germany, resigned from the Berlin Academy of Sciences and campaigned against the National Socialists, one of whose aims was to introduce anti-semitic laws. This was done in April 1933, barely two months after Hitler became Chancellor. Many Jews were dismissed from their jobs, including nearly 2000 university teachers and 20 current or future Nobel prizewinners (see Chapter 7.2). Most of the latter found university posts abroad, mainly in Britain and the USA, and thus were more fortunate than those Jews for whom escape was not possible. It is estimated that about 6 million Jews were killed during the 12 years of National Socialist power in Germany (1933–45).

To return to Einstein, Lenard had this to say while opening a new physics institute in 1935:

I hope that the institute may stand as a battle flag against the Asiatic spirit in science. Our Führer has eliminated this same spirit in politics and national economy, where it is known as Marxism. In natural science, however, with the over-emphasis on Einstein, it still holds sway. We must recognise that it is unworthy of a German to be the intellectual follower of a Jew. Natural science, properly so called, is of completely Aryan origin and Germans must today find their way out into the unknown.
Heil Hitler.[4]

And in 1936 another Nobel prizewinning physicist, Johannes Stark, made this statement:

Fifteen years ago, when Relativity was made high goddess of science ... Lenard pronounced boldly against this general madness and described the theory as nonsense. His brave attack at a Congress at Nauheim in 1920 will always be remembered as honourable to him as it is disgraceful to Professor Planck that he has served as Einstein's lieutenant. ... There are in Germany even now adherents of Einstein who continue to work in his spirit. His main supporter, Planck, still retains his position at the head of the Kaiser Wilhelm Society. His interpreter and friend, von Laue, is still allowed to play a part as expert in physics in the Berlin Academy of Sciences. The theoretical formalist, Heisenberg, who works in the very spirit of Einstein, is actually to be honoured by a call to a university chair. With such regrettable evidence of opposition to the Nazi attitude, Lenard's struggle against Einstein sets an example. It is surely desirable that the competent officials in the Ministry of Culture should seek Lenard's advice before appointing professors of physics.[5]

A final quotation to illustrate this appalling situation comes from a text-book called *German Physics* published by Lenard in 1936–7:

> 'German Physics?' one asks. I might rather have said Aryan Physics or the Physics of the Nordic Species of Man, Physics of those who have fathomed the depths of Reality, of seekers after Truth, Physics of the very founders of Science. But it will be replied to me 'Science is and remains international'. It is false. In reality Science, like every other human product, is racial and conditioned by blood.[6]

Einstein's work was also strongly attacked in the Soviet Union during the period 1948–55. Here is one published statement from this period:

> The unmasking of reactionary Einsteinism in the area of physical science is one of the most pressing tasks of Soviet physicists and philosophers.

Now we return to the 1930s—to discuss one of the predictions of Einstein's special theory of relativity which was being confirmed at just about the same time as Einstein left Germany for the last time. This was the interconvertibility of mass and energy according to the equation $E = mc^2$. This equation was confirmed during the detailed study of nuclear reactions. The neutron was discovered in 1933 and this made it possible for many more such reactions to be studied. However, all these reactions could only be carried out on a very small scale and there seemed little likelihood of making nuclear energy available in large quantities. In all the early experiments far more energy was used in getting the nuclear reaction going than was released by the reactions themselves. Indeed, in 1933, Ernest Rutherford, the man who did more than anyone else to create nuclear physics, told a meeting of the British Association for the Advancement of Science that:

> These transmutations of the atom are of extraordinary interest to scientists, but we cannot control atomic energy to an extent which would be of any value commercially, and I believe we are not likely ever to be able to do so.[7]

And in 1937, the year of his death, Rutherford repeated this view:

> The outlook for gaining useful energy from the atoms by artificial processes of transformation does not look very promising.[8]

Yet within two years the prospects for obtaining nuclear energy on a large scale had been transformed. It was the discovery of uranium fission in early 1939 that made the difference. After this the large-scale release of nuclear energy, using a chain reaction in uranium, was almost certain to be feasible and a nuclear bomb became a distinct possibility. Some nuclear physicists were afraid that Hitler might get such a bomb first and so they drafted a letter and asked Einstein to send it to the American President Roosevelt. This he did on 2nd August 1939, just a month before World War 2 started. Here is the text of that momentous letter.

Albert Einstein,
Old Grove Road,
Nassau Point,
Peconic, Long Island.
August 2nd, 1939

F. D. Roosevelt,
President of the United States,
White House,
Washington, D.C.

Sir:

Some recent work by E. Fermi and L. Szilard, which has been communicated to me in manuscript, leads me to expect that the element uranium may be turned into a new and important source of energy in the immediate future. Certain aspects of the situation which has arisen seem to call for watchfulness and, if necessary, quick action on the part of the Administration. I believe therefore that it is my duty to bring to your attention the following facts and recommendations:

In the course of the last four months it has been made probable—through the work of Joliot in France as well as Fermi and Szilard in America—that it may become possible to set up a nuclear chain reaction in a large mass of uranium, by which vast amounts of power and large quantities of new radium-like elements would be generated. Now it appears almost certain that this could be achieved in the immediate future.

This new phenomenon would also lead to the construction of bombs, and it is conceivable—though much less certain—that extremely powerful bombs of a new type may thus be constructed. A single bomb of this type, carried by boat and exploded in a port, might very well destroy the whole port together with some of the surrounding territory. However, such bombs might very well prove to be too heavy for transportation by air.

The United States has only very poor ores of uranium in moderate quantities. There is some good ore in Canada and the former Czechoslovakia, while the most important source of uranium is Belgian Congo.

In view of this situation you may think it desirable to have some permanent contact maintained between the Administration and the group of physicists working on chain reactions in America. One possible way of achieving this might be for you to entrust with this task a person who has your confidence and who would perhaps serve in an unofficial capacity. His task might comprise the following:

(a) to approach Government Departments, keep them informed of the further development, and put forward recommendations for Government action, giving particular attention to the problem of securing a supply of uranium ore for the United States:

(b) to speed up the experimental work, which is at present being carried on within the limits of the budgets of University laboratories, by providing funds, if such funds are required, through his contacts with private persons who are willing to make contributions for this cause, and perhaps also by obtaining the co-operation of industrial laboratories which have the necessary equipment.

I understand that Germany has actually stopped the sale of uranium from the Czechoslovakian mines which she has taken over. That she should have taken such early action might perhaps be understood on the ground that the son of the German Under-Secretary of State, von Weizsacker, is attached to the Kaiser-Wilhelm-Institut in Berlin where some of the American work on uranium is now being repeated.

Yours very truly,
(Sgd.) A. EINSTEIN.[9]

Einstein had thus completely abandoned his former pacifism in the face of the threat from Hitler and National Socialism. And many of the eminent physicists who had been expelled from Germany in the 1930s worked on the development of nuclear weapons during the early 1940s. Roosevelt and Churchill agreed to set up a joint American/British project, called the Manhattan Project, and this incorporated the earlier British project which was code-named Tube Alloys. The Manhattan Project grew into the largest co-operative scientific and technological project ever known up till then. The first controlled release of large amounts of nuclear energy was achieved in a graphite moderated reactor which was working in 1942 (see Fig. 6.3.3). It was built in a disused squash court at Chicago University under the direction of Enrico Fermi, who had carried out many of the early experiments on neutron-induced nuclear reactions before he fled from Mussolini's Italy.

After this success a bomb was now almost certainly possible, although many difficult and novel technical problems still needed to be solved. After much intensive work, the first nuclear explosion took place in the United States on 16th July 1945, in a desert region in New Mexico (see Fig. 6.1.1, page 241). By this time the war in Europe was over but the war in the Pacific was far from won. Less than a month later two nuclear bombs were dropped on the Japanese towns of Hiroshima and Nagasaki on August 6th and 9th 1945, respectively. Many people were killed and even more were injured, not only by blast and fires similar to those produced by any large bomb but also by the insidious effects of nuclear radiation, a totally new kind of hazard. The bombs also caused widespread devastation (see Fig. 6.3.4) and their use brought World War 2 to an end in a matter of days with, arguably, less destruction and loss of life than would have resulted from an invasion of Japan.

Einstein himself was greatly disturbed when he heard that the bomb had been used. In the remaining ten years of his life (see Fig. 6.3.5) he devoted much of his time and prestige in trying to warn mankind of the dangers of nuclear war. He deeply resented being called the 'father' of the atomic bomb and often said that, if it hadn't been for the menace of National Socialist Germany, he would have done nothing to hasten its development.

Einstein had little or nothing to do with the way in which nuclear power has

Fig. 6.3.3 Artist's impression of the nuclear reactor in which the first ever controlled chain reaction involving uranium fission was achieved in December 1942 under the direction of Enrico Fermi working at the University of Chicago.

Fig. 6.3.4 The devastation at Hiroshima after the atomic bomb attack on August 6, 1945.

Fig. 6.3.5 Einstein after World War 2.

been developed since the 1950s. So we will postpone discussing later applications of his famous equation $E = mc^2$, such as the hydrogen bomb or large power reactors, and return to these topics in later chapters.

'GOD DOES NOT PLAY DICE WITH THE WORLD'

Let us finish this brief account of Einstein and his life by leaving the world of power and politics and returning to Einstein's first love, his desire to understand and make sense of the physical world.

Before he was 40 Einstein had recast our ideas of the physical world in ways that have stood the test of time, at least until the present. In this chapter and the last, we have tried to outline the essence of Einstein's ideas, and you will remember that two of his famous 1905 papers used statistical methods to describe the behaviour of large numbers of atoms and of light quanta. In the early 1920s Einstein did some more fundamental work on the statistical behaviour of matter. This was the development of the so-called Bose-Einstein statistics which are applicable both to light and to most atoms and molecules.

But after this Einstein ceased to be a leader in research on theoretical physics and even became openly hostile to the developments which took place from 1925 onwards. 1925 was the year in which modern quantum or wave mechanics originated in the work of Schrödinger and Heisenberg. Physicists were just getting used to the idea that light behaved sometimes as a wave, for example in diffraction experiments, and sometimes as a particle, as in the photoelectric effect. Then suddenly it was suggested that particles of matter—electrons, atoms or nuclei—might behave in a similar chameleon-like way. At first Einstein greeted this new work with enthusiasm. He wrote in a letter to Schrödinger: 'The idea of your article shows real genius'.[10] But soon Einstein changed his mind. We are going to say more about wave mechanics in Chapters 6.6 and 8.3, so here we will just outline Einstein's main objection to this new theory which had grown directly out of his own work in the early 1920s. Einstein objected to the way in which the waves of Schrödinger's theory were related to the experimental behaviour of actual particles. Fig. 6.3.6 illustrates the interpretation of Schrödinger's waves which was first put forward by Born in 1926 and which is now accepted by the vast majority of scientists. In essence, this is that the wave represents the average behaviour of large numbers of particles but cannot tell us anything about the detailed behaviour of any individual particle. All the wave can tell us is something about the *probable behaviour* of any individual particle. Where the wave has a large intensity we would expect to find many particles, and vice versa (see Fig. 6.3.6).

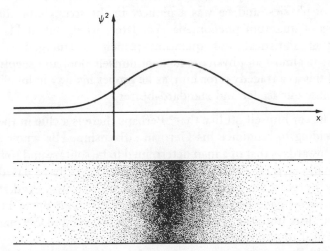

Fig. 6.3.6   Bohr's probabilistic interpretation of wave-mechanics. The intensity (ψ^2) of the wave at any point is proportional to the probability of finding a particle at that point.

Einstein never accepted this statistical interpretation and for nearly 30 years he carried on a one-man fight against it. Much of the debate took place in letters exchanged by Bohr and Einstein and many of these letters have now been published. Throughout Einstein remained convinced that a strictly deterministic theory was possible by which he meant a theory which can describe the detailed path of each individual particle. As he said again and again 'God does not play dice with the world'.

In one letter, to James Franck, Einstein put it like this:

> I can, if the worst comes to the worst, still realize that God may have created a world in which there are no natural laws. In short, a chaos. But that there should be statistical laws with definite solutions, i.e. laws which compel God to throw the dice in each individual case, I find highly disagreeable.[11]

And in 1928 he wrote, in a letter to Schrödinger:

> The Heisenberg-Bohr tranquilizing philosophy—or religion—is so delicately contrived that, for the time being, it provides a gentle pillow for the true believer from which he cannot very easily be roused. So let him lie there.[12]

C. P. Snow described Einstein's lone stand like this:

> Perhaps I can, by an analogy, suggest the effect on Einstein's colleagues of his one-man counter-revolution. It was rather as though Picasso, about 1920, at the height of his powers, had announced that some new kind of representational painting alone could be made to contain the visual truth: and had spent the rest of his life industriously but unavailingly trying to find it.[13]

Most physicists thought that Einstein had wasted the last 25 years of his life. As Max Born, the man who first put forward the statistical interpretation, says:

> He has seen more clearly than anyone before him the statistical background of the laws of physics, and he was a pioneer in the struggle of conquering the wilderness of quantum phenomena. Yet later, when out of his own work a synthesis of statistical and quantum principles emerged which seemed acceptable to almost all physicists he kept himself aloof and sceptical. Many of us regard this as a tragedy—for him, as he gropes his way in loneliness, and for us, who miss our leader and standard-bearer.[14]

Why did he cut himself off like that? Perhaps there is a clue in the 15-year-old Einstein deciding to renounce his German citizenship. His whole life from his schooldays onwards is that of a man determined to be solitary and to make his own decisions. Without these characteristics he would never have dared in his early years to try to recast so much of physics, all on his own. And it is doubtful whether he would have had the inner resources to stand up to Hitler and the National Socialists from the start. Einstein never returned to Germany or forgave those who failed to oppose Hitler in his early days.

After World War 2, when he was invited to rejoin the Bavarian Academy of Science which he had been forced to leave in 1933, Einstein replied:

> The Germans slaughtered my Jewish brethren; I will have nothing further to do with them, not even with a relatively harmless academy.[15]

REFERENCES

1 Popper, K., *Unended Quest—an Intellectual Autobiography*, (London: Fontana Paperbacks, 1976), p. 37; originally published in Schilp, P. A. (ed.), *The Philosophy of Karl Popper* (Illinois: Open Court, 1974).
2 Ibid., p. 38; italics in original.

3　Lenard, P., *Erinnerungen eines Naturforschers*—an unpublished autobiography completed in 1943; quoted in Beyerchen, A. D., *Scientists under Hitler*, (New Haven: Yale University Press, 1977), p. 93.

4　Lenard, P., quoted in Bernstein, J., *Einstein*, (London: Fontana, 1973), pp. 170–1.

5　Stark, J., *Philipp Lenard als deutscher Naturforscher* in *Nationalsocialistiche Monatshefte*, Heft 71, February 1936, p. 109; referred to in Beyerchen, A. D., op. cit., pp. 143–4.

6　Lenard, P., Preface to *Deutsche Physik*, (Munich: J. F. Lehmanns, 1936–7).

7　Rutherford, E., Address to the British Association for the Advancement of Science, 1933; quoted in Snow, C. P., *Variety of Men*, (London: Macmillan, 1969), p. 19. Reproduced by permission of Curtis Brown Ltd., London, and Macmillan.

8　Rutherford, E., *The Newer Alchemy*, (Cambridge: Cambridge University Press, 1937), p. 65.

9　The letter is in the Franklin D. Roosevelt Library; a facsimile reproduction appears in Bronowski, J., *The Ascent of Man*, (London: BBC Publications, 1973), Fig. 184, p. 371. The letter was drafted by Leo Szilard after discussion with Einstein and the nuclear physicists Edward Teller and Eugene Wigner; *see* Weart, S. R. and Szilard, G. W., *Leo Szilard: His Version of the Facts*, (Cambridge, Mass: MIT Press, 1978), pp. 90–100.

10　Postscript to a letter from Einstein to Schrödinger on April 16 1926; quoted in Bernstein J., op. cit., p. 174.

11　Letter from Einstein to James Franck; quoted in Snow, C. P., op. cit., p. 93. Reproduced by permission of Curtis Brown Ltd., London, and Macmillan.

12　Letter from Einstein to Schrödinger, 1928; printed in Klein, M. J., *Letters on Wave Mechanics*, (New York, 1967), p. 31.

13　Snow, C. P., op. cit., p. 94. Reproduced by permission of Curtis Brown Ltd., London, and Macmillan.

14　Born, M., in 'Einstein's Statistical Theories' in Schilpp, P. A. (ed.), *Albert Einstein: Philosopher—Scientist*, (Evanston, Illinois: The Library of Living Philosophers Inc., 1949), pp. 163–4.

15　Quoted in Bernstein, J., op. cit., p. 169.

SUMMARY

1　In 1905, Einstein published three papers—on special relativity, on the photoelectric effect and on Brownian motion—any one of which would have been enough to win him a Nobel prize.

2　Einstein's general theory of relativity of 1916 was successfully tested by experiments carried out during an eclipse of the sun in 1919. This experimental test greatly influenced Karl Popper's philosophy of science.

3　During the 1930s Einstein's work was condemned by the National Socialists in Germany as 'a Jewish fraud, which one could have suspected from the first with more racial knowledge than was then disseminated, since its originator Einstein was a Jew'. Einstein left Germany, his birthplace, in 1933 and never returned.

4　Einstein's famous 1905 equation '$E = mc^2$' was dramatically confirmed on a large scale during World War 2 by the development of the first atomic bombs.

5　For the last 25 years of his life Einstein rejected the probabilistic interpretation of wave mechanics which was accepted by nearly every other physicist. Einstein maintained that 'God does not play dice with the world'.

6.4 Genetics and molecular biology

In this chapter we will focus on three main topics in the development of genetics and modern biology—the work of Mendel and its rediscovery at the start of the 20th century; the development of population genetics in the 1920s and 1930s which unified Mendel's ideas on inheritance with Darwin's on natural selection; and the understanding of the biochemical basis of inheritance culminating in the discovery of the structure of DNA in 1953.

MENDEL AND THE ORIGINS OF GENETICS

Modern genetics has its origins in the 1860s—in the work of Gregor Mendel (1822–84) the Austrian monk whose extensive experiments on inheritance in pea plants were performed at about the same time as Charles Darwin was writing *The Origin of Species*. Mendel carried out his experiments in the gardens of the Augustine monastery at Brno, now part of Czechoslovakia.

In his classic paper, published in 1866 in the Proceedings of the Natural History Society of Brno, Mendel introduces his work in these words:

> Experience of artificial fertilisation, such as is effected with ornamental plants in order to obtain new variations in colour, has led to the experiments which will be here discussed. The striking regularity with which the same hybrid forms always reappeared whenever fertilisation took place between the same species induced further experiments to be undertaken, the object of which was to follow up the developments of the hybrids in their progeny.[1]

Mendel made very careful study of the hybrids produced by recrossing different varieties of pea plant. His experiments involved meticulous attention to detail and they took him eight years to complete. But Mendel did not lose himself in the detail. He had read *The Origin of Species* with great attention and was clearly aware of the wider issues. Again in his own words:

> It requires some courage to undertake a labour of such far-reaching extent; this appears, however, to be the only right way by which we can finally reach the solution of a question the importance of which cannot be overestimated in connection with the evolution of organic forms.[2]

Mendel prepared his grand experiment with great care. He first selected a plant, the common pea, which existed in a number of pure varieties with different characteristics. In preliminary experiments Mendel identified seven pairs of characteristics for more detailed study—differences in the form of the ripe seed (smooth or wrinkled); the colour of the seed albumen (yellow or green); the colour of the seed coat (grey-brown or white—also correlated with coloured or white flowers); the form of the ripe pods (smooth or deeply wrinkled); the colour of the unripe pods (green or yellow); the position of the flowers on the stem (axial or terminal); and finally the length of the stem (long or short). Mendel found that when he crossed one pure form with another, the character of the hybrid

resembles that of one of the parental forms so closely that the other either escapes observation completely or cannot be detected with certainty. This circumstance is of great importance in the determination and classification of the forms under which the offspring of the hybrids appear. Henceforth in this paper those characters which are transmitted entire, or almost unchanged in the hybridization, and therefore in themselves constitute the characters of the hybrid, are termed the *dominant*, and those which become latent in the process *recessive*. The expression 'recessive' has been chosen because the characters thereby designated withdraw or entirely disappear in the hybrids, but nevertheless reappear unchanged in the progeny, as will be demonstrated later on.[3]

Mendel is here introducing for the first time the now commonplace terminology of dominant and recessive traits. In the list given above the first character mentioned in each case is dominant so that, for example, all the hybrids produced by crossing pure varieties with smooth or wrinkled seeds will themselves have smooth seeds (see Fig. 6.4.1).

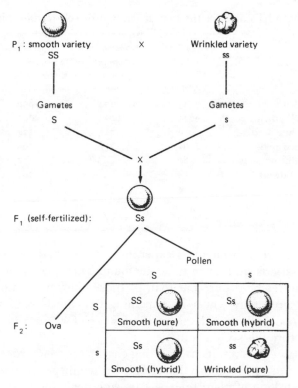

Total: 3 smooth (2 hybrid, 1 pure) : 1 wrinkled

Fig. 6.4.1   Segregation of characteristics (smooth S or wrinkled s seeds) producing the Mendelian ratios; the smooth characteristic S is dominant.

But in the next generation, produced by carefully self-fertilising the hybrids

> ... there reappear, together with the dominant characters, also the *recessive ones* with their peculiarities fully developed, and this occurs in the definitely expressed average proportion of three to one, so that among each four plants of this generation three display the dominant character and one the recessive. This relates without exception to all the characters which were investigated in the experiments. The angular wrinkled form of the seed, the green colour of the albumen, the white colour of the seed coats and the flowers, the constrictions of the pods, the yellow colour of the unripe pod, of the stalk, of the calyx, and of the leaf venation, the umbel-like form of the inflorescence, and the dwarfed stem, all reappear in the numerical proportion given, without any essential alteration. *Transitional forms were not observed in any experiment.*[4]

The other six recessive characters also reappeared in the same proportion, 1:3, as the wrinkled seeds. But to get this round number 3 to 1, Mendel had to use very large numbers of plants because with 'a smaller number of plants ... very considerable fluctuation may occur.'[5] Table 6.4.1 shows the results actually obtained by Mendel for each pair of characteristics in the first generation of self-fertilised hybrids.

Table 6.4.1   Mendel's data for the first generation of self-fertilised hybrid pea plants

Character (dominant first)	Number of plants/ seeds	Number with dominant character	Number with recessive character	Dominant— recessive ratio
Smooth or wrinkled seed	7324	5474	1850	2·96
Yellow or green albumen	8023	6022	2001	3·01
Grey-brown or white seed	929	705	224	3·15
Smooth or wrinkled pods	1181	882	299	2·95
Green or yellow unripe pods	580	428	152	2·82
Axial or terminal flowers	858	651	207	3·14
Tall or short plants	1064	787	277	2·84
Total	19 959	14 949	5010	2·98

Mendel also carried through his experiments to second and subsequent self-fertilised generations and each time he found that plants or seeds showing recessive characters bred pure but that, of those showing dominant characters, only one third bred pure while the remaining two-thirds remained hybrid (see Fig. 6.4.1). Mendel confirmed this ratio of 1:2:1 for pure dominant: hybrid: pure recessive through six generations.

Mendel explained his results by invoking characters which combined randomly from generation to generation so that simple probability theory would tell us the numbers of each kind of plant to expect in each generation.

If A be taken as denoting one of the two constant characters, for instance the dominant, a, the recessive, and Aa the hybrid form in which both are conjoined, the expression

$$A + 2Aa + a$$

shows the terms in the series for the progeny of the hybrids of two differentiating characters.[6]

And Mendel goes on to develop similar but more complex equations for two or three or more pairs of characters each of which is independent of the others. And he showed that:

> All constant combinations which in Peas are possible by the combination of the said seven differentiating characters were actually obtained by repeated crossing. Their number is given by $2^7 = 128$. Thereby is simultaneously given the practical proof *that the constant characters which appear in the several varieties of a group of plants may be obtained in all the associations which are possible according to the (mathematical) laws of combination, by means of repeated artificial fertilization.*[7]

Mendel's work is probably the first example of the detailed application of mathematics to biology. He had no idea of the detailed mechanism by which his characters were transmitted from generation to generation. But he had clearly established the laws underlying the inheritance of discrete, either-or, characters— at least for the pea plant. Yet despite the importance of Mendel's work it was neglected for more than 30 years. In 1868 Mendel became Abbot of his monastery and thereafter had little time for further lengthy experiments. And when he died, in 1884, his successor burnt all the papers which Mendel had left at the monastery.

THE REDISCOVERY OF MENDEL'S WORK

Then in 1900, Mendel's work was rediscovered by no less than three scientists, all within a few months of one another. By this time the main emphasis in studies of inheritance was on biological characters which could vary continuously from individual to individual—such as human physical characteristics like height or weight (see the discussion of the work of Galton, Pearson and Weldon in Chapter 6.7). Discontinuous variations, like those studied by Mendel, seemed to be at odds with this work. So Mendel's laws became the focus for controversy. Mendel's main champion was William Bateson (1861–1926) who, in 1902, published a book called *Mendel's Principles of Heredity: A Defence*. Bateson also had Mendel's original papers translated into English and it is from this translation that the quotations from Mendel, cited above, were taken.

The controversy lasted for nearly 20 years and was made more difficult to resolve by the very real problems involved in carrying out complex series of experiments on inheritance through several generations. But gradually most biologists changed their minds and accepted Mendel's work. This change is exemplified in the attitude of T. H. Morgan (1866–1945) who was initially critical of Mendel's ideas but whose own work later did much to confirm them.

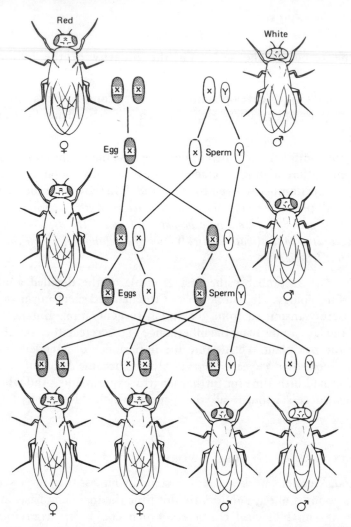

Fig. 6.4.2   Sex-linked inheritance in *Drosophila*. A white-eyed male (right) is crossed with a normal, red-eyed female (left). The first generation are all red-eyed but in the second generation one out of every four flies is white-eyed and male. In 1910 Morgan explained this by assuming that some X-chromosomes (shaded in diagram) contained the gene for red eyes while other X-chromosomes contained the white-eyed gene.

At first Morgan objected to Mendel's ideas on various grounds. They might work for pea plants but there was little evidence that they applied, say, to animals. Mendel's ideas on dominance could not account for the inheritance of sex in a 1:1 ratio. And the categories 'dominant' and 'recessive' were not always clear-cut.

But within the space of only one year, Morgan had changed his mind. In that time he had begun to work on the fruit fly, *Drosophila*, which was to play a major role in the development of genetics. These fruit flies were very suitable for genetic experiments since they were relatively simple organisms and a new generation could be bred every ten days or so. Morgan's first experiments were concerned with eye colour—red was the normal colour for *Drosophila*'s eyes but individuals with white eyes were occasionally observed. Morgan showed that the inheritance of eye colour was sex-linked and he was able to explain his detailed quantitative observations of the numbers of males and females with red or white eyes using the principles of particulate inheritance that Mendel has used in his work on pea plants (see Fig. 6.4.2).

Morgan's group also developed methods of observing changes in *Drosophila* chromosomes under the microscope, of mapping the overall structure of chromosomes, and of relating some features of chromosome structure to the characteristics of adult flies (see Fig. 6.1.2 for a modern photograph of chromosomes). This work led to the publication, in 1915, of an epoch-making book *The Mechanism of Mendelian Heredity* by Morgan, A. H. Sturtevant (1891–1970) and C. B. Bridges (1889–1938). By 1920 the ideas it contained were accepted throughout most of the biological community.

All this work was conducted under fairly cramped conditions in the biology department at Columbia University in New York—much of it in the famous 'fly room' next to Morgan's office which contained a few old tables and thousands of milk bottles for culturing *Drosophila* (see Fig. 6.4.3).

Fig. 6.4.3   The 'fly room' at Columbia University, New York in which T. H. Morgan and his collaborators carried out their famous work on *Drosophila*.

THE ORIGINS OF POPULATION GENETICS

The next major development in genetics took place in the 1920s in the Soviet Union. In 1922 H. J. Muller (1890–1968), who had worked on *Drosophila* at Columbia with Morgan, visited the Soviet Union and took some pure-bred *Drosophila* stocks with him. Muller was later awarded the Nobel prize for his work on the production of mutations in *Drosophila* by X-rays. Muller's visit stimulated great interest in Mendelian genetics and a group of Russian scientists, led by S. S. Chetverikov (1880–1959), were inspired to work on the application of Mendelian genetics to the evolution of natural populations. Chetverikov reasoned that natural populations must possess a large amount of hidden or recessive variation which could be of great importance in evolution. In 1925 Chetverikov studied the genetic properties of a natural population of wild *Drosophila*. By crossing with the pure-bred *Drosophila* stocks brought by Muller, Chetverikov and his co-workers detected 32 masked recessive traits in a wild population of 239 flies.

In 1926 Chetverikov published a paper 'On several aspects of the evolutionary process from the viewpoint of modern genetics' in which he derived simple equations which related gene frequencies to the probability of recessive variations appearing in natural fly populations. The conclusion was that hidden variations would be more likely to become visible in small isolated populations which meant evolution would be expected to occur more rapidly in such isolated populations. Over the period from 1922 to 1929, Chetverikov's group developed simple mathematical and experimental models of inheritance whose validity they tested with both laboratory and field experiments. Chetverikov's group were the first to establish the scale of recessive variation in natural populations and to work with the concept of the gene pool of an entire population.

Later T. Dobzhansky (1900–76), one of Chetverikov's pupils, became one of the world's leading experts on experimental population genetics. Dobzhansky went to the USA in 1927 to work with Morgan at Columbia and later, during the 1930s and 1940s, he carried out a series of classic field and laboratory studies on the evolution of *Drosophila*.

These studies showed that chromosome patterns could be observed to change by natural selection over periods as short as a few months—both in the laboratory and in nature. Dobzhansky was observing evolution right before his eyes.

The pioneering work of Chetverikov and his school was not known to the three men who later developed the detailed statistical theory of population genetics—J. B. S. Haldane (1892–1964) and R. A. Fisher (1890–1962) in Britain and Sewall Wright (1889–) in America. All three were trying to derive quantitative relationships between changes in gene frequencies and the properties of natural populations by introducing new statistical concepts for dealing with factors such as inbreeding, linkage and multiple interactions. Fisher's work was the most mathematically abstract of the three, and dealt with simplified mathematical models of very large populations (see Chapter 6.7). Haldane and Wright were concerned with different aspects of population genetics which had been neglected in Fisher's simplified model. Wright was particularly interested in changes in the gene pools of naturally occurring populations. Between them, these three men laid the foundations for most subsequent developments in population genetics and brought about a synthesis between evolution at the level of the organism and the detailed mechanisms of Mendelian inheritance. This evolutionary synthesis is summed up in the title of Fisher's classical book, published in 1930, *The Genetical Theory of Natural Selection*. And, as we have seen, evolution was not only now being better understood theoretically—it was also being subjected to experimental and quantitative tests. As Dobzhansky put it in 1947 in the introduction to a paper on his *Drosophila* studies:

> Controlled experiments can now take the place of speculation as to what natural selection is or is not able to accomplish. Furthermore, we need no longer be satisfied with mere verification of the existence of natural selection. The mechanics of natural selection in concrete cases can be studied. Hence the genesis of adaptation, which is possibly the central problem of biology, now lies within reach of the experimental method.[8]

MOLECULAR BIOLOGY

Since the 1930s the main focus in genetics has shifted towards an attempt to understand the biochemical basis of inheritance—that is, towards the development of what we now call molecular biology. The key events here were the working out of the structure of DNA in 1953; of the mechanism by which the DNA molecule can duplicate itself during cell division; and of the way in which the detailed structure of DNA controls the structure of the proteins made by cells.

These dramatic discoveries were only possible because of developments over the previous 30 years in many different fields of research. There were three main strands in this work: structural studies, concerned with the architecture of biological molecules; biochemical studies, concerned with how biological molecules interact in cell metabolism and heredity; and information studies concerned with how information is transferred from one generation of organisms to another and how that information is translated into unique biological molecules.

Structural studies were first concerned with the actual chemical structure of biological molecules—which were very difficult to work out as most of the molecules involved are so large. Much of this early work concentrated on the structure of proteins, and it was established early in this century that proteins consisted of long chains of amino-acids. What was more difficult to establish was the exact sequence of these amino-acids in any specific protein. It was not until the mid-1940s that the chemical structure of the first protein, insulin, was worked out by Frederick Sanger (1918–), who showed that insulin from cows consisted of a sequence of 51 amino-acids.

Later structural studies were concerned with the three-dimensional structure of biological molecules rather than just with their chemical composition. Here the key technique was the application of the X-ray diffraction methods which had been so successful in working out crystal structures. The problems involved in applying these techniques to biological molecules were enormous. It was difficult to make good crystals and the molecules were so large that the structures were difficult to establish even if good X-ray pictures could be obtained. However, some progress was made in the 1930s on the structure of keratin—the fibrous protein in wool. This work, by Astbury (1893–1961) at Leeds, inspired several groups to try to find the structures of more complex proteins. Amongst these was Max Perutz (1914–), a refugee from Nazi Germany, who in 1937 began work at Cambridge on the structure of haemoglobin. It wasn't until 29 years later, in the early 1960s, that Perutz and his co-workers finally worked out its structure (see Fig. 6.1.3, page 243).

Biochemical studies were concerned with two main problems—the chemical mechanisms of respiration and the way in which genes controlled the metabolism of cells. The latter problem was significantly advanced by Beadle (1903–) and Tatum (1909–75) working at Stanford University in California. They used a simpler organism than *Drosophila*, a bread mould called *Neurospora*. By studying many mutant strains Beadle and Tatum were able to relate individual mutations to individual stages in biochemical metabolic pathways. They did this by starving the *Neurospora* cultures of different nutrients and seeing which ones the mutant strains were unable to make for themsleves. By the late 1940s and early 1950s this work had established that genes controlled cellular metabolism by controlling the production of specific proteins. The question was—how did they do this? It was this question that was studied by the informationists.

INFORMATIONAL STUDIES

It was originally thought that proteins were the likeliest molecules to carry genetic information, but work based on another very simple type of organism, a bacterial virus or bacteriophage, pointed to the importance of a nucleic acid called deoxyribonucleic acid or DNA. Two of the key workers here were Max Delbrück (1906–81) and Salvador Luria (1912–) working at the Massachusetts Institute of Technology; both were refugees from Fascism, from Hitler's Germany and Mussolini's Italy respectively. These workers and their collaborators studied what happened when the bacteriophage virus infected bacterial cells and their work was greatly aided by new physical techniques such as electron microscopy and radioactive tracers. The climax of this work came in 1952 when radioactive tracer studies clearly showed that it was the DNA in the virus, rather than the protein, which was transferred to the infected bacteria.

It was this result that stimulated the intensive work on the structure of DNA which culminated in the double helix model put forward in 1953 by Francis Crick (1916–) and James Watson (1928–) while working at the Cavendish Laboratory in Cambridge.

THE THREE-DIMENSIONAL STRUCTURE OF DNA

In the early 1950s there were many groups of workers who were trying to establish the three-dimensional structure of DNA. Perhaps the best known were the group of X-ray crystallographers under Maurice Wilkins at King's College, London and the structural chemists at the California Institute of Technology under the Nobel prize winner Linus Pauling.

Wilkin's group were trying to use X-ray diffraction techniques to establish the DNA structure. It had long been known that DNA was a long-chain polymer consisting of three main components—sugars, phosphates and four bases called adenine, thymine, guanine and cytosine. It was also known that the sugars and phosphates were chemically linked in a kind of backbone which ran the whole length of the molecule while the bases were attached at regular intervals to this backbone. But each DNA molecule contained so many thousands of atoms that it was not at all easy to find its detailed structure from X-ray diffraction photographs like Fig. 6.1.7 (page 246). What the photographs did show was that DNA had a spiral structure with a definite repeat distance. But it was not clear from the X-ray work how many spirals there were in each molecule or whether the bases were on the inside or outside of the DNA molecules. Pauling's approach was almost the reverse of that of the X-ray crystallographers. Rather than try to deduce the structure from the X-ray photographs, Pauling built possible models for the DNA molecule using the known structures and bond-lengths of the chemical components of DNA. Then he tested his models against the actual X-ray results.

The structure of DNA was finally found by a combination of these two approaches adopted by two unorthodox scientists—a physicist, Francis Crick, and a biologist, James Watson. These two met in 1951 at the Cavendish Laboratory in Cambridge and soon decided to work together on the structure of DNA. One reason for their success was that they were individualists who were not committed to any particular approach. They were prepared to consider and bring together evidence from many different sources. From the X-ray work of Wilkins and his collaborator Rosalind Franklin, some of it unpublished, they took the fact that DNA was helical and that the phosphate groups were on the outside. This meant that the bases must be on the inside. From theoretical chemistry and biochemistry they took the fact that unlike pairs of bases—say adenine (A) and thymine (T) or guanine (G) and cystosine (C)—attracted one another and that the ratio of A:T or G:C was always one in DNA. This suggested that the bases in DNA might exist in complementary pairs. From further discussions with organic chemists, Watson and Crick discovered that the bases existed in two forms and that if the right form were chosen an adenine-thymine (A-T) pair was almost exactly the same shape and size as a guanine-cytosine (G-C) pair (see Fig. 6.4.4). They then adopted Pauling's model building approach and began to construct three-dimensional models of DNA with two helical strands. As Watson said 'I had decided to build two-chain models. Francis would have to agree. Even though he was a physicist, he knew that important biological objects came in pairs.'[9]

Within a couple of months of deciding to build two-chain models, Watson and

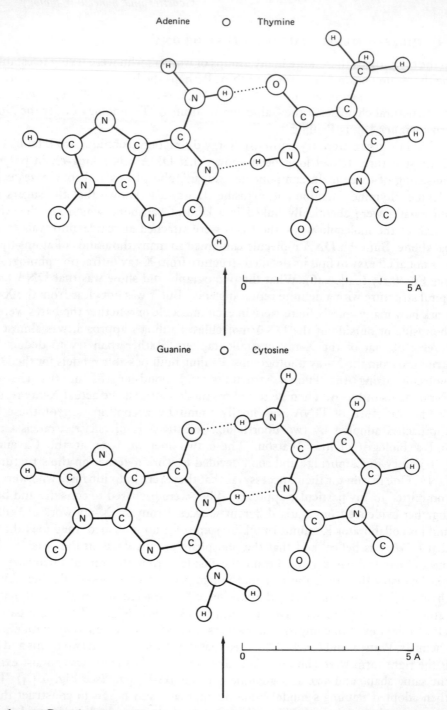

Fig. 6.4.4 Complementary hydrogen bonded base pairs of almost exactly the same shape. *From Watson, J. D. and F. H. C. Crick, 'Genetical Implications of the Structure of Deoxyribonucleic Acid'. Redrawn by permission from Nature, vol. 171, p. 964. Copyright ©️ 1953, Macmillan Journals Limited.*

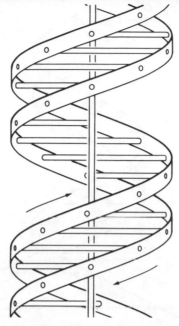

Fig. 6.4.5 Diagrammatic representation of the DNA double helix. The two ribbons symbolise the two phosphate-sugar chains and the horizontal rods the pairs of bases holding the chains together. *From Watson, J. D. and F. H. C. Crick, 'Genetical Implications of the Structure of Deoxyribonucleic Acid'. Redrawn by permission from Nature, vol. 171, p. 965. Copyright © 1953, Macmillan Journals Limited.*

Crick had found the structure of DNA. They announced their momentous discovery in the scientific periodical *Nature* on April 25 1953. Their article was only one page long—surely the shortest scientific paper ever to gain a Nobel prize for its authors.[10] The same issue of *Nature* also contained papers by Wilkins, Franklin and their co-workers at King's College which gave X-ray data for DNA. Some of these data had been known to Watson and Crick from long informal conversations with Wilkins but had not been previously published.

Watson and Crick start by saying that they 'wish to suggest a structure for the salt of deoxyribose nucleic acid (D.N.A.). This structure has novel features which are of considerable biological interest.' They then describe their structure which '. . . has two helical chains each coiled round the same axis . . . both chains follow right-handed helices, but . . . the atoms in the two chains run in opposite directions . . . the bases are on the inside of the helix and the phosphates on the outside' (see Fig. 6.4.5; this diagram is from the original paper in *Nature*).

Watson and Crick continue thus:

> The novel feature of the structure is the manner in which the two-chains are held together by the purine and pyrimidine bases. . . . They are joined together in pairs . . . adenine (purine) with thymine (pyrimidine) and guanine (purine) with cytosine (pyrimidine). (See Fig. 6.4.4.)

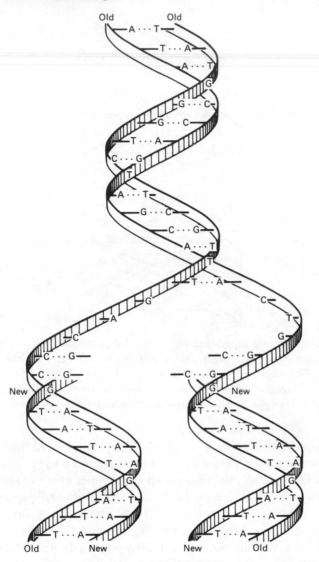

Fig. 6.4.6   Diagram of the copying mechanism for DNA proposed by Watson and Crick.

They then come to the most important feature of their structure:

> ... The sequence of bases on a single chain does not appear to be restricted in any way. However, if only specific pairs of bases can be formed, it follows that if the sequence of bases on one chain is given, then the sequence on the other chain is automatically determined. ...
>
> It has not escaped our notice that the specific pairing we have postulated immediately suggests a possible copying mechanism for the genetic material.

A month later in another paper in *Nature*—two pages this time—Watson and Crick discussed this 'possible copying mechanism' in more detail:[11]

... any sequence of the pairs of bases can fit into the structure. It follows that in a long molecule many different permutations are possible, and it therefore seems likely that the precise sequence of the bases is the code which carries the genetical information ... our model for deoxyribonucleic acid is, in effect, a *pair* of templates, each of which is complementary to the other. We imagine that prior to duplication the hydrogen bonds are broken, and the two chains unwind and separate. Each chain then acts as a template for the formation on to itself of a new companion chain, so that eventually we shall have *two* pairs of chains, where we only had one before. Moreover, the sequence of the pairs of bases will have been duplicated exactly. (emphases in original).

Fig. 6.4.6 shows in principle how this copying mechanism might work. This is how it is now thought that genetic information is transmitted from one generation to the next and from the single fertilised germ cell to all the cells of a multi-cellular organism. And since 1953 details of protein synthesis have also been worked out. We now know that one of the main functions of DNA is to control the synthesis of proteins in the cells of living creatures. Almost without exception the cell nuclei of all living creatures contain long molecules of DNA. The chemical structure of DNA in all these creatures is the same. The differences are primarily in the length of the DNA molecules.

Fig. 6.4.7 shows a DNA molecule from a bacterial virus or bacteriophage. This

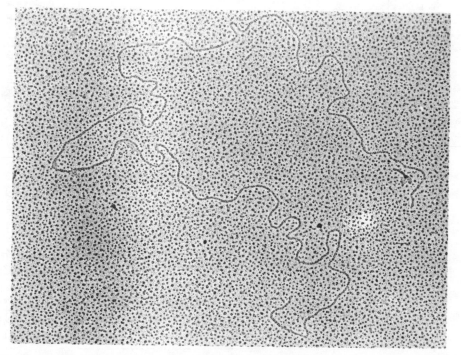

Fig. 6.4.7 Very long single molecule of DNA which has burst from a bacterial virus or bacteriophage.

The nucleotide bases are shown as: U, uracil; C, cytosine; A, adenine; G, guanine.

SECOND LETTER

FIRST LETTER	U	C	A	G	THIRD LETTER
U	UUU } Phe UUC } Phe UUA } Leu UUG } Leu	UCU UCC } Ser UCA UCG	UAU } Tyr UAC } Tyr UAA Ochre (End) UAG Amber (End)	UGU } Cys UGC } Cys UGA ?Cys (?End) UGG Try	U C A G
C	CUU CUC } Leu CUA CUG	CCU CCC } Pro CCA CCG	CAU } His CAC } His CAA } Gln CAG } Gln	CGU CGC } Arg CGA CGG	U C A G
A	AUU AUC } Ile AUA AUG (Start) Met	ACU ACC } Thr ACA ACG	AAU } Asn AAC } Asn AAA } Lys AAG } Lys	AGU } Ser AGC } Ser AGA } Arg AGG } Arg	U C A G
G	GUU GUC } Val GUA GUG (Start)	GCU GCC } Ala GCA GCG	GAU } Asp GAC } Asp GAA } Glu GAG } Glu	GGU GGC } Gly GGA GGG	U C A G

Amino acids: Ala = alanine; Arg = arginine; Asn = asparagine; Asp = aspartic acid; Cys = cysteine; Gln = glutamine; Glu = glutamic acid; Gly = glycine; His = histidine; Ile = isoleucine; Leu = leucine; Lys = lysine; Met = methionine; Phe = phenylalanine; Pro = proline; Ser = serine; Thr = threonine; Try = tryptophan; Tyr = tyrosine; Val = valine.

Fig. 6.4.8 The genetic code. The base T (thymine) in DNA is replaced by the base U (uracil) in the RNA molecules which act as intermediaries in protein synthesis.

DNA molecule is about 50 micrometers long. In a simple bacterium like *Escherichia coli*, DNA molecules are about 2 millimetres long and in man they are about one metre long and each one contains several thousand million base pairs. Since there are about a thousand million million cells in the human body, the total length of the DNA molecules in a man is many thousands of times greater than the distance from the earth to the sun.

The proteins in all living creatures are very similar in structure. They consist of long chains of amino acids. But of all the possible amino-acids, only 20 occur in proteins and it is the *same* twenty in all living organisms from bacteria to mammals. We now know in detail how the sequence of base pairs of the DNA

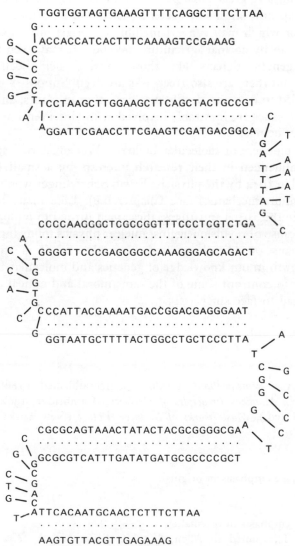

Fig. 6.4.9 Sequence of bases in a gene synthesised in the laboratory. *From Portugal and Cohen, A Century of DNA (Cambridge, Mass: MIT Press, 1977).*

molecules in the cell nucleus controls the sequence of amino acids made in the cells. This genetic code is shown in Fig. 6.4.8. A sequence of three base pairs codes for each amino acid. But since there are 4^3 or 64 possible triplet sequences and only 20 amino acids, the code is degenerate—that is, different triplets of bases may code for the same amino acid (see Fig. 6.4.8).

The operation of the genetic code is now so well understood that genes have been synthesised in the laboratory (see Fig. 6.4.9) and it is now proving possible to transfer genes—either synthetic or natural—into the DNA of living organisms and thus to change the proteins which they can synthesise. This process of genetic engineering is an amazing technical feat but one which, like the development of nuclear power, can have great possibilities both for good and for evil. It may be possible to develop new strains of bacteria which will fix atmospheric nitrogen more efficiently or which may convert inedible materials into food stuffs much more efficiently than by normal agriculture methods. It may even be possible to correct human genetic defects like those causing sickle cell anaemia or phenylketonuria. But there are also great risks involved. Suppose a bacterium was produced which led to new and deadly diseases either in plants, animals or man. And major ethical issues are inevitably involved in any intervention in the genetic makeup of human beings.

Several of the pioneers of molecular biology—Watson, Crick and Wilkins— were all much influenced in their research interests by a small book *What is Life?*[12] published in 1944 by the physicist Erwin Schrödinger who was one of the founders of quantum mechanics (see Chapter 6.6). Like many other talented young scientists in the 1940s and 1950s, these men turned to biology rather than physics partly because of revulsion against nuclear physics following the dropping of nuclear weapons on Hiroshima and Nagasaki. So it is ironic that the extraordinary growth in our knowledge of genetics and molecular biology is now forcing biologists to confront some of the same moral and ethical dilemmas that physicists have had to face since 1945.

REFERENCES

1   Mendel, G., *Experiments in Plant-hybridisation*, first published in 1866 and translated in Bateson, W., *Mendel's Principles of Heredity*, (Cambridge, 1902); reprinted in Knedler, J. S. (ed.), *Masterworks of Science—Vol. 3* (New York: McGraw-Hill, 1973), p. 179.
2   Ibid., p. 179.
3   Ibid., p. 183.
4   Ibid., pp. 183–4; emphasis in original.
5   Ibid., p. 185.
6   Ibid., p. 187.
7   Ibid., p. 192; emphasis in original.
8   Dobzhansky, T., quoted in Allen, G., *Life Science in the Twentieth Century*, (Cambridge: Cambridge University Press, 1978), p. 142.
9   Watson, J. D., *The Double Helix*, (Harmondsworth: Penguin, 1970), p. 134.

10 Watson, J. D., and Crick, F. H. C., 'Molecular Structure of Nucleic Acids', *Nature*, Vol. 171, pp. 737–8. Copyright © 1953 Macmillan Journals Limited. Reprinted by permission.

11 Watson, J. D., and Crick, F. H. C., 'Genetical Implications of the Structure of Deoxyribonucleic Acid', *Nature*, Vol. 171, pp. 964–7. Copyright © 1953 Macmillan Journals Limited. Reprinted by permission.

12 Schrödinger, E., *What is Life?*, (Cambridge: Cambridge University Press, 1944).

SUMMARY

1 Genetics has its origins in the 1860s in Mendel's classic work on the inheritance of seven different either/or characteristics in pea plants; Mendel explained his simple numerical ratios for plants of different types by using probability theory and introducing the idea of dominant and recessive characters.

2 T. H. Morgan, like many biologists, was originally opposed to Mendel's ideas but changed his mind when he found that his experimental results with the fruit fly *Drosophila* could be explained using Mendel's ideas on particulate inheritance.

3 In the 1920s and 1930s the development of population genetics unified Mendel's ideas on particulate inheritance with Darwin's on natural selection; this was largely due to the experimental work of the Russians Chetverikov and Dobzhansky, much of it with *Drosophila*, and the theoretical work of Fisher and Haldane in Britain and Sewall Wright in the USA.

4 Advances in chemical structure analysis, biochemistry, X-ray diffraction and many other fields led, in 1953, to Watson and Crick discovering the three-dimensional double helical structure of DNA, the key genetic material.

5 Details are now known of how the DNA molecule can duplicate itself during cell division and of how the detailed structure of DNA controls the structure of the proteins made by cells.

6.5   Developments in the medical sciences

We are so familiar with modern developments in medicine and health care that it is difficult for us not to take them for granted. But if we go back in time, we can see the achievements of the 20th century in a better perspective—the perspective of history. This journey in time will also show us how other sciences have contributed to medical knowledge and give us some understanding of the interconnectedness of human endeavours in different fields.

ORIGINS IN ANCIENT GREECE

The origins of western medicine are found in the classical era of ancient Greece and Rome. Greek civilisation was built on ideas of health which were expressed by thinkers such as Plato and Hippocrates—a healthy mind in a healthy body. These words from the famous Hippocratic writings are as timely now as they were over 2000 years ago:

> Life is short, science is long: opportunity is elusive, experiment dangerous, judgement is difficult. It is not enough for the physician to do what is necessary, but the patient and attendants must do their part as well, and circumstances must be favourable.[1]

The Greek idea of health was also embodied in their religious worship and can still be appreciated today by visiting such famous shrines as Delphi, Delos and Olympia. These shrines emphasise the idea of a person's complete well-being—spiritual, social and physical. They all contain temples symbolising man's relationship to the gods; a theatre for acting out man's place in the universe and his relationships with his fellow men and women; and a sports stadium where athletes offered their achievements as part of their worship. Hence, the enormous importance of the Olympic Games, held at Olympia every four years for over 1000 years. Fig. 6.5.1 shows the original stadium in which the games always took place. If you look carefully you can see the actual starting line.

It was in this kind of world that the first great physician Hippocrates began to develop a science of medicine. Doctors and students came from far and wide to study with him at the island of Kos and many of the ideas which were developed there have remained influential across the centuries. Examples include the need to use careful observation, the concept of internal harmony and the need to study the influence of the environment on the individual. Indeed, Hippocrates was the first sociologist for he advocated the study of comparative ethnography, or studies of conditions in different societies which might be conducive to good health or to illness. He also laid the foundations of professional ethics in the famous Hippocratic oath:

Fig. 6.5.1 The stadium at Olympia where the ancient Olympic Games took place for more than 1000 years.

THE OATH

I swear by Apollo the healer, by Aesculapius, by Health and all the powers of healing, and call to witness all the gods and goddesses that I may keep this Oath and Promise to the best of my ability and judgement.

I will pay the same respect to my master in the Science as to my parents and share my life with him and pay all my debts to him. I will regard his sons as my brothers and teach them the Science, if they desire to learn it, without fee or contract. I will hand on precepts, lectures and all other learning to my sons, to those of my master and to those pupils duly apprenticed and sworn, and to none other.

I will use my power to help the sick to the best of my ability and judgement; I will abstain from harming or wronging any man by it.

I will not give a fatal draught to anyone if I am asked, nor will I suggest any such thing. Neither will I give a woman means to procure an abortion.

I will be chaste and religious in my life and in my practice.

I will not cut, even for the stone, but I will leave such procedures to the practitioners of that craft.

Whenever I go into a house, I will go to help the sick and never with the intention of doing harm or injury. I will not abuse my position to indulge in sexual contacts with the bodies of women or of men, whether they be freemen or slaves.

Whatever I see or hear, professionally or privately, which ought not to be divulged, I will keep secret and tell no one.

If, therefore, I observe this Oath and do not violate it, may I prosper both in my life and in my profession, earning good repute among all men for all time. If I transgress and forswear this Oath, may my lot be otherwise.[2]

Another significant Greek name in the history of ancient medicine is Galen who wrote his famous books on *Hygiene* about AD 180. Galen used many of Hippocrates' ideas of health and illness and remained the dominant medical authority until the Renaissance (see Chapter 3.2 and Fig. 3.2.8, page 54). During these intervening centuries, however, the Christian Church added a dimension of compassion to health care with religious orders caring for the sick and infirm in hospitals and monasteries while the prevailing religious faith gave meaning to human problems of suffering and death, which were accepted as God's will. And throughout this time women, as well as men, were respected as healers and their contributions in caring for those in need were highly valued.

DEVELOPMENTS IN RENAISSANCE EUROPE IN ANATOMY, PUBLIC HEALTH AND PHYSIOLOGY

Galen's work provided the basis of medical knowledge and health care for nearly 14 centuries. During those long years, medical research in Europe made no significant progress. Europe had descended into the Dark Ages. Then the universities emerged, Europe experienced the Renaissance, and the natural sciences started to burgeon. Man began to feel master of the natural world and of his own destiny. The foundations of religious belief were shaken and many practices which had been forbidden by religion were now tolerated. For example, in 1315 at the University of Bologna, an Italian, Mondino, began to teach anatomy by dissection of the human body. It was the first time this had been done for over 1500 years. Even the mighty Galen had had to restrict himself to dissecting animals and so could not avoid errors in human anatomy. But it took a further 200 years for this pioneering work to reach maturity in Vesalius' epoch-making book *On the Structure of the Human Body* in 1543 (see Chapter 3.2 and Figs. 3.2.9–11, pages 54–5). This was the first major work of modern medicine, paving the way for later breakthroughs.

We now enter a time which was relatively optimistic. Advances were occurring in other areas of knowledge and people were beginning to feel that they could become masters of their own fate. Life could still be 'nasty, brutish and short', but for many people it was already becoming more enjoyable and more worth living. So, man became interested in prolonging life—an interest reflected in another famous book: Paracelsus' *Book on Long Life* of 1560. Still following Hippocrates, Paracelsus stressed the importance of public health and of man's control over the physical and social environment. He put this into practice by draining the swamps around Venice and clearing the Venetian harbour of pollution—a great

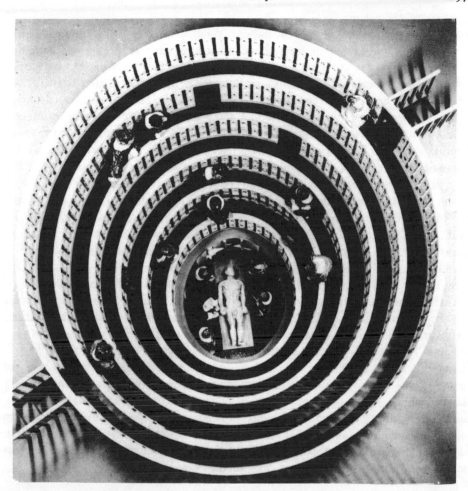

Fig. 6.5.2 Model of the anatomy theatre at Padua where Harvey watched Fabricius' anatomical demonstrations.

achievement! Paracelsus also looked after his personal health by adopting a rigid low-calorie diet—and to good effect for he lived to 98!

The next key advance came in the field of physiology. Modern physiology has its origins in the work of William Harvey and his discovery of the circulation of the blood. His book *De Motu Cordis et Sanguinis in Animalibus* (*On the Motion of the Heart and Blood in Animals*) published in 1628, was as important for medicine as Newton's *Principia* was for the physical sciences. Harvey's ideas grew from his studies at Padua where his teacher, Fabricius, was the first person to demonstrate the operation of the valves in veins—giving Harvey clues for his own theory (see Chapter 3.2 and Fig. 3.2.12, page 56). Fig. 6.5.2 shows a model of the anatomy theatre in Padua where Harvey stood and watched Fabricius' anatomical demonstrations. So, by the 17th century, we have advances in knowledge of anatomy, in public health and in physiology.

The care of the patient
Now let us turn to the plight of the patient. What was it like at this time to be ill or injured? How were patients treated and what kind of health care was available?

Very little was available in the way of drugs or medicines. Surgery was a crude and desperate remedy, full of terror and danger for the unfortunate patient. It was thus usually reserved for the wounds of the battlefield and undertaken to avoid gangrene. Because of the conditions of the time, wounds often became infected and amputation was necessary to prevent the dreaded gangrene spreading. Operations were carried out with a butcher's skill and boiling oil was used as a means of cautery to stop bleeding. Progress was made by a French surgeon, Paré, in 1536, when he ran out of boiling oil. He had to use a simple dressing instead and he found that the wound healed better. Paré also developed methods of tying off blood vessels to stop bleeding, and he made use of simple dressings and bandages. Fig. 6.5.3, from a fresco in the Sorbonne, shows Paré doing an operation. There has been a battle and Paré is treating a wounded soldier. The dangers are emphasised by the presence of the bishop giving a blessing, and being ready to administer the Last Rites in the all-too likely event of death.

At this time, hospitals were often relatively advanced and pleasant, symbolising the Hippocratic ideas of treating the whole patient in an appropriate, healing environment. Fig. 6.5.4 shows the hospital of Santa Cruz in Toledo in the 16th century. The equally ancient hospital of the Knights Templar in Rhodes still stands and is just as clean, airy, restful and pleasant as this. So we can see that some of these early hospitals were in many ways more therapeutic and sanitary than those which were to develop centuries later during the Industrial Revolution, some of which are still with us today. There are few wards, even in modern hospitals, which look as good for nursing patients as this hospital in Santa Cruz.

As we move on now through time the number of significant developments increases and only some of them can be mentioned here.

Chemistry and physiology
In the 18th century the famous chemist, Lavoisier, made a great contribution to physiology when he separated ordinary air into two gases and developed his theory of respiration. Fig. 4.5.4 (page 150) is a drawing by Madame Lavoisier showing Lavoisier undertaking an experiment in his laboratory in Paris in 1789. He is analysing gases which are being inhaled and exhaled, studying the relationship of respiration to exercise (note the treadle) and observing the effects on the subject's pulse rate. Lavoisier's work and life were cut short by the French Revolution, when he was sent to the guillotine in 1794 (see Chapter 4.3). However, his work paved the way for later progress in physiology, pharmacology, biochemistry and experimental medicine.

The rise of clinical medicine
Despite many such advances in knowledge, doctors still could not get inside patients' bodies to study what was happening there. The risks of internal surgery were far too great. However, a breakthrough occurred with the development of

Fig. 6.5.3 Paré operating on a wounded soldier in the 16th century.

Fig. 6.5.4 The hospital of Santa Cruz at Toledo in the 16th century.

Fig. 6.5.5   Laennec using his newly invented stethoscope early in the 19th century.

the stethoscope, which enabled doctors to listen to changes within the body—for example, to hear fluid in the lungs. The stethoscope was invented by a Frenchman, Laennec, and its use was publicised in 1819. Fig. 6.5.5, also from a wall painting in the Sorbonne, shows Laennec listening to chest sounds. He was particularly interested in tuberculosis and died of it at the age of 45.

Another important pioneer of modern clinical medicine is Edward Jenner who was born in a Gloucestershire village and became a country doctor. During the course of his work Jenner noticed that cows sometimes developed ulcers on their udders and became generally unwell. Farmers used to say that they were suffering from 'cowpox'. It was noticed that milkmaids who milked these cows tended to develop symptoms of inflammation on their wrists and hands together with a fever. Even more interestingly, it was also noticed that those who had once suffered from cowpox never succumbed to the dreaded disease of smallpox. On the basis of such observations, Jenner undertook some experiments. In 1796 he took pus from a milkmaid's sore and transferred it to some scratches he had made on the arm of a healthy boy. The boy subsequently developed the signs and symptoms of cowpox. Jenner then proceeded with the experiment to a stage which was exceedingly dangerous—he introduced pus from the spots of a smallpox victim into further scratches on the boy's arm. Mercifully, the boy did not succumb to smallpox. The experiment was repeated with other patients, and so the process of vaccination was established.

MAJOR ADVANCES IN THE 19TH CENTURY

Coming into the 19th century, there are three other major advances we need to describe if we are to understand our own era—the rise of the germ theory of disease, and twin developments in surgery: anaesthetics and antisepsis.

The germ theory of disease
Bacteria had first been observed in the 17th century by a Dutch microscope maker—Anton van Leeuwenhoek (see Chapter 3.2 and Figs. 3.2.4–5, pages 49–50). But it was not until the middle of the 19th century that Louis Pasteur linked bacteria with the theory of contagion and demonstrated that germs caused particular diseases such as anthrax, chicken pox and cholera.

Pasteur was born in France in 1822 and graduated as a chemist at the age of 25. His research activities were very diverse, but he was particularly interested in microbes. Many contemporary scientists believed that these organisms emerged spontaneously and Pasteur set out to challenge this theory. In order to do this, he set up an experiment with meat broth, showing that it would go bad if exposed to microbes in polluted air. He took 20 boiled broths to the top of Mont Blanc, opened the containers in the pure mountain air and resealed them; 20 others were exposed to the polluted air of Paris. Only four of the flasks open on Mont Blanc subsequently fermented, in contrast to all of those exposed to the Parisian air. He thus demonstrated the probable existence of microbes in the air and undermined the theory of spontaneous generation. Pasteur also experimented with wine, having been asked by French wine manufacturers to study reasons for wine going sour. He recommended that the wine should be heated to 50°C–60°C, suspecting that microbes were the cause of souring and that this treatment would destroy them. The method was successful and is familiar to us now as the process of *pasteurization*, used to render milk safe for human consumption. Applying his ideas on microbes to disease, Pasteur discovered that hens could be made immune to chicken cholera by the administration of a mild dose of cholera bacteria. Similarly, he gave a public demonstration of vaccination against anthrax in sheep. He injected 24 sheep with a mild culture of anthrax and later injected these same sheep, plus another 24, with a fatal dose of anthrax microbes. Two days later, the 24 sheep who had been vaccinated were alive and well; the other 24 lay dead. Pasteur then, like Jenner, took the great risk of experimenting with a human being. He chose that most deadly disease, rabies, and injected into a nine-year-old boy, who had been bitten by a rabid dog, an extract made from the central nervous system of an animal suffering from rabies. Happily, the boy developed no symptoms and returned home in good health.

Pasteur's work was of immense importance because it encouraged attempts both to prevent and to cure disease. Prevention could be achieved by public health measures—such as vaccination for smallpox, following Jenner's and Pasteur's work, or by sanitation and control of water supplies. The rising standard of living in the 19th century (see Chapter 4.11) was also important in improving health. As for cures, the isolation of pathogenic bacteria paved the way for the identification of substances which would destroy them—and thus for the revolution in therapeutics of this century.

New horizons in surgery: anaesthesia and antisepsis
Any surgeon faces three major problems: stopping bleeding, suppressing pain and preventing infection of wounds. The first problem, as we have seen, had been solved with the discovery of methods of ligature of blood vessels, but we have to wait until the latter part of the 19th century for solutions to the other two. Until this time, surgeons could do little to help patients to withstand the pain of surgery. Patients had to be held or tied down and surgeons had to work as quickly as possible. Some surgeons developed skills to the point where they could amputate a leg and tie off the arteries in less than 30 seconds. But such speed made refinement difficult; nor was sophisticated internal surgery possible.

Anaesthesia   The idea of anaesthesia began with Humphry Davy (see Chapter 4.5). In 1797 Davy suggested the use of nitrous oxide, or laughing gas, as an aid to surgeons. However, his ideas were not followed up for over 40 years until, on 16th October 1846, the first surgical operation was performed with a general anaesthetic. Fig. 6.5.6 shows a contemporary drawing of this epoch-making event, which took place in Boston, USA. The patient is having tumour removed from his neck. Note that the surgeons and assistants are still wearing everyday clothes and there is no attempt to establish an aseptic environment.

Fig. 6.5.6   The first surgical operation using a general anaesthetic, Boston, 1846.

Clearly, anaesthesia was an enormous blessing in the alleviation of suffering. It also allowed for great increases in the scope and practice of surgery. But for a while it led to increased hazards, because of the large numbers of patients who died from subsequent infection of wounds. Hence the great significance of that other breakthrough—antisepsis.

Some feeling for what surgery was like in the era preceding antiseptic techniques can be obtained by visiting the operating room preserved from those days which can be found in Southwark, near London Bridge.[3]

Antisepsis In 1867 the *Lancet* reported a successful operation for a compound fracture of the leg, carried out by a surgeon called Joseph Lister. What was significant was not the operation but the manner in which it was done. Lister used a new technique which was to reduce dramatically the death rate for such a procedure, which up till then would have been higher than 50 per cent. Lister had been thinking about Pasteur's ideas on germs. On 12th August 1865, an eleven-year-old boy was run over by a cart and had multiple injuries to both legs. Lister operated (see Fig. 6.5.7) using carbolic acid as a disinfectant; the wound healed rapidly and cleanly. And so began an era of antiseptic, and later of aseptic, techniques which has enabled surgeons to undertake operations that would previously have been unthinkable.

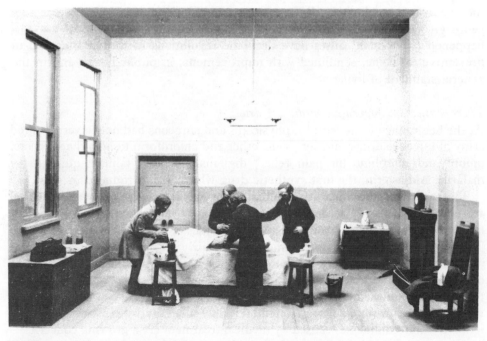

Fig. 6.5.7 Lister operating in 1865 using an antiseptic spray of carbolic acid. Note that there is no attempt to exclude germs and bacteria but merely to kill those that were present. Aseptic techniques came later.

LANDMARKS IN 20TH-CENTURY HEALTH CARE[4]

We now have some idea of the state of medical knowledge and health care at the beginning of this century. But much of the pay-off from the increase in knowledge only came in later decades. At the turn of the century life was still 'nasty, brutish and short' for many people. For example, in 1900, a married couple who were planning to have a child had to reckon with an appreciable risk of the mother dying in childbirth. The maternal mortality rate was nearly 5 per 1000 live births compared with 2 in 10,000 nowadays. Also, there was a probability of one child out of every seven dying in the first year of life compared with less than one in 50 now. And infectious diseases took an enormous toll of life. For every person who dies from them now, 60 would have done so 40 years ago—especially from whooping cough, measles, diphtheria and tuberculosis. Think of the contrast. Nowadays in the western world the possibility of a mother dying in childbirth or of a child not surviving to adult life, except perhaps of being killed in a road accident, hardly crosses parents' minds. And people do not expect to catch infectious diseases. Or, if they do, they expect to be cured. The situation has changed very fast—so fast that many people have completely forgotten what things were like only 40 years ago: the wards of fever hospitals; the sentence of lifelong invalidism that a diagnosis of pulmonary tuberculosis entailed; young children blue and exhausted from paroxysms of whooping cough; a child of 12 dying after an attack of diphtheria; the virtually inevitable death from meningococcal meningitis. All these were commonplace phenomena of our own century. Yet now they are things of the past in our society. How has this happened? The main causes have been the revolutions in therapeutics and in preventive medicine, combined with improvements in public health and in the general standard of living.

Therapeutics and preventive medicine: drugs

At the beginning of this century, physicians and surgeons had only a very limited array of effective drugs: nitrous oxide, ether and chloroform to help the surgeon; opium and morphine for pain relief; digitalis for heart failure; quinine for malaria; and aspirin, the first synthetic drug which was introduced in 1899.

The explosion since then has resulted in the vast array of drugs with which we are familiar. A glance at the pharmacopoeia will bring home dramatically how much progress has been made in just a few decades of human history. The search for new drugs continues. For example, until fairly recently the American drug industry was bringing on to the market about 400 new drugs, or combinations of old drugs, every year. It is this context that we need to consider in discussions on the pharmaceutical industry or the issues involved in tragedies such as those which occurred with the drug thalidomide. Side-effects of drugs clearly cannot be ignored—indeed illnesses caused by medical treatment are a growing concern[5]—but they always need to be set against the benefits of the modern pharmacopoeia such as its contribution to the treatment of infectious diseases (see Fig. 6.5.8).

Remember, too, that this fall in the number of deaths particularly affects those

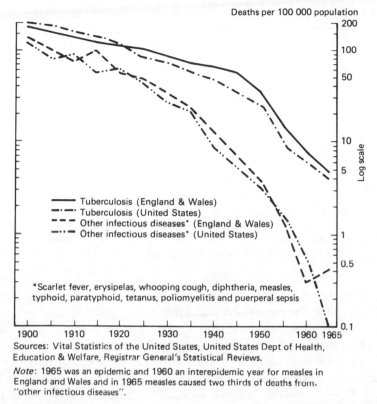

Deaths per 100 000 population

Tuberculosis (England & Wales)
Tuberculosis (United States)
Other infectious diseases* (England & Wales)
Other infectious diseases* (United States)

*Scarlet fever, erysipelas, whooping cough, diphtheria, measles, typhoid, paratyphoid, tetanus, poliomyelitis and puerperal sepsis

Sources: Vital Statistics of the United States, United States Dept of Health, Education & Welfare, Registrar General's Statistical Reviews.

Note: 1965 was an epidemic and 1960 an interepidemic year for measles in England and Wales and in 1965 measles caused two thirds of deaths from "other infectious diseases".

Fig. 6.5.8   Death rates from tuberculosis and other infectious diseases in the USA and England and Wales, 1900–65. *Redrawn from Norton, A., The New Dimensions of Medicine (London: Hodder and Stoughton, 1969).*

in the first half of life and needs to be seen in the context of overall population growth (see Fig. 4.11.1, page 213). It is important to emphasise that other factors were also at work. Social factors such as a rise in the general standard of living and improvements in diet, housing, sanitation and work conditions were all important and had contributed to the fall in death rates preceding the introduction of drugs. However, drugs were clearly highly significant. Table 6.5.1 shows that deaths from streptococcal infections (scarlet fever, erysipelas, puerperal sepsis) fell by 80 per cent following the introduction of sulphonamides in 1935; deaths from diphtheria fell dramatically after the immunisation programme began in 1940; the largest drop in deaths from tuberculosis occurred in the decade 1945–55 after the discovery of streptomycin; and deaths from poliomyelitis fell significantly in the years following the mass vaccination campaign beginning in 1956.

Therefore, the twin effects of rising standards of living and the discovery of so many effective drugs have revolutionised the patterns of disease and death in western societies. Two examples from personal experience may help to bring this home. My father, in the early days of his career as a surgeon, worked in the East End of London. He remembered a situation where rickets was so frequent that

Table 6.5.1　Deaths from certain infectious diseases: England and Wales

	1925	*1945*	*1965*
Scarlet fever and erysipelas	1838	201	12
Whooping cough	6058	689	21
Diphtheria	2774	694	0
Measles	5337	728	115
Puerperal sepsis	1110	76	4
Meningococcal infection	354	527	112
Typhoid and paratyphoid	388	44	8
Tetanus	159	79	21
Tuberculosis—all forms	40 392	23 468	2282
Poliomyelitis	156	126	3
Population	38·9 M	42·6 M	47·9 M

(*Source: Registrar General's Statistical Review for various years*)

when children came to hospital for operations for other causes, it was not uncommon to take the opportunity to break their legs while they were under anaesthetic, so as to straighten them. Rickets has now almost disappeared in Britain—the exceptions being a small, but possibly increasing, number of cases in deprived inner city areas and among the children of some ethnic minority groups. For example, some Asians have difficulty in adjusting their young children's diet to the kinds of food available in this country. The second example comes from my own experience of nursing. I remember nursing patients in the 1950s with pulmonary tuberculosis who were dying because the available treatments were very limited—surgery, rest, diet, fresh air. I myself later developed tuberculosis but was fortunate to do so after the anti-tubercular drugs had been discovered. A combination of these and six months in hospital cured my tuberculosis. Only ten years earlier it might have been a different story.

In medicine and health care, however, success often breeds new problems. We all have to die of something. Cures for some diseases mean that we may survive but, in due course, we will inevitably suffer from something else instead. Figure 6.5.9 shows how, with the fall in tuberculosis, there has been a dramatic rise in the incidence of lung cancer. Some of the most pressing problems now facing us are not infectious diseases but the malignant diseases, cardio-vascular diseases, the diseases of old age, and certain psychiatric problems. We will be looking in later chapters at some of the implications of these changing patterns of disease and death and at the problems they pose.

Diagnostic procedures
Remember that, until the last century, it was very difficult to have any idea of what was going on *inside* the body. This made accurate diagnosis very difficult. As we come into the 20th century, a whole range of diagnostic procedures has developed. For example, X-rays, discovered by Röntgen in 1896, were very soon

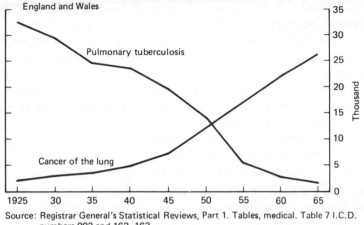

Source: Registrar General's Statistical Reviews, Part 1. Tables, medical. Table 7 I.C.D.
numbers 002 and 162, 163.

Fig. 6.5.9 Deaths from pulmonary tuberculosis and lung cancer, England and Wales, 1925–65. *Redrawn from Norton, A., The New Dimensions of Medicine (London: Hodder and Stoughton, 1969).*

being used to diagnose fractures and internal lesions. Now we are so familiar with the repertoire of investigations undertaken in pathological laboratories, analysing cell tissues and body products, that we may take them for granted. However, it is not so long ago that doctors had to taste patients' urine to see if it was sweet and so likely to contain sugar, indicating diabetes. Some experienced physicians used to teach students by dipping in one finger and licking another!

Surgery
This is probably the area which most often hits the headlines as it tends to be the most 'heroic' and dramatic. As we know, advances have been multitudinous—all made possible by the 19th-century breakthroughs in anaesthesia, antisepsis and asepsis. Surgical techniques are improving all the time. One tragic illustration: it was recently reported that in Northern Ireland, terrorists have changed their policy of shooting people in the knees. They now shoot them in the ankles— apparently because surgeons have become so good at replacing kneecaps that shooting in the knees was not seen to cause sufficient damage. As is often the case, advances in technology may stem from situations of conflict and 'man's inhumanity to man'.

Radiotherapy
This is another 20th-century development, used when there is a need for the selective destruction of cells. Also, there have been recent developments in the use of laser-beam therapy—for example, to destroy malignant cells in the treatment of some conditions of the cervix, or to 'burn on' a retina which has become detached.

Psychiatry
Here, there have been twin developments: in physical treatments and in psychological methods. The latter lie more in the realm of behavioural science and

we will be discussing some aspects of these in Chapter 8.6. In the realm of physical treatments, there is first and foremost the revolution in the development of drugs used for the treatment of the psychoses such as schizophrenia, and for neuroses—such as depression or acute anxiety states. In addition, there have been other developments such as the use of electro-convulsive therapy (ECT). All these represent as dramatic a revolution in the care of the mentally ill as has occurred in the care of physical illness. Overall, the revolution in psychiatric treatment has meant that official policy has been able to change to what is called an 'open door' policy. Many people who, in early days, would have had to remain incarcerated in mental hospitals all their lives, can now live in the community for at least part of the time. Also, life within mental hospitals has changed greatly, with the virtual disappearance of the need for physical restraints, locked doors and padded cells. All this has happened in living memory—largely since World War 2.

Health care and the sciences
It is important to emphasise that most of these advances in medicine and health care have involved contributions from many fields of science. If we list but a few of these, we can more easily appreciate how interdisciplinary medical knowledge has now become. As a matter of routine, modern health care draws on:

Biochemistry—in the analysis of body fluids and the light this subject has shed on a myriad of diseases such as anaemia and diabetes;

Immunology—in preventive health care and transplant surgery;

Electrical engineering—in the development of machines such as electro-cardiographs and electroencephalograph machines;

Polymer science—with the development of materials for use in orthopaedic and arterial surgery, such as artificial heart valves and hip joints;

Physics—with X-rays and radiography, laser-beam therapy and the use of radioisotopes;

Psychology—in psychoanalysis and behaviour therapy;

Statistics—in epidemiology and the development of mathematical research techniques;

Physiology—in the understanding of respiration and developments in anaesthesia.

CURRENT ISSUES IN HEALTH CARE

Let us finish this historical overview by reminding ourselves of the dilemmas posed by advances in health care and the ethical issues they raise.

Changing patterns of disease and death
As we have already noted, achievements bring new problems in their wake. But we must all die of something. Now that many of the diseases which used to kill people in infancy or childhood have been eradicated, more and more people are

surviving to middle and old age. They then succumb to other disabilities or diseases and thus create new demands for health care. Also, more effective ways of caring for people who suffer disabilities such as spina bifida, or who are injured in serious accidents, means that there are now more people who are chronically handicapped. Therefore our population now contains many more elderly people, who are prone to the diseases of old age, and many more handicapped people. Many of these people cannot be cured but they need care—and this raises all sorts of problems of how best to use scarce resources. High technology medicine is immensely expensive and we only have a limited budget available for health care. There is therefore a real conflict between different groups who urgently require money and staff for various worthwhile purposes. Who can balance the need for a kidney dialysis unit which can save a number of lives, against the care of the elderly or mentally ill in the community? These are tough, difficult questions of priorities which, in a free democratic society, should be the concern of us all.

The quality of life and its termination

Given modern technology, it is now possible to keep people alive to an extent undreamed of before in human history. By using antibiotics, or heroic surgery, or life-support systems in intensive care units, it is now possible to keep someone alive when 'nature's way out' would have been death. Agonising questions confront perhaps the patient, often the relatives, and inevitably the medical and nursing staff, about the quality of life which the patient may expect. Arthur Hugh Clough made the famous statement: 'Thou must not kill, but should not strive officiously to keep alive.' However, reality can be heartbreakingly complicated and uncertain. The enormous advances in health care have now put people, increasingly, in the almost god-like position of having the power of decision of life and death over their fellows. And in such decisions there are no easy answers.

These are ultimate issues of great human significance. If we are to play a responsible role as citizens in our own society, as well as in our professional capacity as scientists, we need to think about them seriously and to work out our own moral position. For there can be nothing of greater importance than the life and the quality of life—and death—of our fellow human beings.

REFERENCES

1 Lloyd, G. E. R. (ed.), *Hippocratic Writings*, (Harmondsworth: Pelican, 1978), p. 206.
2 Ibid., p. 67.
3 The Wellcome galleries in the Science Museum, London, also show many examples of the conditions under which health care was practised in earlier times.
4 Much of the information in this section is taken from Norton, A., *The New Dimensions of Medicine*, (London: Hodder and Stoughton, 1969).
5 See Illich, I., *Medical Nemesis*, (London: Calder and Boyars, 1975) for a discussion of illness caused by medicine or iatrogenic illness, as it is called.

SUMMARY

1 Ancient Greek writers on medicine like Hippocrates emphasised the idea of man's complete well-being—social, spiritual and physical.

2 Later developments in Europe, like those of Vesalius in anatomy and Harvey on the circulation of the blood, treated man more mechanically—as a collection of parts rather than as a whole.

3 The range of effective medical and surgical treatments was very limited until the latter part of the 19th century. Then the rise of the germ theory of disease and the use of vaccination, combined with advances in surgery such as anaesthesia and antisepsis, began to extend the scope of medical practice.

4 In this century substantial improvements in mortality rates and life expectancy have been achieved by a combination of increased medical and scientific knowledge, the availability of many new drugs and vaccines, advances in surgical techniques and improvements in the standard of living.

5 As health care improves, many more people survive to middle and old age who would have died in infancy or childhood at any earlier period in history. These changing patterns of disease and death will require a shift in emphasis from curing specific diseases to the more comprehensive provision of health care.

6.6 The unification of chemistry and physics: Mendeleev, Bohr and Schrödinger

Many of the basic facts about atoms and their structure are now taken for granted in all the sciences. They form the framework that all scientists use when they think about the natural world. These facts are now so well accepted that most people have no knowledge of the long trail of argument and experiment by which they were established. So in this chapter we will focus on part of the story. We will describe the work of Mendeleev on the periodic table and of Bohr on the structure of the atom. Then we will briefly discuss Schrödinger's ideas on wave mechanics which superseded Bohr's work in the 1920s. The aim will be to show how vast areas of physics and chemistry were unified—at least in principle—as a result of all this work.

In Chapter 4.5 we saw how chemistry emerged in the late 18th and early 19th centuries as a separate subject of study in something like its modern form. The key names in this development were Priestley, Lavoisier, Dalton and Davy. Physics too took on much of its modern shape in the 19th century and became clearly distinguished from the more general natural philosophy of earlier centuries. As the 19th century proceeded, chemistry and physics became more specialised and separate, and the story of how they were unified in the 20th century is one of the most extraordinary and unexpected in the history of science.

MENDELEEV'S PERIODIC LAW

To understand how this happened we must start with chemistry and with the work of the Russian chemist Dmitry Mendeleev (1834–1907) (see Fig. 6.6.1). During the early 19th century Dalton's atomic theory formed the basis of most chemical work and much attention was concentrated on making it more quantitative by the experimental measurement of atomic weights. However, progress was slow. This was not because the experiments weren't accurate but rather that the differences between atoms and molecules were not clearly understood. Consequently by the 1850s many different atomic weights for the same element were in use and similarly there were large numbers of possible chemical formulae for the same compound. Kekulé, in his famous textbook of organic chemistry, published in 1861, lists no fewer than 19 different formulae for the relatively simple compound acetic acid for which the modern formula is CH_3COOH.

To try to clear up the confusion Kekulé organised the first ever International Chemical Congress at Karlsruhe in 1860 and the 26-year-old Mendeleev was one of the participants. After this there was certainly more agreement on atomic weights and in the 1860s various people tried to look for relationships between the chemical properties of the elements and their atomic weights. Mendeleev's was

Fig. 6.6.1 Dmitri Mendeleev (1834–1907), the originator of the Periodic Table of the Elements.

the most successful of these attempts but he was by no means the only one working along the same lines. What he did was to arrange the elements in a table in order of increasing atomic weight and when he came to an element which was similar to an earlier element, he started a new row. To help him in his arrangement and rearrangement of the elements, Mendeleev wrote the elements and their properties out on cards which he shuffled in an inspired version of his favourite card game, patience. Sometimes Mendeleev had to alter the atomic weights or to leave gaps in order to bring similar elements together in the columns of his table.

Mendeleev first published his Periodic Law in 1869 in a textbook called *Osnovy Khimii* or Principles of Chemistry. He expressed it like this—'If all the elements be arranged in the order of their atomic weight, a periodic repetition of properties is obtained'. Initially the idea of arranging the elements in the order of their atomic weights was not well received. One facetious critic even asked if anyone had ever tried arranging the elements in the order of the initial letters of their names! But the experimental evidence which changed opinion decisively in Mendeleev's favour concerned the gaps we mentioned earlier.

Fig. 6.6.2 shows Mendeleev's Periodic Table of 1871; it only contains 63 elements—all that were known at the time—and the atomic weights given are very approximate.

Note in particular the gaps in the third and fourth columns at the approximate atomic weights of 44, 68 and 72. Mendeleev left these gaps because, if he didn't, later elements fell in the 'wrong' places—that is they didn't lie in the same column as elements with similar properties. In doing this you could say that Mendeleev was fiddling his data so as to fit in with his ideas. But he was so confident in his periodic law that he went even further. He predicted the existence of three new, as yet undiscovered, elements—eka-boron (44), eka-aluminium (68) and eka-silicon (72). From his extensive knowledge of the chemistry of the known elements, Mendeleev made some very detailed predictions about the properties of these unknown elements. That was in 1871. In 1875 eka-aluminium or gallium was discovered, followed in 1879 by eka-boron or scandium and in 1886 by eka-silicon or germanium. Their properties agreed extraordinarily closely with Mendeleev's predictions. Look at Table 6.6.1 which compares his 1871 predictions for eka-silicon with the actual properties of germanium when it was discovered, 15 years later, in 1886. The agreement is extraordinarily close.

Table 6.6.1

Property	Eka-silicon (*1871 prediction*)	Germanium (*1886*)
Relative atomic mass	72	72·32
Specific gravity	5·5	5·47
Specific heat	0·073	0·076
Atomic volume	13 cm³	13·22 cm³
Colour	dark grey	greyish white
Specific gravity of dioxide	4·7	4·703
Boiling point of tetrachloride	100°C	86°C
Specific gravity of tetrachloride	1·9	1·887
Boiling point of tetraethyl derivative	160°C	160°C

This was striking enough but the discovery of the first of Mendeleev's new elements—gallium in 1875—had been in some ways even more dramatic. Gallium was discovered spectroscopically by a Frenchman, Lecoq de Boisbaudran, who measured some of its properties including its specific gravity. Most of the properties of gallium fitted those of eka-aluminium but its specific gravity was too low. At Mendeleev's suggestion, Lecoq de Boisbaudran remeasured the specific gravity more carefully and confirmed that it was very nearly equal to Mendeleev's prediction. This made a great impact on the scientific world for Mendeleev seemed to know more about the new element than the man who had discovered it!

Series	Group I R_2O	Group II RO	Group III R_2O_2	Group IV RH_4 RO_2	Group V RH_3 R_2O_5	Group VI RH_2 RO_3	Group VII RH R_2O_7	Group VIII RO_4
1	H=1							
2	Li=7	Be=9·4	B=11	C=12	N=14	O=16	F=19	—
3	Na=23	Mg=24	Al=27·3	Si=28	P=31	S=32	Cl=35·5	—
4	K=39	Ca=40	−=44	Ti=48	V=51	Cr=52	Mn=55	Fe=56 Co=59 Ni=59 Cu=63
5	(Cu=63)	Zn=65	−=68	−=72	As=75	Se=78	Br=80	—
6	Rb=85	Sr=87	?Yt=88	Zr=90	Nb=94	Mo=96	−=100	Ru=104 Rh=104 Pd=106 Ag=108
7	(Ag=108)	Cd=112	In=113	Sn=118	Sb=122	Te=125	I=127	—
8	Cs=133	Ba=137	?Di=138	?Ce=140				—
9	(−)							
10	—	—	?Er=178	?La=180	Ta=182	W=184	—	Os=195 Ir=197 Pt=198 Au=199
11	(Au=199)	Hg=200	Tl=204	Pb=207	Bi=208			—
12	—	—	—	Th=231	—	U=240	—	—

Fig. 6.6.2 Mendeleev's Periodic Table of 1871.

In 1894 the first rare or inert gas, argon, was discovered by Ramsay and this seemed to upset Mendeleev's work. Ramsay predicted that if Mendeleev was right, there should be a whole family of such gases and over the next four years he found them—helium in 1895, which had first been discovered spectroscopically in the sun in 1868, and neon, krypton and zenon in 1898. These discoveries were another dramatic confirmation of Mendeleev's ideas. His periodic table had unified and systematised chemistry in a way which is still valid. Today, more than 100 years later, Mendeleev's periodic table is on the wall of nearly every chemistry lecture room in the world. Mendeleev was acclaimed all over the scientific world—except in Tsarist Russia. There his liberal political views were held against him and he was never given the recognition he deserved.

But, although Mendeleev's table worked, at that time no-one had any idea why it worked. Listing the elements in the order of their atomic weights seemed to have a mystical aspect, possibly reminiscent of Kepler or the ancient Pythagoreans, rather than one based on rational explanation. The underlying reason why Mendeleev's idea worked was not understood until the 20th century when the detailed structure of atoms began to be established.

It is interesting that Mendeleev never accepted the first steps in this process— the discovery of the electron in 1897 by J. J. Thomson and the explanation of radioactivity, discovered in 1896, in terms of the transmutation of elements. Mendeleev thought that these discoveries would undermine the very foundations of his periodic law but in a curious way they were eventually to confirm it. And the man who played the major part in this process was a Dane, Niels Bohr (1885–1962).

BOHR AND THE STRUCTURE OF THE ATOM

Bohr's original interests had nothing to do with chemistry. He was concerned with the new discoveries in physics—and particularly with those dramatic discoveries in atomic physics which, at that time, were changing all our ideas on the structure of matter.

In 1900, there were still those who denied the existence of atoms. By 1910 no-one of any stature could maintain such a view. In those ten years, electrons— discovered in 1897 by J. J. Thomson at Cambridge—had been shown to be negatively charged particles which were constituents of every kind of matter. Since ordinary matter is electrically neutral, there were presumed to be positive charges in the atom to balance the negative charges of the electrons. However, little was known about these positive charges until the work of Rutherford in 1910. Rutherford was extremely interested in radioactivity—which had been discovered by a Frenchman, Becquerel in 1896—and he spent many years studying the properties of the penetrating radiations emitted. The three main types of radiation—α, β, and γ-rays—were all first identified and named by Rutherford. So it was fitting that it was with the α-rays, which he had identified and named, that Rutherford made his greatest discovery—the nucleus of the atom. Fig. 6.6.3 shows the sort of experiment that he did.

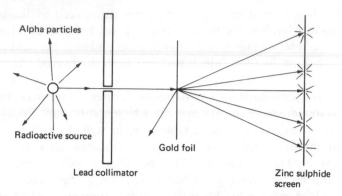

Fig. 6.6.3   Rutherford's α-particle scattering experiment; the α-particles were detected by observing the scintillations they produced when they hit the zinc sulphide screen.

What Rutherford expected was that most α-particles would go straight through the thin gold foil and that some would be deflected a little to one side or the other. What he found was that some α-particles bounced right back off the foil, almost back the way they had come. This was totally unexpected. As Rutherford put it— it was almost as if you had fired a fifteen inch shell at a sheet of tissue paper and it came back and hit you. This result led Rutherford to propose his nuclear model of the atom—a small positively charged heavy nucleus with the much lighter electrons orbiting like planets around it, rather like Kepler's model of the solar system. If a positively charged α-particle makes a direct hit on a similarly charged heavy nucleus, you would expect it to be repelled straight back. Rutherford's nuclear model was able to explain all the details of his experimental results and it also allowed him to estimate the approximate size of the nucleus. The answer came out surprisingly small—about 10^{-15} metres which is about a hundred thousand times smaller than the size of a typical atom (around 10^{-10} metres). Since it is so difficult to grasp such figures, it may be helpful to illustrate their meaning in another way. If a million atoms were arranged neatly in a line they would only extend for about a tenth of a millimetre. And within each atom the nucleus is far smaller. If we represent the nucleus by a tennis ball then the planetary electrons would be circling in orbits about five miles away. So Rutherford's atom is largely empty space.

In 1912, soon after Rutherford had proposed the nuclear model, Bohr came to work with him at Manchester. At that time neither Rutherford nor anyone else had any idea how the electrons in the atom were arranged around the nucleus and it was on this problem that Bohr worked. Bohr saw, almost at once, that the nucleus was the key to the radioactive properties of atoms and that it was the planetary electrons which determined their ordinary chemical and physical properties. The question was—how did they do this? Bohr found the clues he needed in the nature of the light which atoms emit. This light provides a window into the interior of the atom. That this was roughly true had been known for many years. Fifty years earlier, Bunsen and Kirchhoff had discovered that different elements emitted light of definite colours—the so-called line spectra. These characteristic spectra had been used to identify elements in analysis and to discover new ones. Remember that helium and Mendeleev's eka-aluminium or gallium had first been discovered spectroscopically. But although it was known that each element emitted only some wavelengths or frequencies and not others, no–one knew why this was so. What Bohr did was to find a connection between the frequencies of the emitted spectral lines and the arrangement of the electrons in the atom. He started with the simplest atom, hydrogen, which has one planetary electron and a nucleus carrying a single positive charge. Its spectrum is fairly simple too (see Fig. 6.6.4); note the various series in the ultraviolet, visible and

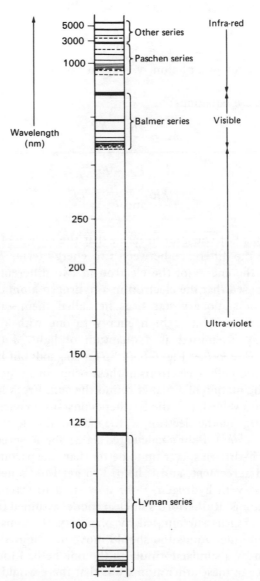

Fig. 6.6.4 The complete hydrogen spectrum; note the different series of lines in the ultra-violet, visible and infra-red.

infra-red. The frequencies of these lines had been measured with great precision and it had been found these very precise frequencies were given by a simple formula:

$$v = R \left(\frac{1}{n^2} - \frac{1}{m^2} \right)$$

in which v is the frequency of a line; R is a constant called the Rydberg constant; and n and m are integers. If $n = 1$ we get the Lyman series; if $n = 2$, the Balmer series and if $n = 3$, the Paschen series and so on.

In 1913 Bohr combined this formula with Planck and Einstein's famous quantum formula:

$$E = hv$$

in which E is the energy of a photon, v is again the frequency and h is Planck's universal constant.

This gave him these equations:

$$hv = \frac{hR}{n^2} - \frac{hR}{m^2}$$

or

$$E = E_1 - E_2$$

in which

$$E_1 = \frac{hR}{n^2} \text{ and } E_2 = \frac{hR}{m^2}.$$

Bohr interpreted this last equation to mean that the energy of the emitted light photon is given by the difference between two energy terms E_1 and E_2 each of which represented the energy of the electron in two different orbits round the nucleus. Bohr supposed that the electron in a hydrogen atom could only exist in particular orbits—or stationary states, as he called them—and that when an electron jumps from an orbit of high energy to one with a lower energy the difference in energy is emitted as a quantum of light of the corresponding frequency given by $E = hv$ (see Fig. 6.6.5). So far so good, but Bohr went further. He analysed the motion of the electrons in these orbits in a way which was similar to the analysis of the motion of a planet round the sun. From his equations Bohr was able to calculate a value for R, the Rydberg constant, in terms of the measured values for e, the charge on the electron, m, the mass of the electron and h, Planck's constant ($R = 2\pi^2 me^4/h^3$). Bohr's calculated value for R agreed with the value measured from the hydrogen spectrum to better than one part in a thousand. This was strikingly good agreement, and it helped to get Bohr's new ideas accepted.

After this success with hydrogen, Bohr now tried to extend his ideas to the structure of other atoms. Rutherford's nuclear model assumed that all atoms had a positively charged nucleus surrounded by planetary electrons. So Bohr argued that the electrons in other atoms too should move in a limited number of orbits with energies given by a similar formula to the one he had found for hydrogen. The main difference in these other atoms was that there would be more than one electron and that the nucleus would have a larger charge. We can't go into all the

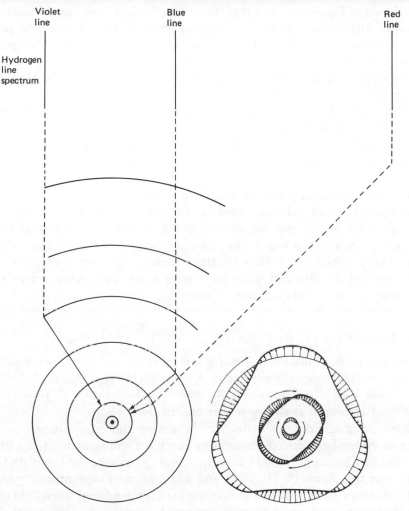

Fig. 6.6.5 Diagram illustrating Bohr's theory of the hydrogen spectrum; the electron jumps from a high energy to a low energy orbit and emits a light quantum of frequency equivalent ($E = h\nu$) to the energy difference. The inset shows diagrammatically how Bohr's electron orbits can be interpreted using Schrödinger's wave mechanics. *Adapted from Bronowski, J., The Ascent of Man, BBC Publications.*

details of the work which followed, but two striking developments must be mentioned.

First, by accurate measurements of the wavelengths of X-rays emitted by various elements, it proved possible to use Bohr's theory to calculate the charge on the nucleus for each element—what we now call its atomic number—and therefore to find the number of planetary electrons for each kind of atom.

Secondly, by detailed analysis of spectra, Bohr was able to build up a picture of how the electrons were arranged in other atoms more complicated than hydrogen. Bohr's analysis showed that these electrons were arranged in orbits very similar to those shown in Fig. 6.6.5—and that there were definite limits to the numbers of electrons which could be in any one of these orbits, two in the first orbit or shell, eight in the next, and so on. In other words the electrons were arranged in a regular series of layers which Bohr was able to show could explain the regular periodic similarities of chemical properties which were displayed in Mendeleev's periodic table of the elements.

Bohr's work had thus provided an underlying explanation for Mendeleev's periodic table which was based on atomic numbers—the number of electrons in the atom or the charge in the nucleus—rather than atomic weights. So our state of knowledge after Bohr's work was more satisfactory and systematic than it had been in 1900.

But there were major problems with Bohr's theory. For example, he could not really explain why only a limited number of orbits were possible for his planetary electrons. The spectra pointed in this direction and once you assumed the existence of these orbits much else fell into place, as we have seen. But why couldn't other orbits exist? After all, there were no such limits on the orbits of planets around the sun and the solar system is in many ways a very similar dynamical system to Bohr's planetary atom.

SCHRÖDINGER'S WAVE MECHANICS

The answer to this question was not given by Bohr but by the work of Erwin Schrödinger (1887–1961) in the 1920s. And, in giving this answer, Schrödinger had to break even more fundamentally than Bohr had done with the traditional methods of physics in analysing matter and motion.

What Schrödinger did was to develop the idea mentioned in Chapter 6.3—that matter as well as light behaves sometimes as a wave and sometimes as a particle. This idea had been put forward as a theoretical possibility by Louis de Broglie (1892–) in 1924 in his Ph.D. thesis, and was confirmed experimentally in 1927. Fig. 6.6.6 shows the very similar diffraction patterns produced when light (below) and electrons (above) are diffracted through by a pair of parallel slits. These patterns, for both light and electron beams, are average effects produced by large numbers of individual particles. Fig. 6.6.7 shows diagrammatically how a double-slit pattern is built up from the accumulated effect of many individual electrons or photons. Electron diffraction patterns are of course now very well known—for example, in the widespread use of electron microscopes.

In 1926, even before electron diffraction had been observed experimentally, Schrödinger had developed a new theory, wave mechanics, which enabled detailed calculations to be made of the behaviour of electron waves inside the hydrogen atom. For our purposes the details of how he did this are not important. In principle what he did was to show that the electron waves inside the hydrogen atom formed three-dimensional standing wave-patterns which are directly analogous to the one-dimensional standing wave patterns for vibrating stretched strings which had first been studied by the ancient Pythagoreans (see Fig. 6.6.8). The inset to Fig. 6.6.5 attempts to show, diagrammatically, the possible standing

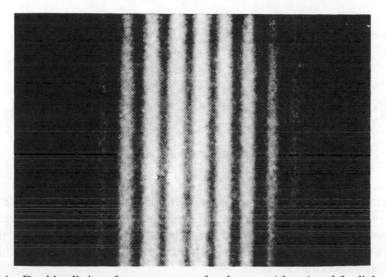

Fig. 6.6.6 Double-slit interference patterns for electrons (above) and for light (below). *From Open University Science Foundation Course S100, Quantum Theory, Unit 29, Copyright © The Open University 1971. (The electron pattern was obtained by C. Jönsson at the University of Tubingen.)*

1   2   3

Fig. 6.6.7 Build-up of a double-slit diffraction pattern (electrons or light quanta).

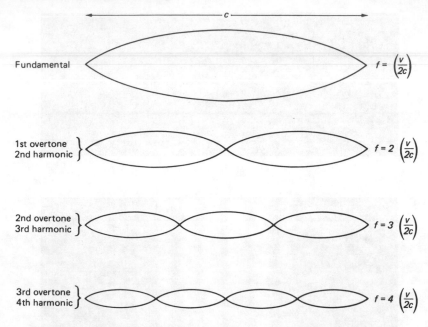

Fig. 6.6.8 Standing wave patterns on a vibrating stretched string.

wave patterns for the electron in a hydrogen atom. Schrödinger's three-dimensional standing wave patterns thus explained the existence of Bohr's limited number of orbits and when he did the detailed calculations he found that the energies associated with each of these patterns were given by a formula which was just the same as Bohr's.

So Schrödinger's wave mechanics gave a theoretical justification for one of Bohr's rather arbitrary assumptions. And in the next few years wave mechanics was used to solve satisfactorily many other problems in atomic physics. One of these involved the study of the motion of electrons in solids—first in metals and then in insulators and semiconductors. At the time, in the 1930s, this was a very academic field of study with little practical significance. But the understanding gained of how semiconductors work was a key factor in the discovery of the transistor in 1948 and in the mushroom growth of semiconductor electronics in the decades since then—the latest product of which is the silicon chip microprocessor (see Chapters 6.7 and 8.3).

In this chapter we have seen how Mendeleev's work resulted in the incorporation of an enormous amount of diverse chemical data into a systematic framework and that this framework predicted the existence of several new elements—predictions which were later confirmed in detail. We have also seen how Mendeleev's work survived intact the 20th-century redefinition of a chemical element in terms of atomic number rather than atomic weight, and how his framework was given additional support and explanation by Bohr's early quantum theory of the atom which was developed in response to work in spectroscopy rather than chemistry. And we have seen how Bohr's theory itself has been incorporated in the much more generally applicable wave mechanics of Schrödinger and is now a particularly successful application of that more general theory.

We now have a general physical theory, wave mechanics, which explains the greatest chemical generalisation ever made, Mendeleev's periodic table. So we can justly claim that, in principle, chemistry and physics have been unified by the work of Mendeleev, Bohr and Schrödinger.

SUMMARY

1   By arranging the elements in order of ascending atomic weight, Mendeleev devised his periodic table of the elements which incorporated an enormous amount of diverse chemical data into a systematic framework.
2   Mendeleev predicted very accurately the properties of three previously unknown elements—gallium, scandium and germanium; within 15 years these elements had all been discovered and were found to have just the properties which Mendeleev predicted.
3   Bohr developed the theory of the planetary electrons in Rutherford's nuclear atom so as to explain in detail the spectrum of hydrogen and many other elements.
4   Bohr's theory postulated that the electrons in different atoms were arranged in a regular series of layers or shells which could explain the regular periodic similarities in chemical properties shown in Mendeleev's periodic table.
5   Schrödinger's wave mechanics gave a more satisfactory explanation of Bohr's electron orbits in terms of three-dimensional standing electron waves.

6.7 The origins and development of statistics and computing

Statistics and computing are probably growing more rapidly in modern industrialised societies than any other area of applied science. Many people are now predicting that silicon microprocessors (see Fig. 6.7.1), or silicon chips as they are popularly called, will have very far-reaching effects. It is claimed that we are on the verge of a second industrial revolution which will change our lives as much as the first industrial revolution changed people's lives in the 18th and 19th centuries. So it is both interesting and useful to see how statistics and computing originated and have developed. In this chapter we will deal first with statistics, tracing its development from the 17th to the 20th century, and then outline the main stages in the growth of computing and computers over the same period.

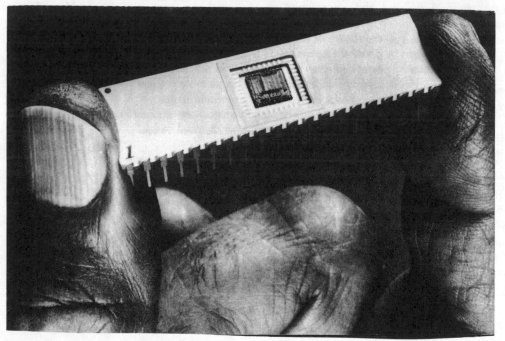

Fig. 6.7.1 Silicon microprocessor by Ferranti, produced in 1977.

STATISTICS

R. A. Fisher, one of the founders of modern statistics, defined the subject as 'mathematics applied to observational data' and this fundamental link with observational data can be seen throughout the history of statistics.

Statistics had its origin in three very different areas of practical activity—censuses and other inventories; the study of games of chance, such as dice-throwing or card games; and the quantitative estimation of errors of measurement.

Early censuses, such as the Domesday Book compiled by the Normans after their conquest of England in 1066, were inventories which sought to give a complete picture of an existing, and usually fairly static, situation. Then in the late 17th century several works appeared which made rounded estimates of population and national wealth rather than precise counts. See, for example, Fig. 6.7.2 which is taken from *Natural and Political Conclusions Upon the State and Conditions of England* (1696) by Gregory King. This table gives the 'income and expense of the several families of England calculated for the year 1688'. But note that most of the figures given are clearly approximate, rounded estimates rather than precise enumerations. However, research has shown that 'King's figures are surprisingly accurate whenever we have been able to provide independent checks on them' even though King offered 'very little evidence to justify any of the statistical work'.[1] Other similar works which appeared about this time including *Natural and Political Observations on the London Bills of Mortality* (1662) by John Graunt and *Political Arithmetic* (1690) by John Petty. These books dealt with a society which was beginning to change fairly rapidly and the authors reasoned about their data in ways which would be recognisable to modern statisticians and demographers.

In the 18th century this work was developed as actuaries began to construct life expectancy tables for insurance purposes. These tables were, in effect, tables of probabilities and this work drew on the theory of probability which had developed in the 17th century out of interest in gambling—the throwing of dice, the tossing of coins and various card games. Many of the individuals who contributed to this development are already familiar to us from their work in other fields. They include Galileo, Pascal, Fermat and Christian Huyghens whose book *De Ratiociniis in ludo aleae* (On Reasoning in Games of Dice) appeared in 1657. An interesting paper by Halley in 1692 was one of the earliest attempts to apply probability theory to mortality tables; it was called *An estimate of the degrees of the mortality of mankind, drawn from the curious tables of the births and funerals at the City of Breslau; with an attempt to ascertain the price of annuities upon lives*. In the 18th century this work was continued by men such as De Moivre (*The Doctrine of Chances or A Method of Calculating the Probability of Events in Play*, 1718) and Jacob Bernoulli (*The Art of Conjecturing*, 1713) and it culminated in the *Théorie Analytique des Probabilités* (*Analytical Theory of Probability*) of Laplace which was published in 1812.

The third area of practical interest from which statistics arose is scientifically rather more respectable than the rather dubious subjects of political arithmetic and gambling. This was the study of errors of measurement which had originated with Tycho Brahe's estimates of errors in his measurements of astronomical positions in the late 16th century (see Chapter 2.2). It was Carl Friedrich Gauss (1777–1855; see Fig. 6.7.3) who put this subject on a sound mathematical footing and Gauss' interest in it also came from astronomical observations of the positions of stars. As in nearly all his work Gauss' mathematical derivation of the error

Number of Families	Ranks, Degrees, Titles and Qualifications	Heads per Family	Number of Persons	Yearly income per Family (£ s.)	Yearly income in general (£)	Yearly income per Head (£ s. d.)	Yearly expense per Head (£ s. d.)	Yearly increase per Head (£ s. d.)	Yearly increase in general (£)
160	Temporal Lords	40	6,400	3,200	512,000	80 0 0	70 0 0	10 0 0	64,000
26	Spiritual Lords	20	520	1,300	33,800	65 0 0	45 0 0	20 0 0	10,400
800	Baronets	16	12,800	800	704,000	55 0 0	49 0 0	6 0 0	76,800
600	Knights	13	7,800	650	390,000	50 0 0	45 0 0	5 0 0	39,000
3,000	Esquires	10	30,000	450	1,200,000	45 0 0	41 0 0	4 0 0	120,000
12,000	Gentlemen	8	96,000	280	2,880,000	35 0 0	32 0 0	3 0 0	288,000
5,000	Persons in greater Offices and Places	8	40,000	240	1,200,000	30 0 0	26 0 0	4 0 0	160,000
5,000	Persons in lesser Offices and Places	6	30,000	120	600,000	20 0 0	17 0 0	3 0 0	90,000
2,000	Eminent Merchants and Traders by Sea	8	16,000	400	800,000	50 0 0	37 0 0	13 0 0	208,000
8,000	Lesser Merchants and Traders by Sea	6	48,000	198	1,600,000	33 0 0	27 0 0	6 0 0	288,000
10,000	Persons in the Law	7	70,000	154	1,540,000	22 0 0	18 0 0	4 0 0	280,000
2,000	Eminent Clergymen	6	12,000	72	144,000	12 0 0	10 0 0	2 0 0	24,000
8,000	Lesser Clergymen	5	40,000	50	400,000	10 0 0	9 4 0	0 16 0	32,000
40,000	Freeholders of the better sort	7	280,000	91	3,640,000	13 0 0	11 15 0	1 5 0	350,000
120,000	Freeholders of the lesser sort	5½	660,000	55	6,600,000	10 0 0	9 10 0	0 10 0	330,000
150,000	Farmers	5	750,000	42 10	6,375,000	8 10 0	8 5 0	0 5 0	187,500
15,000	Persons in Liberal Arts and Sciences	5	75,000	60	900,000	12 0 0	11 0 0	1 0 0	75,000
50,000	Shopkeepers and Tradesmen	4½	225,000	45	2,250,000	10 0 0	9 0 0	1 0 0	225,000
60,000	Artizans and Handicrafts	4	240,000	38	2,280,000	9 10 0	9 0 0	0 10 0	120,000
5,000	Naval Officers	4	20,000	80	400,000	20 0 0	18 0 0	2 0 0	40,000
4,000	Military Officers	4	16,000	60	240,000	15 0 0	14 0 0	1 0 0	16,000
500,586		5⅓	2,675,520	68 18	34,488,800	12 18 0	11 15 4	1 2 8	3,023,700
								Decrease	*Decrease*
50,000	Common Seamen	3	150,000	20	1,000,000	7 0 0	7 10 0	0 10 0	75,000
364,000	Labouring People and Out Servants	3½	1,275,000	15	5,460,000	4 10 0	4 12 0	0 2 0	127,500
400,000	Cottagers and Paupers	3¼	1,300,000	6 10	2,000,000	2 0 0	2 5 0	0 5 0	325,000
35,000	Common Soldiers	2	70,000	14	490,000	7 0 0	7 10 0	0 10 0	35,000
849,000	Vagrants; as Gipsies, Thieves, Beggars, &c.	3½	2,795,000	10 10	8,950,000	3 5 0	3 9 0	0 4 0	562,500
			30,000		60,000	2 0 0	4 0 0	2 0 0	60,000
	So the general Account is								
500,586	Increasing the Wealth of the Kingdom	5⅓	2,675,520	68 18	34,488,800	12 18 0	11 15 4	1 2 8	3,023,700
849,000	Decreasing the Wealth of the Kingdom	3½	2,825,000	10 10	9,010,000	3 3 0	3 7 6	0 4 6	622,500
1,349,586	Neat Totals	4 1/12	5,500,520	32 5	43,491,800	7 18 0	7 9 3	0 8 9	2,401,200

Fig. 6.7.2 'Gregory King's scheme of the income and expense of the several families of England calculated for the year 1688.' From Laslett, P., *The World We Have Lost* (Methuen and Co., 1971).

Fig. 6.7.3   Karl Friedrich Gauss (1777–1855).

distribution curve was preceded by a large number of practical measurements in which such curves were obtained empirically. Since Gauss' time this error curve has been found to be applicable in very many different experimental situations; indeed whenever the overall error is due to a combination of a large number of 'elementary errors', then we obtain a Gaussian error curve. (See Fig. 6.7.4.)

The main focus of Gauss' work on errors was aimed at estimating the size of errors with a view to making these as small as possible—although he realised that some area of uncertainty would always remain even in very precise physical measurements. However, many later developments in statistics, in both the 19th and 20th century arose from biological experiments in which it was the error or variation itself which was the main point of interest.

We have already mentioned one application of statistics in biology—the work of Mendel in the 1860s on the inheritance of pairs of opposite characteristics in pea plants, tallness versus shortness or wrinkled versus smooth seeds (see Chapter 6.4). The variations that Mendel studied were discontinuous. His pea-plants were *either* tall *or* short. So the mathematics he needed was similar to that used in analysing experiments in coin tossing or the throwing of dice.

However, the main biological event of the second half of the 19th century was Darwin's work on evolution. Here the emphasis was much more on continuous,

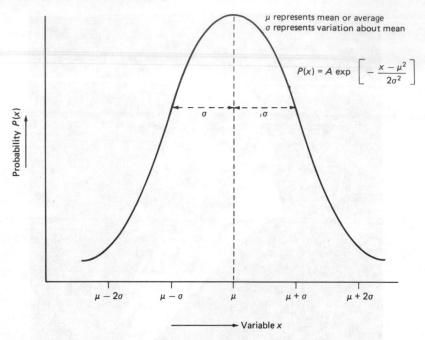

Fig. 6.7.4   Gauss' Error Curve.

rather than discontinuous variation and it was studies of continuous variation that stimulated the further development of statistics.

A pioneer in this field was Francis Galton (1822–1911). Galton believed that virtually anything could be quantified. He even devised a numerical scale for womanly beauty and conducted statistical experiments on the effectiveness of prayer! His most famous experiments were measurements of human physical characteristics like height or weight. He found that many of these variables were distributed according to Gauss' error curve. But unlike Gauss, the main focus of Galton's interest was on the differences between individuals—that is, on the size of the variation (σ) rather than on the average or mean (μ) of the curve. As he wrote:

> The primary objects of the Gaussian Law of Errors were exactly opposed, in one sense, to those to which I applied them. They were to get rid of, or to provide a just allowance for, errors. But these errors or deviations were the very things I wanted to preserve or know about.[2]

Galton was also very interested in the extent to which variations were inherited and he carried out experimental studies on the correlations between the heights of parents and their children. He found that a positive correlation did exist but that the heights of the children regressed towards the mean—that is, they were nearer the overall average than those of their parents had been. This work of Galton's marked the beginning of the study of correlation and regression which has since become a major branch of statistics.

Much of Galton's work was summarised in his book *Natural Inheritance* which he published in 1889. The book inspired a large amount of work in the new field of biometry and this soon led to further advances in statistics, most of which were due to Karl Pearson (1857–1936), a Professor of Applied Mathematics at University College, London. Pearson was greatly excited by Galton's book and together with a colleague W. F. Weldon, Professor of Zoology at University College, Pearson started on a programme aimed at the 'development and application of statistical methods for the study of problems of heredity and evolution.'[3]

In the course of the 15 years from 1890 to 1905, Pearson created many of the basic tools of modern statistics. He was the first to use the term *standard deviation* for σ, and it was he who established the modern practice of calling the Gaussian error curve a *normal distribution*. Pearson also derived mathematical expressions for many other kinds of distribution which have since found application in statistical theory and he started to give a mathematical basis to the theory of correlation. Pearson, in 1902, also derived the chi-squared (χ^2) distribution, that measure of the 'goodness of fit' between a theoretical distribution and a set of experimental observations which is now used in very many tests of statistical significance.

Pearson's statistical techniques were widely used by Weldon and others in biometric experiments. According to Weldon in 1893 'It cannot be too strongly urged that the problem of animal evolution is essentially a statistical problem: that before we can properly estimate the changes going on in a race or species we must know accurately'[4] four things. These were (a) the frequency distribution of a character; (b) the correlation between characters in an individual; (c) the link between death rate and the value of a character; and (d) the inheritance of characters.

These very practical aims were difficult to achieve, particularly when the effects sought were small. So the early biometricians, as they were called, usually tried to work with as large an experimental population as they could find. The use of small samples, and the development of appropriate statistical methods for handling these, arose from work in another practical area but a very different one—the brewing of Guinness.

The early statistical theory of small samples was developed in the Guinness Breweries in Dublin as a byproduct of attempts to find relations between the quality of the raw materials, barley and hops, the conditions of production and the quality of the beer produced. The man responsible was W. S. Gosset, or Student as he was also known, and it was in this very practical environment that the widely used Student's t-test for small samples was first worked out.

Finally, in this very rapid review of the development of statistics, let us briefly mention the work of R. A. Fisher (1890–1962), which was stimulated by the problems of another very practical area—the improvement of agriculture. During the most formative intellectual period of his life, Fisher worked at the Agricultural Research Station at Rothamsted. He derived inspiration from almost every branch of the Experimental Station's work-crop yields, rainfall, bacterial counts, genetics and, above all, field trials—and he developed advanced statistical methods to deal with the complex correlations which these experiments sought to establish. Fisher's books *Statistical Methods for Research Workers* (1925) and *The Design of Experiments* (1935) have remained standard texts for more than 40 years—and he turned Rothamsted into one of the great statistical training centres of the world.

We can't possibly discuss all of Fisher's work but one facet of it is worth mentioning because of its intrinsic importance for genetics and evolution. In 1930 Fisher published a book called *The Genetical Theory of Natural Selection* which is generally regarded as laying the foundations of modern population genetics (see Chapter 6.4). In this book Fisher developed a rigorous mathematical model which showed how changes in survival rates of individual organisms could be related to overall population growth rates *and* to changes in individual gene frequencies in the population. In doing this he used his statistical methods to show how to separate the three main sources of·observed variability—genetic variation, environmental differences and the observer's sampling errors.

This work of Fisher's resolved, at least in principle, certain conflicts within the theory of evolution which had been going on ever since the rediscovery of Mendel's work in 1900. The problem concerned the relationship between individual genes—and the characteristics associated with them—and the overall characteristics measured by the biometricians which were determined by the effects of many individual genes. Fisher's work pointed to ways in which the apparent conflict between 'particulate' and 'blending' theories of inheritance could be resolved. It was an important step in creating the framework of ideas within which modern genetics and molecular biology has developed (see Chapter 6.4).

Statistics has clearly come a long way from the relatively simple problems of census enumeration and coin-tossing from which it started. And the data with which it deals have progressively become both more detailed and more extensive. This means, of course, that developments in computing are of direct relevance to those in statistics, and it is to these developments that we will now turn.

COMPUTING

Simple calculators are almost as old as civilisation itself—the abacus, for example, was in use more than 2000 years ago. However, like so much else in modern science, modern computers derive from developments during the 17th-century scientific revolution. Early in the 17th century John Napier (1550–1617) invented logarithms. His book *Mirifici Logarithmorum Canonis Descriptio* (A Description of the Marvellous Rule of Logarithms) appeared in 1614 and contained the first logarithm tables. The effect was to reduce all arithmetical calculations to addition and subtraction and the principle involved was soon applied in one of the earliest analogue computers, the slide rule, which appeared around 1630. Napier also invented a new digital calculating aid which is still known as Napier's rods.

The 17th century, the era of the mechanical philosophy, also saw the

Fig. 6.7.5   Leibniz mechanical calculator of 1671.

development of mechanical calculators using trains of gear wheels. Pascal (1623–62) made such a calculator in 1642.

Pascal's calculator could only add and subtract, but in 1671 Leibniz (1646–1716) made a calculator which could multiply as well (see Fig. 6.7.5). The crucial feature of its design, the Leibniz stepped wheel, remained in widespread use until mechanical calculators ceased to be manufactured in the late 1960s.

Now let us jump to the 19th century to discuss the ideas of Charles Babbage (1791–1871). Babbage devoted much of his life to making new kinds of mechanical calculators. He made the first general purpose analogue computer or differential analyser as he called it; part of this machine is still in the Science Museum, London. But it was Babbage's digital calculators which are of most interest today. He built a Difference Engine in order to calculate and print tables of mathematical functions using the method of differences. This machine was a special purpose automatic digital computer. And Fig. 6.7.6 shows part of his Analytical Engine which was conceived as a completely automatic general-purpose digital computer. Babbage's Analytical Engine had all the main features of a modern digital computer—a store or memory for holding numbers, an arithmetic unit for performing arithmetical operations, a unit which controls the sequence of these operations, and a method of automatically supplying numbers and operating instructions to the machine. Babbage proposed to control his machine by using punched cards similar to those used to control automatic looms which were capable of weaving fabrics containing very complex designs. Fig. 6.7.7 shows an early version of such a loom. As Babbage wrote:

The Jacquard loom weaves any design which the imagination of man can

Fig. 6.7.6 Part of Babbage's Analytical Engine designed as the first completely automatic general purpose digital computer.

conceive. The patterns designed by artists are punched by a special machine on pasteboard cards and when these cards are placed within the loom, it will weave the desired pattern.[5]

On the wall of his study Babbage had a portrait of Jacquard—woven on one of his own looms—which was so good that most people thought it was an engraving. 24,000 cards were needed to produce it.

Inspired by this complex feat of automatic control, Babbage devised a system of punched cards which controlled both the action of his arithmetic unit and the transfer of numbers to and from his store. His cards operated in exactly the same way as the punched program cards which are still used to control many of today's electronic computers. They provided a very flexible means of making the machine perform any number of different tasks so that it was a true general purpose computer. Babbage's design even included 'the possibility of bringing any particular card or set of cards in use *any number of times successively* in the solution of one problem'[6]—or what we would now call program loops.

So Babbage's Analytical Engine had all the main features of a modern electronic digital computer. But, unfortunately, he never got it to work! This was

Fig. 6.7.7 Pattern-weaving loom from 1728; the pattern is controlled by the punched card loop. Jacquard's control system was much more complex.

because it was very difficult to realise his ambitious schemes with the bulky mechanical components he had to use. His machines were too bulky and too slow to be practicable. But all the key ideas were there and he also clearly realised the major limitation of a computer, as this quotation shows:

> The Analytical Engine has no pretension whatever to originate anything. It can do whatever we *know how to order it to perform*.[7]

Babbage's ideas were forgotten for nearly a century although, from about 1880 onwards, a number of special purpose electromechanical machines were built. These machines used data on punched cards which could be made to control electromagnetic counters or relays. They were widely used in complex tabulations like national censuses, starting with the US National Census in 1890.

The first fully automatic general purpose computer was not built until 1944, and most of the people who worked on the early machines had little or no knowledge of Babbage's ideas. The 1944 machine was still electromechanical. Electronic circuits were considered but were thought to be too unreliable. The machine was the result of collaboration between Harvard University and IBM, then a major manufacturer of punched card machines. This computer was slow by today's standards—0·3 seconds were needed for a single addition—but it remained in use at Harvard for about 15 years.

The next major change in design was the switch from electromechanical to electronic circuits. This change was stimulated by the rapid development of electronic radar and radio equipment during World War 2. These early electronic computers used large numbers of thermionic valves and were fairly bulky machines. For example, the EDSAC machine, completed at Cambridge University in 1949, contained more than 3000 valves. The EDSAC was the first practical electronic computer which could store programs. These early valve computers were about a thousand times faster than machines using electro-mechanical relays—a single addition taking about one five-thousandth of a second.

Since this time, around 1950, the principles governing the organisation of general purpose digital computers have not changed very much. Indeed most of the ideas involved would be familiar to Babbage. What has changed dramatically since 1950 is the physical hardware used to implement these principles. Due to the switch first from valves to transistors and then from transistors to integrated circuits, computers are now much smaller, cheaper and more reliable. Just how much the hardware has changed can be seen if we compare a modern micro-

© 1982 Ferranti Electronics Limited

Fig. 6.7.8 Prototype at Manchester University in the 1940s of one of the earliest commercially available electronic computers; it contained 4000 valves, six miles of wire and 100,000 soldered joints.

computer with an early valve machine. Fig. 6.7.8 shows the prototype at Manchester University of one of the earliest commercially available computers, made by Ferranti in 1950. It filled a small room—two bays each sixteen feet long, eight feet high and four feet deep. It contained about 4000 valves, six miles of wire and 100,000 soldered joints. It produced 27 kilowatts of heat and was very temperamental. Air conditioning was essential. The cost was around a million pounds in today's prices.

By contrast, Fig. 6.7.1 shows a modern microprocessor, the F100, again made by Ferranti but in 1977. This device contains about 7000 components and about six feet of metallic interconnection. It produces only five milliwatts of heat— about five million times less than the valve machine. It is millions of times more reliable and will operate at temperatures ranging from $-55°C$ to $+125°C$. It has about a hundred times the computing capacity of the earlier machine but is around ten thousand times cheaper—the cost being around a hundred rather than a million pounds.

These improvements in computing power and reliability, combined with massive reductions in size and cost, have taken place almost continuously over the last 20 years or so and there is no reason to suppose that they have yet reached any limit. At the heart of a modern micro-computer are one or more silicon chips. Fig. 6.7.9 shows a circular slice of silicon, a few centimetres in diameter, which is large enough to make a thousand or more chips; this slice is being cut up into individual silicon chips.

But in our amazement at these dramatic changes in computer hardware it is as well to realise two things. In order to make a computer do anything useful,

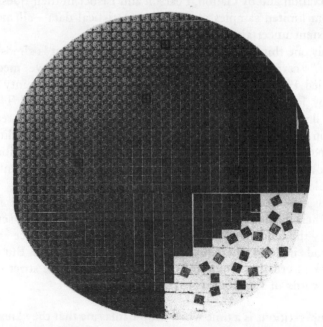

Fig. 6.7.9   Slice of silicon large enough to make about a thousand silicon chips.

however fast or reliable or cheap it may be, we need first to be able to program it and then to supply it with reliable data.

Remember Fig. 6.1.3 which shows a model of the structure of haemoglobin. This structure was worked out using the methods of X-ray crystallography. The detailed calculations involved in this process are well understood but they required many man-years of programming effort before they could be made intelligible to the central processor of a computer. And, even when a suitable program has been successfully developed, we still need accurate experimental data. In this case the original data are the positions and intensities of the spots on photographs like Fig. 6.1.7 (page 246) which is an X-ray photograph of DNA. If these data are seriously in error then nothing that the computer or programmer can do can make the output meaningful.

So perhaps we should look sceptically at some of the more sweeping claims being made for the microprocessor revolution. There is little doubt that microprocessors will change many aspects of our lives. But the ways in which they will do so are no easier to predict than were the changes which flowed from the invention of printing in the 15th century or of the steam engine in the 18th century.

In the words of Lady Lovelace, Babbage's chronicler, the computer 'can do whatever *we know how to order it to perform*'. If a task can be broken down into a sequence of simple arithmetical or logical steps, then the computer can do it—and do it much faster and more reliably than we can. But much of human life does not consist of such tasks. And even when we are dealing with tasks which can be analysed in this way, we still have the problems faced by Gauss in his theory of the errors of observation and by Galton, Pearson and Fisher in their quest for reliable inferences from limited samples of extensive statistical data—all measurements are to some extent uncertain.

There is only one thing we can be certain of in science—it is almost impossible to be *absolutely* certain about anything. Schrödinger's wave mechanics (see Chapter 6.6) led Heisenberg to formulate his famous Uncertainty Principle—rechristened by Jacob Bronowski as the Principle of Tolerance.[8] This principle sets a natural limit to the precision of even the simplest of measurements.

So if, even in science, we are centrally confronted with our fallibility—with the ever-present possibility of error—how much more does this possibility exist in more complex areas of human life. And how much more important then is the principle of tolerance.

Some enthusiasts for a computerised society fail to recognise the fundamental limitations of their machines. In their writings they slip easily from science proper into scientism—the tendency to overestimate the powers of science and to seek to apply its methods in areas where these are of doubtful relevance. But if we do this we run the risk of creating a new age of superstition—in the strict sense of that word. In the words of Friedrich von Hayek:

An age of superstitions is a time when people imagine that they know more than they do. In this sense the twentieth century was certainly an outstanding age of superstition, and the cause of this is an overestimation of what science has achieved—not in the field of the comparatively simple phenomena, where it has of course been extraordinarily successful, but in the field of complex phenomena, where the application of the techniques which proved so helpful with essentially simple phenomena has proved to be very misleading.[9]

REFERENCES

1 Laslett, P., *The World We Have Lost*, (London: Methuen, 1971), p. 267.
2 Quoted in Taton, R., *Francis Galton* in Gillispie, C. C. (ed.), *Dictionary of Scientific Biography*, (New York: Charles Scribner, 1974), Vol. V, p. 266.
3 Pearson, E., 'Karl Pearson, An Appreciation' in *Biometrika*, Vol. 28, p. 218.
4 Weldon, W. F., quoted in Pearson, E. S., *Some Reflections on Continuity in the Development of Mathematical Statistics* in Pearson, E. S. and Kendall, M. G., *Studies in the History of Statistics and Probability*, (London: Charles Griffin, 1970), p. 344.
5 Babbage, C., quoted in *Mathematics and Man*, Unit 11 of Open University Course AM289 *History of Mathematics*, p. 27.
6 This quotation is from Lady Lovelace, daughter of the poet Lord Byron. Lady Lovelace worked with Babbage and wrote extensively about his ideas; quoted in *Mathematics and Man*, op. cit., p. 34 (italics in original).
7 Lady Lovelace, *Mathematics and Man*, op. cit., p. 33 (italics in original).
8 Bronowski, J., *The Ascent of Man*, (London: BBC Publications, 1973), pp. 365–7.
9 von Hayek, F. A., *Law, Legislation and Liberty, Vol. III, The Political Order of a Free People*, (London: Routledge and Kegan Paul; Chicago: University of Chicago Press, 1979), p. 176.

SUMMARY

1 Statistics had its origins in three very practical activities—censuses; the study of games of chance like dice-throwing; and the quantitative estimation of errors of measurement—primarily in the work of Gauss.
2 Modern statistics also grew from practical problems—primarily in biology. Attempts to quantify Darwin's theory of evolution led Karl Pearson to create many of the basic tools of modern statistics, such as correlation theory and the χ^2 distribution, while work on the improvement of agriculture inspired R. A. Fisher to develop tests of statistical significance and correlation methods for experiments with many variables.
3 In the 19th century Charles Babbage devised a mechanical automatic general-purpose digital computer—his Analytical Engine—which had all the main features of a modern electronic digital computer—a memory, an arithmetic unit, a unit for controlling the arithmetic operations, and automatic control by punched cards.
4 Modern electronic computers now use integrated circuits made from silicon and are far smaller, cheaper and more reliable than the early valve machines.
5 Microcomputers will undoubtedly change many aspects of our lives but we should always remember that—in the words of Lady Lovelace, Babbage's chronicler—the computer can only 'do whatever *we know how to order it to perform*'.

6.8 The physical evolution of the universe: the modern world picture

Let us now look again at Fig. 1.1 (page 2) which shows the medieval European view of the universe as it was depicted by Dante in his *Divine Comedy* only 600 years ago—with the earth firmly at the centre, the planets and stars revolving about it and Paradise beyond.

Contrast this picture with the spiral galaxy shown in Fig. 1.2. Our sun is now thought to be a minor star in such a galaxy, which itself is just one amongst the myriads of galaxies now known to exist.

These figures show one way in which our view of the universe has changed since the 14th century. But our world picture, as it is sometimes called, has changed in another very important way. Dante's universe was unchanging. Even though the spheres were moving, for the men of the 14th century they had always been moving in just the same way, ever since the creation. Their world picture was essentially static. By contrast our modern world picture is dynamic. We now have a fairly clear idea of how the earth, the elements and the universe itself have all changed over vast aeons of time from structures which were once very different from those we observe today. Our universe and all its parts has a history, and the story of its physical evolution has been largely pieced together in the 20th century. By contrast the main ideas on biological evolution were the work of the 19th century (see Chapter 4.8).

In this chapter we will outline our modern ideas on the structure and evolution of the universe, of the elements, and of the earth itself—and give some account of the experimental and observational basis for these ideas.

We obviously can't do controlled experiments on the universe or the earth any more than we can in most areas of biological evolution. What we can do is to try to correlate all the observational evidence and to bring to its interpretation all of our modern scientific knowledge, much of which has been checked by controlled experiments.

THE STRUCTURE AND EVOLUTION OF THE UNIVERSE

The first tentative steps in creating the modern world picture were taken by Copernicus and Galileo. Remember Galileo's telescopic observations of the moon (see Fig. 3.1.3, page 35) and his crude sketches of Jupiter and its four moons which he published in 1610 in his book *The Starry Messenger*. Today we have colour photographs of the moon and planets taken from spacecraft and even, for the moon, photographs taken from the surface of the planet itself. See Fig. 6.8.1 which shows a modern photograph of the famous red spot on the surface of Jupiter and two of its four main moons—the ones seen by Galileo.

Yet, despite these great advances in detailed observation of the planets, our

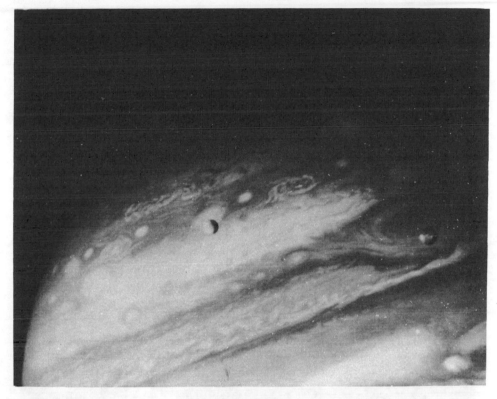

Fig. 6.8.1   Close-up of Jupiter showing the red spot and two of its four main moons first observed by Galileo in 1610.

modern picture of the solar system is not, in any major way, different from that developed by Copernicus, Galileo, Kepler and Newton.

Where our ideas have changed dramatically since the 17th century is in our view of the universe beyond the solar system. Stars emit the whole range of electromagnetic radiations (see Fig. 6.2.2, page 254), but the earth's atmosphere absorbs most of them. It so happens that there are only two windows in the atmosphere through which radiation can reach the earth's surface so that we can obtain information about the stars—one window is in the optical or visible region and the other is in the radio region of the electromagnetic spectrum. The optical window has always been open, but it was only after the development of radar in World War 2 that the radio window opened and radio-astronomy became possible.

What can we learn by looking through the optical window? For 200 years or so after the time of Galileo and Newton, optical astronomy was primarily involved in making more accurate and detailed maps of the sky. As telescopes increased in power, many more stars were discovered and classified, and many new objects were discovered—new galaxies like the spiral galaxy shown in Fig. 1.2 (page 3) and nebulae like the Andromeda nebula shown in Fig. 6.8.2.

Fig. 6.8.2   Andromeda nebula.

Yet, though the observations piled up, for a long time the stars were very little understood. The breakthrough in understanding came from the study of spectra—just as it had in chemistry and in atomic physics (see Chapter 6.6).

In astronomy the most useful spectra are the absorption spectra first observed by Fraunhofer in the 19th century. Any gas or vapour will absorb certain characteristic frequencies from white light and absorptions show up as dark lines in the spectrum, which are known as Fraunhofer lines. For example, the sun's spectrum has many dark Fraunhofer lines corresponding to the absorption spectra of elements in its outer atmosphere such as hydrogen, helium and calcium. Indeed, helium, as its name implies, was first discovered in the sun's atmosphere from observations of these Fraunhofer lines, some of which can be seen in the part of the sun's spectrum shown in Fig. 6.8.3 (fifth spectrum from the bottom). These

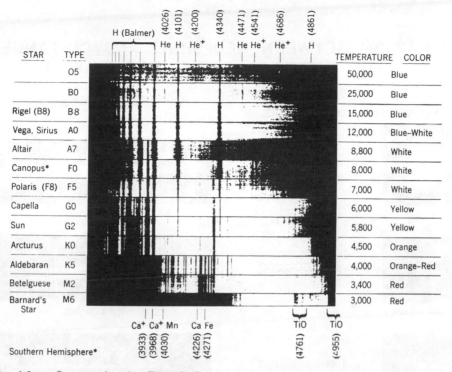

Fig. 6.8.3 Spectra showing Fraunhofer absorption lines from the sun and other stars of differing types, brightnesses and surface temperatures; all the stars are from our galaxy.

dark lines in the sun's absorption spectrum show that the same elements exist there as on earth.

Now look at the other spectra in Fig. 6.8.3 which are taken from a large number of stars of different brightnesses and types in our galaxy. Again these spectra all contain Fraunhofer lines and, most important, many of the lines, although they differ in intensity, appear in exactly the same place in all the spectra. Perhaps the most extraordinary thing that these spectra tell us is that the stars are made of exactly the same elements as our sun and our earth and that we can make sense of all these observations using exactly the same laws of physics as have been established here on earth. That would have amazed Aristotle and Dante but not, I think, Galileo or Newton.

But there is even more we can learn from these spectra—and in particular from the spectra of the light from other galaxies rather than from stars in our own galaxy. All the spectra in Fig. 6.8.3 are from stars in our own galaxy, and the spectral lines are all in the same places. However, when we look at light spectra from other galaxies we don't find the Fraunhofer lines in exactly the same places as in our sun's spectrum. At first sight this might seem to indicate that these galaxies were made of different elements. But on closer inspection an even more

RELATION BETWEEN RED-SHIFT AND DISTANCE FOR EXTRAGALACTIC NEBULAE

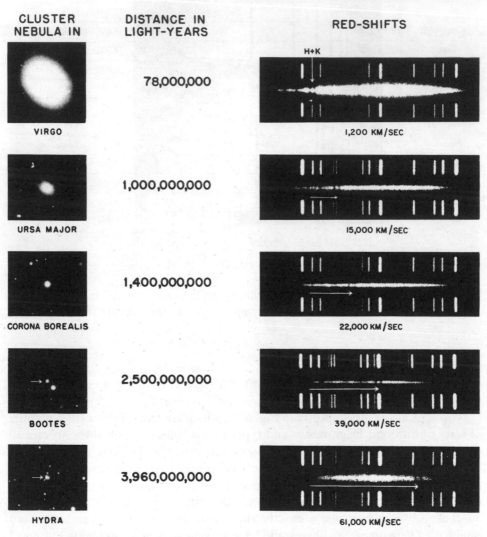

CLUSTER NEBULA IN	DISTANCE IN LIGHT-YEARS	RED-SHIFTS
VIRGO	78,000,000	1,200 KM/SEC
URSA MAJOR	1,000,000,000	15,000 KM/SEC
CORONA BOREALIS	1,400,000,000	22,000 KM/SEC
BOOTES	2,500,000,000	39,000 KM/SEC
HYDRA	3,960,000,000	61,000 KM/SEC

Red-shifts are expressed as velocities, $c\,d\lambda/\lambda$. Arrows indicate shift for calcium lines H and K. One light-year equals about 9.5 trillion kilometers, or 9.5×10^{12} kilometers.

Distances are based on an expansion rate of 50 km/sec per million parsecs.

Fig. 6.8.4 Spectra from a number of galaxies showing differing red-shifts for two Fraunhofer lines in the calcium spectrum.

remarkable fact emerges. The dark lines in the spectrum of a galaxy *do* correspond to those in the sun's spectrum—except that they are all shifted by a definite amount towards the lower frequency or red end of the spectrum. The red-shift is the same for all the lines in the spectrum of any one galaxy but is different for different galaxies. Fig. 6.8.4 shows the different red-shifts observed for two Fraunhofer lines in the calcium spectrum for a number of different galaxies.

Red shifts are almost certainly due to the Doppler effect—the shift in frequency observed in either light or sound waves when there is relative motion between the source and the observer. A common example of the Doppler effect is the change in frequency of the note emitted by a racing car exhaust when it is approaching or receding. So it is very likely that the spectra of galaxies are red-shifted *because* these galaxies are moving away from us, and from the size of the red-shift we can calculate the speed of recession of the galaxy. These red-shifts were first extensively studied in the 1920s by Edwin Hubble (1889–1953) working with the 100-inch telescope at Mount Wilson in California. In 1929 Hubble put forward evidence which suggested that the red-shift of a galaxy—and thus its recession velocity—was directly proportional to its distance from us. So the faster a galaxy is receding, the further away it is. If Hubble is right, we live in an expanding universe, and Hubble also gave an estimate for what has since been called Hubble's constant H—the constant of proportionality in the equation $V = HD$ relating recession velocity V and distance D.

Hubble's original paper on red-shifts was based on very little observational evidence; he had made measurements on only 18 galaxies. But his ideas stimulated an enormous amount of observation, and today Hubble's velocity-distance law is very well supported by experiment.

It is important to remember that our cosmological distance scale is derived indirectly by a fairly long and circumstantial chain of reasoning, and it may need to be revised as more evidence is acquired. With that reservation in mind, let us now quote some figures which give some idea of the scale of the universe which these astronomical measurements have revealed. Hubble's earliest measurements were on galaxies with recession velocities of about 1000 kilometres per second which correspond to distances of about 50 million light years (1 light year is approximately 10 million million kilometres, and stars in our galaxy range from 4 to about 50,000 light years away from us). More recently measurements have been made out to much greater distances—up to about 5000 million light years.

Now let us look at these results from another angle. If all these galaxies are moving apart at speeds proportional to their distances from us, then they could all have originated a long time ago at the same place in some kind of massive explosion. This is the so-called Big Bang theory of the universe. By measuring Hubble's constant and taking its reciprocal, we can estimate how long ago the big bang occurred. When we do this, the answer we get is about 20,000 million years (Ma), which is the modern estimate of the age of the universe.

There is yet another way of looking at observations of light from distant stars. When today we look at stars or galaxies which are, say, 300 million light years away, *we are looking at the universe as it was 300 million years ago*. We are looking

Fig. 6.8.5 Part of the Cambridge 1-mile radio telescope; a number of steerable dishes can be moved along a stretch of disused railway line, thus giving an effective telescope diameter of one mile.

back in time. If someone in such a galaxy could see the earth, what he would see, now, would not be us but the earth in the age of the dinosaurs (about 300 Ma ago).

Unfortunately it is difficult to make good measurements on the light from very distant galaxies. If they are more than about 3000 million light years away, the light is far too faint when it reaches us, having been absorbed and scattered by all the intervening dust and debris. So our optical window mists up when we try to look back towards the origin of the universe, but fortunately the other window through our atmosphere—the radio window—remains open.

Radio-astronomy is a very young science. The first radio-star, the sun, was only observed in 1931—and modern radio-telescopes like the Cambridge 1-mile telescope shown in Fig. 6.8.5 are the direct descendants of the radar aerials built during World War 2. Radio-telescopes give us an alternative view of many of the familiar visible objects in the sky. For example, Fig. 6.8.2 shows an ordinary light photograph of the Andromeda nebula, while Fig. 6.8.6 shows a radio map of the same region of the sky; the overall shape is the same but the detail is different.

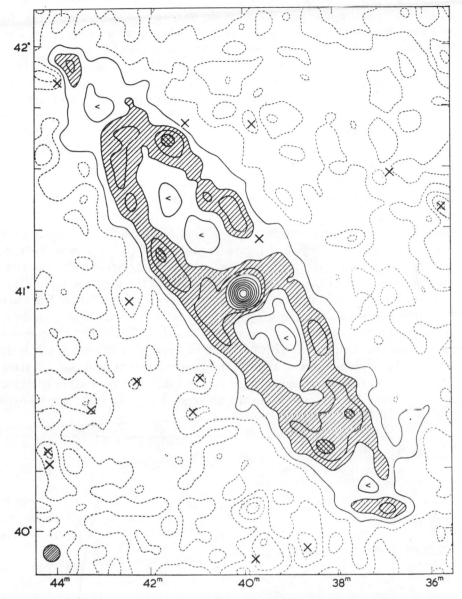

Fig. 6.8.6 Radio map of the Andromeda nebula measured at a wavelength of 73 cm; compare with Fig. 6.8.2, which is an optical photograph of Andromeda.

Radio-telescopes have also detected the radio-signals from the remnants of the supernova seen by Tycho Brahe which in 1572 was a new star bright enough to be seen by day but which is no longer visible. However, it is the fact that radio-waves are much less easily scattered than light that makes radio-telescopes so useful for looking at very distant objects. This means that with radio-telescopes we can look back much further in time towards the original big bang than is possible with optical telescopes. When we do this we find exactly what the Big Bang theory predicts—galaxies were closer together then than they are now and the further back we look, the closer together they get.

That is one interesting piece of evidence provided by radio-astronomy in favour of the Big Bang theory. The next piece of evidence from radio-astronomy was not initially based on observations but grew directly out of modern cosmological theory, which itself is derived from Einstein's General Theory of Relativity. When Einstein's equations are applied to the behaviour of very large amounts of matter, like the universe, it is found that solutions do exist which make the universe isotropic (the same in all directions), homogeneous (the same everywhere), and either expanding or contracting. So Einstein's theory is compatible with a big bang and it forms the basis of all modern cosmological theories.

One of these theories in the 1960s predicted that the initial flash of the big bang should still be detectable—not in the visible region of the spectrum but red-shifted all the way into the microwave or radio region of the electromagnetic spectrum. The theory even predicted what this echo of the big bang should look like; it should be isotropic and its strength should vary with wavelength like black body radiation from a source at a temperature of about three degrees Kelvin. Radio-astronomers looked for this radiation and soon found it. Fig. 6.8.7 shows some of the early results; the radiation was isotropic and fitted a black body distribution for a temperature of 2.7 K. Later results have amply confirmed these early findings. So radio-astronomers have detected this echo or whisper from the original big bang—an echo which fills the universe. Yet it is a very faint echo, as

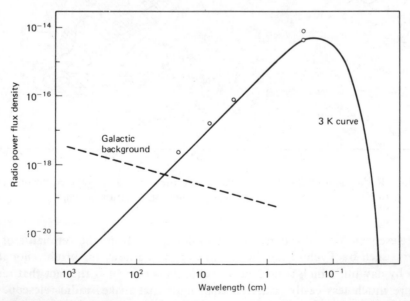

Fig. 6.8.7 Observations by radio-astronomy of the cosmic microwave spectrum—the echo of the original Big Bang.

are all the signals received in radio-telescopes. In fact, if we collected together all the radio energy which has ever been received by all the radio-telescopes in the world, it would only be enough to raise the temperature of one gram of water by about one-tenth of a degree. This should give you some idea of the enormous experimental skill which is involved in making many of the measurements which underpin our present view of the universe.

So far we have been discussing the large-scale astronomical evidence which has led us to this view of an evolving universe. Now let us look at some other kinds of evidence which have made the overall picture more coherent. This evidence is derived from laboratory work here on earth, particularly in nuclear and high energy physics. The conditions in the early stages of the big bang, and in the interiors of stars today, are in many ways comparable to those artificially created in man-made particle accelerators. So in the last few decades the physics of the very large—cosmological structures are measured in light years and one light year is about 10^{16} metres—and of the very small—a typical nucleus is about 10^{-15} metres in diameter—have come together in ways that would have seemed impossible only 50 years ago.

We don't know how the tiny, dense and very hot universe came into existence about 20,000 Ma ago. But we can give a very plausible description, based on the laws of nuclear and particle physics, of how it evolved from being tiny, dense and hot to being huge, almost empty and very cold.

On this picture, the universe has evolved from a soup of matter and radiation at a temperature of around 10^{12} K, 10^{-5} seconds after the big bang, to a plasma of electrons and photons at 10^{11} K after 0.1 seconds. After 14 seconds the temperature drops to about 3×10^9 K and the plasma is mainly electrons, protons and neutrons, while after 4 minutes all the neutrons have fused with protons to form helium nuclei. 700,000 years later the universe has cooled to about 5000 K and the first atoms are formed. Another 1000 Ma and gravity has become the dominating force leading to the concentration of matter into dense clouds and then into even denser stars.

It is at this stage that there begins another kind of evolution—the evolution of the elements in the interiors of the stars.

THE EVOLUTION OF THE ELEMENTS

The modern version of Mendeleev's Periodic Table contains 90 naturally occurring elements; technetium and prometheum do not occur in nature. The nuclei of these elements exist in various isotopic forms so that about 270 stable and more than 50 radioactive nuclei occur naturally on the earth. These 300 or more nuclei exist in varying abundances on earth and in rather different amounts in other parts of the solar system. The task for any theory of the evolution of the elements is to show how all these nuclei could be synthesised with about the same abundances as occur naturally. The raw materials are protons, helium nuclei and the variety of conditions of high temperature and pressure which are known to exist in the interiors of the different types of star. The other key factor in the

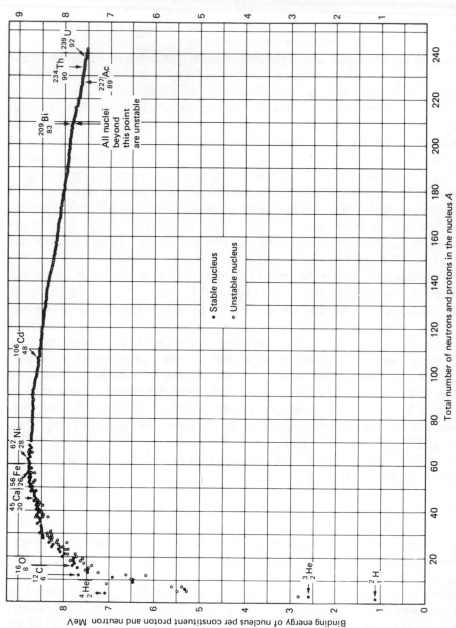

Fig. 6.8.8 Binding energy curve for stable and unstable nuclei; the energy release per neutron or proton vs. atomic number. *Redrawn from Open University Science Foundation Course S101, Units 29–30, Quantum Theory, copyright © The Open University Press.*

theoretical calculations is the vast amount of data concerning nuclear reactions which has been established since the 1920s. Fig. 6.8.8 shows perhaps the most important of these data—the amounts of energy released when complex nuclei are built up from protons and neutrons; most energy is released when elements near the middle of the periodic table are formed; fission of heavier elements (as in atomic bombs or fission reactors) or fusion of lighter elements (as in hydrogen bombs or the stars) leads to the emission of large amounts of energy.

We obviously can't go into the detailed mechanisms by which all these elements can be synthesised, and the estimates of the rates at which the reactions proceed. However, we can state the conclusion. A perfectly plausible set of processes, based on known laws of nuclear physics and known data, can be postulated, which explains both the formation of all known naturally occurring nuclei *and* their relative abundances. One of these processes is particularly interesting. It now seems very likely that all the carbon atoms in every living thing—including us— were produced in the interiors of stars, over very long periods of time, by collisions in which two helium nuclei collided to form a beryllium nucleus, which then in turn collided with a third helium nucleus to form a carbon nucleus.

Thus the evolution of the elements as well as the evolution of the universe can now be explained in detail by a combination of direct observations and known physical laws established here on earth.

Now let us turn to the third example of physical evolution—the more parochial one of the history or evolution of the earth.

THE EVOLUTION OF THE EARTH

In the late 18th and early 19th century, ideas about the earth and how it was formed changed drastically (see Chapter 4.6). The traditional view was that the earth had been formed relatively recently as a result of one or more sudden catastrophic events, possibly of divine origin; an age of about 6000 years was often quoted. This view gradually gave way to a very different one, called uniformitarianism, in which the earth was seen as having changed gradually by means of the same physical processes which we can observe going on today. For uniformitarianism to be plausible, the age of the earth needed to be very much longer than 6000 years. So estimates of the earth's age grew—to around 100,000 years by the French naturalist Buffon (1707–88) in the middle of the 18th century; to a million years by the geologist Werner (1750–1817) in the late 18th century; and to about 300 million years by Charles Darwin (1809–82). Charles Lyell (1797–1875) tried to apply uniformitarian methods directly by measuring modern deposition rates, in both sedimentary and volcanic regions, and then working out how long it must have taken for strata of known thickness to be deposited (see Chapter 4.6). The results were not very accurate but they gave figures for the age of the earth which ranged from a few million to around a thousand million years. So Lyell's rough estimates were compatible with the hundreds of millions of years Darwin needed for his ideas on the evolution of life to be plausible.

Then in the later part of the 19th century a new line of physical reasoning was applied to the problem of the age of the earth. The man who did this was Lord Kelvin (1824–1907), one of the founders of modern thermodynamics; our absolute temperature scale—degrees Kelvin—is named after him.

Kelvin was one of the leading authorities on heat, and he used his knowledge to try to calculate the age of the earth. By working out the rate at which the earth gains heat from the sun's radiation and loses heat to outer space, Kelvin showed that the earth was cooling—and cooling fairly rapidly compared with the long times required by Darwin and the uniformitarian geologists. Whatever favourable assumptions Kelvin made about the temperature of the earth when it was formed, his calculations showed that, if the earth was as old as Lyell and Darwin estimated it to be, it would long ago have cooled to temperatures far below those which would support life. Kelvin's estimates of the age of the earth were of the order of tens of millions of years only.

Kelvin was also worried about the sources of the sun's energy; he doubted whether they were sufficient to maintain the sun's enormous output of heat and light for hundreds of millions of years.

This conflict between the physical evidence and that from biology and geology was not resolved until the 20th century. What Kelvin had left out of his calculations about the earth was the enormous amount of heat produced inside the earth by the disintegration of naturally occurring radioactive elements. Kelvin couldn't take this factor into account because he didn't know about it. Radioactivity was not discovered until 1896 and the quantity of radioactive material in the earth could not be estimated until much later. However, when this was done, the discrepancy was cleared up. The earth could be old enough to allow for the evolution of life and warm enough to sustain it. Later Kelvin's second problem was also cleared up when it was established that the primary source of the sun's energy came from nuclear reactions in the sun's interior (see Chapter 6.2). What the controversy did finally show was that the existence of life on earth is only possible because the sun is fuelled by nuclear energy and because enormous quantities of radioactive minerals lie buried deep within the earth. Without nuclear energy and this radioactivity, we wouldn't be here.

It is another extraordinary fact that it is by using this natural radioactivity that nearly all our modern information about the age of the earth and of rocks has been obtained. Radioactive isotopes have the property that they go on emitting their radiation at a definite rate, no matter what we do to them. This rate is different for different isotopes, but for each isotope it is an unvarying quantity—which is usually expressed as the half-life of the isotope. The half-lives of many of these isotopes are very short—sometimes only fractions of a second or a few minutes, hours or days. But others have half-lives in the ranges of thousands, millions or even hundreds of millions of years. As early as 1913, it was suggested that these long-lived isotopes might be used to date rocks, but it was not until the 1950s and 1960s that the technique became reliable enough to be widely applied; it had to await considerable developments in both chemical separation methods and in the electronic detection of very weak sources of radioactivity. Since then radioactive

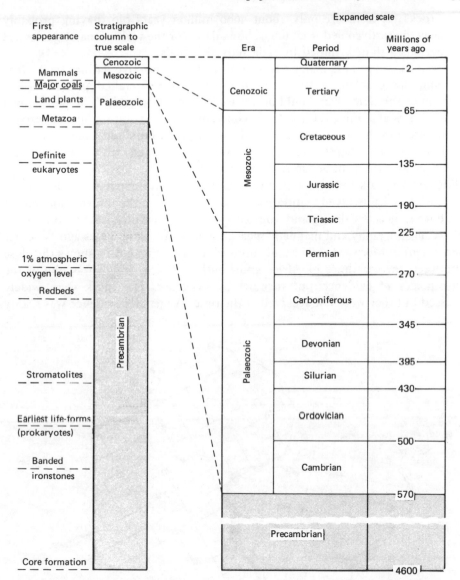

Fig. 6.8.9 Stratigraphic column with dates. *Redrawn from Open University Science Foundation Course S101, Unit 28, Earth History, copyright © The Open University Press.*

dating has been widely applied and it is the source of our modern estimate for the age of the earth as well as for the ages of many of the rocks which have been formed subsequently as the earth has evolved. And radioactive dating, using radioactive carbon, is also the source of most of the accurate dates we have from man's pre-history—the period from about 50,000 years ago up to the time of the earliest written records, perhaps 5000 years ago.

The best modern estimate of the age of the earth is about 4600 million years— and this is also the age of the oldest rocks on the moon and in meteorites. The oldest rocks on earth are only about 3600 million years old, having probably formed as the earth cooled. Fig. 6.8.9 shows dates for the various major geological time-periods—all determined by radioactive methods.

As well as fixing these major dates so precisely, radioactive dating, by its very precision, has enabled us to build up a much more detailed picture of the evolution of the earth than would otherwise have been possible. The data shown in Fig. 6.8.9 and Fig. 1.4 (page 7) concerning the evolution of the earth's atmosphere and the rocks, together with the information about the evolution of life described in Chapter 1, have all been established with much greater confidence as a result of accurate radioactive dating.

The last 20 years has also seen a dramatic change in our view of the way the earth's surface has evolved—and this too has been due to the information gained from both radioactive dating and from many other precise physical measurements such as seismography and magnetic measurements. About 50 years ago Wegener (1880–1930) put forward the idea of continental drift. Fig. 6.8.10 shows the close fit between the coastlines of Africa and South America, which was one of the major pieces of evidence put forward by Wegener. His ideas were widely discussed but they were not accepted at the time. Wegener's evidence was scanty

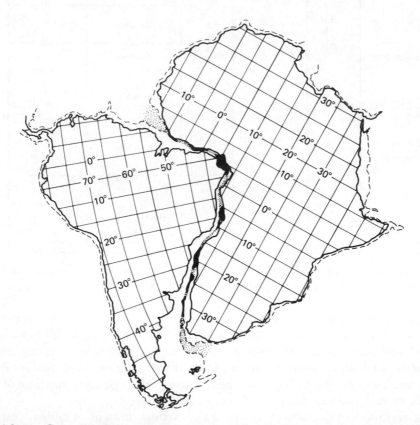

Fig. 6.8.10   Computer fit of the 1000 metre contour between South America and Africa.

and he was unable to suggest any plausible mechanism to explain how the continents could move.

Then, in the 1960s, an enormous amount of geophysical and oceanographic observation was carried out—partly to prospect for minerals, partly to advance pure research as in the International Geophysical Year, and partly because of the arms race and the spread of nuclear submarines. The result was that the earth's surface—all of it, including the sea-bed—was mapped in a much more coherent way than ever before. Fig. 6.8.11 is a diagram of the ocean floor in the Atlantic which shows some of the results of all that work. The discovery of the major features of the ocean floor—mountains, canyons, deep trenches and the prominent mid-oceanic ridges—was dramatic enough. But even more dramatic was the discovery of the very different ages of the different parts of the earth's surface. The youngest regions are the ocean floors, most of which were less than 200 million years old. The major mountain belts—the Andes, the Rockies, the Alps and the Himalayas—are also very young, ranging from less than 10 million to around 200 million years old. The oldest rocks, ranging from 1000 million to nearly 4000 million years in age, are found in regions like the great plains of North America, Europe, Siberia and low-lying regions like the Amazon basin and much of Africa. But it was the detailed dating of the ocean floor that proved to be of most interest. It was found that the mid-oceanic ridges (see Fig. 6.8.11) were the youngest, less than 5 million years old, and that, moving away from these ridges on either side, the ocean floor got progressively older. It was just as if new oceanic crust was all the time being formed at the ridges and then moving outwards. This picture was amply confirmed by measurement of the magnetism of the rocks of the ocean floor. It is known from magnetic measurements on continental rocks that the earth's magnetic field has reversed in direction at least 20 times over the last five million years. Radioactive dating has pinpointed the exact times when the field has reversed—the north magnetic pole became a south pole 690,000 years ago and then a north pole again 870,000 years ago, and so on. When measurements were done on the magnetism of the ocean floors, it was found that, moving outwards from the ridges, successive strips were magnetised in opposite directions and that the ages of the rocks corresponded to the known times of field reversal. All the time, molten rock from the earth's interior is welling up at the ridges and, as it solidifies, it becomes magnetised along the direction of the prevailing magnetic field. Then the new sea floor moves outwards as new molten rock emerges at the ridges. And this spreading of the sea floor has, in effect, been recorded for us on an enormous terrestrial magnetic tape recorder. We can even measure the rate of sea-floor spreading which in the South Atlantic turns out to be about two centimetres per year on each ridge flank. So the width of the Atlantic is increasing by about four centimetres per year (or four metres per century).

As a result of all this work a whole new dynamic picture of the evolution of the earth's surface began to emerge—in which the land-masses are carried about by the movement of enormous plates floating on the fluid core of the earth. As these enormous plates collide with one another we get earthquakes, volcanic activity and even new mountain ranges being formed. For example, it is the collision of

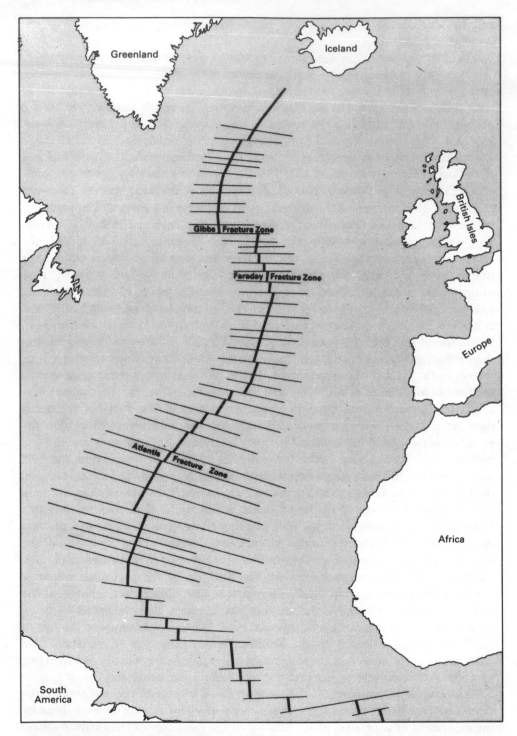

Fig. 6.8.11 Diagram showing the ocean floor in the North Atlantic: the bold line shows the mid-oceanic ridge and the lines across it represent transverse faults.

Fig. 6.8.12 Positions of the continents about 200 million years ago; they are fitted together along their continental shelves rather than their coastlines. *Redrawn from Open University Science Foundation Course S101, Units 6–7, Plate Tectonics, copyright © The Open University Press.*

the African plate with Europe which has created the Alps and made the Mediterranean such a region of volcanoes and earthquakes. And it is the collision of the Indian and Eurasian plates which gave rise to the Himalayas. Fig. 6.8.12 shows how the present continents were probably situated around 200 million years ago in the age of the dinosaurs. Before this the earth's land masses were probably in totally different locations. This whole new subject is called plate tectonics and it visualises the earth's surface as being the arena of a gigantic game of terrestrial billiards in which plates continually collide and recollide, forming and reforming the surface of the earth. And all this movement was going on at the same time as the evolution of the earth's atmosphere and of life (see Fig. 6.8.9 and Fig. 1.4, page 7).

In this chapter we have reviewed the modern world picture which has been established by work in many branches of science. Much of it is fairly recent and it is a rather more coherent picture than it would have been even 20 years ago. The universe it describes is not static like Dante's universe—it is dynamic and continually changing. Perhaps the most extraordinary thing about it is just how much we have been able to deduce about the distant past. And remember the key to most of what is described in this chapter is found in an enormous number of precise and difficult physical measurements interpreted in the light of physical laws established here on earth. Lyell said: 'The present is the key to the past'. This is still true, but now 'the present' includes not only direct observations of physical processes which are going on today but also the vast body of interlocking theoretical and experimental knowledge which has been built up since the 17th-century scientific revolution. It is as a result of all this that we have our modern world picture.

SUMMARY

1   By contrast with the static and earth-centred medieval view of the universe, our modern world picture is dynamic; it describes how the earth, the elements and the universe itself have all evolved from structures which were once very different from those we observe today.

2   Evidence from optical and radio-astronomy shows that the sun and all the stars are made of the same elements as the earth; that the universe is expanding; and that all the matter in the stars and galaxies was once, about 20,000 million years ago, compressed into a very much smaller space that it is today.

3   The experimentally established laws of nuclear physics can give a very plausible account of how the universe evolved from being tiny, dense and hot to being huge, almost empty and very cold; and of how the elements evolved in the interiors of the stars.

4   In the 19th century Lord Kelvin argued that, if the earth was hundreds of millions of years old, as Darwin estimated, it would long ago have become too cold to support life; we now know that radioactive minerals produce enough heat for the earth to be old enough for the evolution of life and warm enough to sustain life today.

5   Radioactive dating, using natural radioactivity, has shown that the earth is about 4600 million years old and, in conjunction with other precise physical measurements, it has established an accurate chronology for the intertwined evolution of the earth's atmosphere, of life on earth and of the earth's surface rocks.

The philosophy of science, political systems and the development of technology in the 20th century

The first chapter in this section will outline some key ideas about how science and technology develop and how their development is related to political and social systems of different kinds.

Subsequent chapters will then describe how science and technology have developed in this century in a number of very different societies—National Socialist Germany, the liberal capitalist countries of Europe and North America, the Soviet Union, Republican and Communist China and Japan since the Meiji restoration.

These chapters—together with the discussion in earlier chapters about the growth of science and technology in Europe since the 17th century—will provide some evidence against which we can test the general ideas set out in Chapter 7.1. For it is only by comparing different societies that we can see whether the benefits and the problems associated with science and technology are universal, or specific to a particular kind of society, or more pronounced in one kind of society than in another.

7.1 The philosophy of science, the scientific community and political systems

So far in this book we have looked at the development of scientific and technological knowledge without explicitly thinking about *how* that knowledge has grown. We have primarily been *describing* its growth rather than *analysing* the process. At the same time we have also mentioned some of the institutions of the scientific community—journals, scientific societies, conferences—but again our account of these institutions has been mainly descriptive rather than analytical. This has been deliberate for it is important to have a good general knowledge of a subject before attempting to analyse it. Without such knowledge, analysis is unlikely to be reliable and may even ignore some of the most important points.

But by now we should have a sufficiently good knowledge of science and its development to be able to give some tentative answers to questions about how scientific knowledge is gained and how scientific institutions function—that is, to questions asked by the *philosophy of science* and the *sociology of science*. I suspect that we have a more substantial basis for answering these questions than would a philosopher or sociologist who lacked a sound appreciation of science and its workings. At the very least, we do have some appreciation of what science is like from the inside.

The philosophy of science—or, for that matter, much of philosophy proper— centres on two questions. What is the world like? (ontology) and How do we know what the world is like? (epistemology).

It is, of course, impossible to answer these questions on their own. Before we can discuss what the world is like, we need to *know* something about it. And before we can ask *how* we know what it is like, we again need to be fairly sure we do know something about it.

In other words it is difficult—or even impossible—to discuss these questions in an abstract way. It is much easier to begin the discussion if we start from a well-established body of knowledge—like natural science—and then ask how it was established. This has been one of the major approaches adopted by philosophers of science in the 20th century.

THE LOGICAL POSITIVISTS AND KARL POPPER

Early in the century the philosophy of science was dominated by a group called the Logical Positivists. These philosophers thought that the only valid form of knowledge was the scientific, together with some extensions of the scientific which used similar methods. They tried to devise precise logical criteria of 'verifiability' for scientific knowledge which primarily rested on direct observations. As three of them put it, 'The representatives of the scientific world-conception resolutely stand on the ground of simple human experience. They confidently approach the task of removing the metaphysical and theological debris of millenia.'[1] But the verification criteria proposed by the logical positivists sometimes excluded too much—including most of the theoretical side of science; and sometimes too little—re-admitting much that was thought to be metaphysical or unscientific nonsense.

So logical positivism was gradually superseded as the prevailing philosophy of science by the ideas of Karl Popper (1902–), some of which we mentioned when discussing Einstein's work (see Chapter 6.3). Popper's ideas differ from those of the logical positivists in two major ways. First he does not regard the scientific as the only valid form of knowledge. Consequently one of his main concerns has been to establish what he calls a *demarcation criterion* which would distinguish between scientific and other forms of knowledge.

Popper finds his demarcation criterion by stressing the importance in science of *falsification* and this is his second main difference from the logical positivists with their emphasis on verification. The key process in Popper's own words:

... is the testing of the theory by way of empirical applications of the conclusions which can be derived from it ... certain singular statements—which we may call 'predictions'—are deduced from the theory. ... Next we seek a decision as regards these ... derived statements by comparing them with the results of practical applications and experiments. If this decision is positive, that is, if the singular conclusions turn out to be acceptable, or *verified*, then the theory has, for the time being, passed its test: we have found no reason to discard it. But if the decision is negative, or in other words, if the conclusions have been *falsified*, then their falsification also falsified the theory from which they were logically deduced.

It should be noticed that a positive decision can only temporarily support the theory, for subsequent negative decisions may always overthrow it.[2]

This is the classic statement of what is often called the hypothetico–deductive model of science and it is also the source of Popper's demarcation criterion. Again in his own words:

... I shall certainly admit a system as empirical or scientific only if is capable of being *tested* by experience. These considerations suggest that not the *verifiability* but the *falsifiability* of a system is to be taken as a criterion of demarcation. ... I shall not require of a scientific system that it shall be capable of being singled out, once and for all, in a positive sense, but I shall require that its logical form shall be such that it can be singled out, by means of empirical tests, in a negative sense: *it must be possible for an empirical scientific system to be refuted by experience.*[3]

In other words, for a theory to be scientific it must be capable of disproof; and experiments can disprove a theory but they can never establish it with certainty.

Popper's view of science is thus much more tentative than the earlier confident statements of the logical positivists. This is illustrated by one of Popper's favourite quotations, from the ancient Greek philosopher Xenophanes ($\sim$570–480 BC):

The gods did not reveal, from the beginning,
All things to us, but in the course of time
Through seeking we may learn and know things better.
But as for certain truth, no man has known it,
Nor shall he know it, neither of the gods
Nor yet of all the things of which I speak.
For even if by chance he were to utter
The final truth, he would himself not know it:
For all is but a woven web of guesses.[4]

There is much of great value in Popper's view of science and his work is certainly an improvement on that of the logical positivists. But it has at least two major weaknesses.

First it is most applicable to science viewed on the small scale—a particular new theory or a possible new application of an old one—these are the areas where Popper's ideas fit best with the actual practice of science. It is much less clear how we can apply them to much larger bodies of scientific knowledge.

Secondly, Popper's work in the philosophy of science overemphasises the logical structure of science and plays down the importance of studying the social institutions of the scientific community and the interactions between science, the scientific community and the rest of society: that is, he overemphasises the philosophy of science, as traditionally understood, at the expense of the sociology of science.

Recent work in the history, sociology and philosophy of science has indicated that the nature of scientific activity is more complex than was supposed either by Popper or the logical positivists.

HISTORY AND THE PHILOSOPHY OF SCIENCE

In his book *The Structure of Scientific Revolutions*,[5] the historian of science, Thomas Kuhn, put forward a view about how science develops which has been extremely influential. Kuhn suggested that the history of science shows that science develops through a series of stages. Initially in any field of study there are a number of competing general ideas or theoretical frameworks. Then one of these conceptions emerges as superior to all the others in its ability to explain a wide range of phenomena and so becomes what Kuhn calls a durable 'paradigm' for the field. A period of 'normal science' follows in which scientists solve puzzles within the framework of ideas of the 'paradigm'. Gradually more and more anomalies emerge which the 'paradigm' cannot explain; it thus becomes fatally undermined and once again a number of competing general ideas emerge. Eventually one of these different and competing conceptions becomes a new 'paradigm' for the field and a new period of 'normal science' begins. What Kuhn calls a 'scientific revolution' has taken place. The new period of 'normal science' goes on for some time until another 'scientific revolution' occurs. And so the cycle continues.

Kuhn's ideas about how science develops have been widely read and accepted in recent years. So it is worth emphasising that his work has a number of significant weaknesses. First Kuhn nowhere clearly defines what he means by a 'paradigm'—indeed one writer has estimated he uses the word in more than 20 different ways.[6] So since Kuhn's main idea is so vague, it is difficult to see whether the historical examples he uses actually support his views or not. Secondly, Kuhn does not always distinguish clearly between the whole scientific community and the small number of workers, possibly as few as half-a-dozen, in a specialised field. Nor does he analyse how the scientific community has changed over the long period since what we and many others have called the scientific revolution of the 17th century. Thirdly, Kuhn does not emphasise the increasingly quantitative, interlinked and cumulative nature of much of modern science, particularly in subjects like astronomy, physics and chemistry from which he draws his main examples. For example, Kuhn says that Newton's and Einstein's dynamical theories are 'fundamentally incompatible' and he rejects the view which regards Einstein's work as a generalisation of Newton's which is valid for a wider range of experimental conditions (see Chapter 6.2). Fourthly, Kuhn plays down some of the major predictive successes of science, as for example, when he dismisses Mendeleev's prediction of the existence and properties of several new elements (see Chapter 6.6) as being 'discoveries predicted by theory in advance' which 'result in no *new sort* of fact'.[7] Kuhn makes no mention of the way in which Mendeleev's work has, in this century, played such a central role in the unification of chemistry and physics (see Chapter 6.6).

By emphasising changes in theoretical concepts rather than the increasing degree of agreement between theory and experiment, Kuhn underestimates the substantial element of continuity between older and new bodies of knowledge. Indeed he sometimes suggests that science is essentially non-cumulative and that successive accepted bodies of knowledge have no points of contact with their predecessors or successors. He even suggests that the natural world is in some way shaped or determined by our own ideas. For example Kuhn states that after Dalton's atomic theory (see Chapter 4.5) was accepted 'chemists came to live in a world where reactions behaved quite differently from the way they had before' and 'even the percentage composition of well-known compounds was different'.[8] And more generally 'the historian of science may be tempted to exclaim that when paradigms change, the world changes with them'.[9]

These unfortunate exaggerations have deflected attention from some important and valid contributions made by Kuhn, such as the importance of studying the detailed history of how theories actually change and the role of scientific communities, both large and small. Yet even here Kuhn often overstates his case when he writes 'inevitably these remarks will suggest that the member of a mature scientific community is, like the typical character of Orwell's *1984*, the victim of history rewritten by the powers that be. Furthermore that suggestion is not altogether inappropriate.'[10]

However, the work of Kuhn and other historians of science does clearly show that there is much more to science than its logical structure. In stressing this, they are challenging not only Popper and the logical positivists but also the whole philosophical tradition stemming from Descartes (see Chapter 3.3) with its overwhelming emphasis on logical and mathematical deduction as the cornerstone of scientific method.

THEORY AND OBSERVATION

Logical analyses of the nature of science, like those of Popper and the logical positivists, have recently been undermined in another way. Such analyses have traditionally assumed that theories and observations are quite distinct and that observations can be described in a clear and unambiguous way—in terms of what was called an independent observation language. But most of the terms we use to describe observations are defined with reference to a whole network of other terms, many of which may be theoretical.

Consider, for example, a simple measurement of the speed of a car in kilometres per hour. Such a measurement involves practical definitions of units of distance—kilometres, and time—hours. It involves theoretical discussions about the nature of space and time like those made by Einstein (see Chapter 6.2). And it involves other laws of physics—both theoretical and experimental—that are involved in understanding the operation of the instruments used to make the actual measurement. Or consider some of the observations of X-ray diffraction patterns used in establishing the structure of DNA (see Chapter 6.4). These observations only have their full meaning when they are considered in relation to the theory of X-ray diffraction, the theory and practice of the instruments used in making reliable X-ray measurements and, perhaps most important, the whole range of knowledge in physics, chemistry, biochemistry and biology which was used in finding the DNA structure.

So each observation does not stand in isolation and there is no clear distinction between the languages used to describe observations and theories. It therefore becomes much more difficult to decide in any given test case if a prediction has been unambiguously falsified as Popper's ideas seem to require. Each experimental test result has to be seen in terms of a much wider network of theories and earlier experiments.

It is considerations such as these which have led some philosophers of science to emphasise that scientific knowledge—like our knowledge of the everyday commonsense world described in ordinary language—is a vast and interconnected network of concepts, regularities, laws, observations and measurements.[11] In this vast network the distinction between observation and theory is not always clear-cut.

Another fundamental problem arises because, in principle, more than one theory can sometimes fit the observed facts reasonably well. Consider again the incorporation of an enormous amount of diverse chemical data into the systematic framework provided by Mendeleev's periodic table. Mendeleev's great generalisation was put forward when even the very existence of atoms was in doubt. But it has survived intact the 20th-century redefinition of a chemical element in terms of atomic number rather than atomic weight, and has been given additional support by Bohr's early quantum theory of the atom, which itself has been incorporated in the more generally applicable wave mechanics of Schrödinger and is now a particularly successful application of that more general theory (see Chapter 6.6). The basic facts of chemistry and Mendeleev's generalisation remain but the underlying concepts and theories have changed almost beyond recognition.

Or consider the observed facts about the motion of bodies on earth and in the heavens which were first given a coherent theoretical explanation by Newton in the 17th century (see Chapter 3.4). These same observations fit just as well with Einstein's theory of special relativity which starts from very different definitions of space and time (see Chapter 6.2)—and they also fit with the equations of Einstein's theory of general relativity which in some ways is a generalisation of his theory of special relativity but which again starts from different definitions.

Once again the basic observations are the same but the underlying concepts and theories have changed almost beyond recognition. These examples illustrate that there can be, in principle, an indefinite number of theories that fit the observed facts—a phenomenon called by philosophers of science the *underdetermination* of theory by empirical data.

THE CUMULATIVE NATURE OF SCIENTIFIC KNOWLEDGE

Science is clearly cumulative in continuing to incorporate more and more data and in providing increasing practical control over our environment. Yet it is also clear that our present theories are unlikely to be the last word. They may change in the future as dramatically as they have changed in the past. But, if history is any guide, it seems likely that there will be a substantial element of continuity between older and newer theories and that scientists will continue to use such continuity as a guide in the development of new ideas—just as Bohr, Einstein and others have done so successfully in the past.

We can hope that new theories will continue to incorporate more data and to synthesise work in different branches of science in the way, for example, that Einstein's theory of general relativity is the starting point for modern cosmology—some of the results of which were described in Chapter 6.8.

But we cannot be sure that this will happen, any more than we can be *absolutely* sure of anything. Recent work by historians, philosophers and sociologists of science has clearly shown that, despite its increasing power and influence, modern science is definitely not as completely logical and certain as once it was thought to be. Of course this does not mean that most of our knowledge is doubtful or that logic has no place in science. We should be aware that 'Logic is itself an ideal to which real theories never perfectly attain, and we should beware of drawing consequences of importance to the philosophy of real science from the presupposition of this ideal'.[12] Likewise in any appraisal of science as it really is, we need to take into account the problems about separating observations unambiguously from theories; the underdetermination of theories by empirical data; the network model of science; and the associated network of individuals and institutions which make up the international scientific community.

THE PHILOSOPHY AND SOCIOLOGY OF SCIENCE AND SOCIETY

We, therefore, need to take a broad view of the scope of the philosophy of science which will incorporate many of the main points of Popper's philosophy but will also try to relate these to the history and to the sociology of science—both the internal sociology of the scientific community and the external sociology of the rest of society. Ideas which attempt to cover such an enormous area must clearly be tentative, provisional and open to revision—as Popper himself would recommend. However, it is necessary to try to establish some unifying ideas if we are not to be overwhelmed by the very size and complexity of the problems involved in assessing the interactions between science and society.

Any coherent philosophy and sociology of science and society needs to deal with four main topics:

the body of *scientific knowledge*;
the *criteria and values* involved in establishing this knowledge;
the *internal structures* of the scientific community;
the *external structures* of the rest of society.

Scientific knowledge

The starting point is the vast body of interlocking theoretical and empirical scientific knowledge which has been established since the 17th century. In Parts 1–4 of this book, we have outlined the main features of this knowledge and its development up to the end of the 19th century. Then in Part 6 we described some of the major scientific developments of the 20th century.

Scientific knowledge has clearly grown extremely rapidly since the 17th century and this growth is still continuing.

Criteria and values

So far in this book we haven't stressed the criteria which are used to assess claims to knowledge and the underlying values which are accepted as vital to the growth of scientific knowledge.

These criteria—*logical coherence* and the *use of all available relevant evidence*—and the associated values—*open criticism, pluralism, tentativeness*—are similar to those involved in all academic work. But their effectiveness is frequently enhanced by the special nature of science. The widespread use of mathematics in science often makes it easier to answer the question—Is this claim coherent and logical? Mathematics does this by forcing scientists to make their concepts more precise and by revealing more easily the logical implications of these concepts.

The evidence available to scientists from *observation* is greatly increased in many areas by the use of controlled and repeatable experiments. In some areas of science—the history of the earth, cosmology, large-scale evolution—controlled experiments are not possible. But even here the knowledge gained with the aid of experiments greatly enhances our understanding of these processes.

These two fundamental criteria—logic and evidence—also combine together in a powerful way in science—as in the testing of detailed logical predictions by observation and controlled experiment, followed by the modification of concepts, if necessary, and their further testing by experiment. This combination provides one of the most effective motors for the growth of science. It does this by making *criticism* much more effective—both self-criticism and criticism by other scientists.

The other key values—*pluralism* and *tentativeness*—are partly associated with the necessity for public criticism, and partly with a recognition of the impossibility of any claim to total knowledge of the complex real world. This is why emotional arguments or moral pressure are normally avoided by scientists in presenting their work. They recognise the enormous problems involved in establishing even a part of the truth and so they try to avoid anything which would deflect attention from the logical coherence of their claims to knowledge and from the experimental evidence which supports these claims.

Structures of the scientific community

These are the structures we discussed in Chapter 3.3—scientific societies, learned journals, and institutions of higher education. These structures base their operations on the criteria and values mentioned above and they are all in continual interaction both with each other and with each individual scientist. All the scientific work of individuals is done in the knowledge that it will have to be subjected to *open, public criticism* by the scientific community and that it will stand or fall on the criteria of logic and evidence. It is in this way that the whole scientific community can generate and use a vastly greater stock of collectively validated knowledge than any individual or centralised group could possibly do. As we have seen in earlier chapters the structures of the scientific community have grown and evolved in a spontaneous way from their rudimentary origins in the 17th century—in parallel with the growth in the body of scientific knowledge over the same period.

External structures of society

Science and the scientific community are generally recognised to be the most international of all human activities or institutions, transcending many differences of language, culture or belief. Yet it is also true that science—both pure and applied—has prospered more in some societies and cultures than in others. However, because the interactions involved are so complex, it is very difficult to make any definite statements about the connections between the development of science and the structures of different kinds of society.

So I now want to put forward some tentative hypotheses about these connections. Then, in later chapters, we will discuss some relevant evidence, drawing examples from as many different kinds of society as possible. Then we will return to these hypotheses and try to assess their validity in the light of the evidence.

These, then, are the hypotheses we will try to test:

(1) For science to prosper in any society, it is necessary for that society to recognise that the scientific community needs to be at least partially autonomous. In scientific matters, scientists must be left to govern themselves by means of the criteria, values and institutions described above.

(2) Societies whose institutions are based on values which are similar to those of the scientific community—tolerance, individual freedoms and the free flow of information—are more likely to have been associated with significant scientific activity, both historically and at present. The compatibility of values makes it likely—but not certain—that some degree of sympathetic understanding will exist between society and the scientific community. Therefore, such societies are more likely *both* to provide the partial autonomy needed for science

and to make use of science and technology in a creative way in the development of society.

(3) Societies whose basic values are incompatible with those of science are less likely—both historically and now—to have fostered the development of science, however much they may have wanted or claimed to be doing so.

(4) The future development of science and technology, and its fruitful application world-wide, may depend on the continuing existence of societies whose basic values are compatible with those of the scientific community.

REFERENCES

1   Hahn, H., Neurath, O., and Carnap, R., *The Scientific Conception of the World: the Vienna Circle*, (Dordrecht: D. Reidel, 1973), p. 19, originally published in 1929.
2   Popper, K., *The Logic of Scientific Discovery*, (London: Hutchinson, 1972), p. 33, emphases in original; originally published in Vienna in 1934 as *Logik der Forschung*.
3   Ibid., pp. 40–1, emphases in original.
4   Quoted in Magee, B., *Popper*, (London: Fontana Paperbacks, 1973), p. 28.
5   Kuhn, T. S., *The Structure of Scientific Revolutions*, (Chicago: Chicago University Press, 1962).
6   Masterman, M., 'The Nature of a Paradigm' in Lakatos, I., and Musgrave, A., *Criticism and the Growth of Knowledge*, (Cambridge: Cambridge University Press, 1970).
7   Kuhn, T. S., op. cit., p. 61, italics in original. Copyright © T. S. Kuhn, 1971.
8   Ibid., pp. 134–5.
9   Ibid., p. 111.
10 Ibid., p. 167.
11 See Hesse, M. B., *The Structure of Scientific Inference*, (London: Macmillan, 1974) for a detailed exposition of the network model of science; this model has also been used by Duhem, P. (*The Aim and Structure of Physical Theory* [Princeton: Princeton University Press, 1954]; first published in 1906) and Quine, W. V. O. (*From a Logical Point of View* [Cambridge, Mass: Harvard University Press, 1953]).
12 Hesse, M. B., *Revolution and Reconstructions in the Philosophy of Science*, (Brighton: Harvester Press, 1980), p. xiii.

SUMMARY

1   The philosophy of science centres on two questions—What is the world like? (ontology) and How do we know what the world is like? (epistemology).
2   Logical positivists thought that science was the only valid form of knowledge and tried to devise precise logical criteria of *verifiability* for scientific knowledge based on direct observations.
3   By contrast, Karl Popper tried to demarcate scientific from other forms of knowledge by emphasising the importance of *falsifiability*: a theory is only scientific if its predictions can be tested by experience and so are potentially falsifiable.

4 Recent work in the history and philosophy of science has shown the importance of studying the detailed history of how theories actually change, of the relationship between theories and observations and of the development of scientific communities, both large and small.

5 Any coherent philosophy and sociology of science and society needs to consider: the body of *scientific knowledge*; the *criteria and values* used in establishing this knowledge; the *internal structures* of the scientific community, and the *external structures* of the rest of society.

7.2 Science and National Socialism in Nazi Germany

'... Science, like every other human project, is racial and conditioned by blood' (Lenard, 1937).

Relativity 'was a Jewish fraud, which we could have suspected from the first with more racial knowledge than was then disseminated, since its originator Einstein was a Jew' (Lenard, 1942).

'Fifteen years ago, when Relativity was made high goddess of science ... Lenard pronounced boldly against this general madness and described the theory as nonsense ... it is disgraceful to Professor Planck that he has served as Einstein's lieutenant ...' (Stark, 1935).[1]

We have already seen these quotations from two German physicists, both winners of the Nobel prize, when we were discussing Einstein's life and work in Chapter 6.3. They illustrate how severely Einstein and his work were attacked before and after the National Socialists came to power in Germany in 1933. In this chapter we will consider what led up to these extraordinary statements and discuss in more detail the nature and consequences of National Socialism.

THE BACKGROUND—SCIENCE AND THE GERMAN UNIVERSITIES

In the 19th century Germany developed rapidly into a major industrial and scientific power (see Chapter 4.11). Many universities were founded and they played a great part in establishing the principles of academic freedom—for both professors and students—which are very much taken for granted in liberal societies today. But 19th-century Germany was far from being a liberal society. Unlike Britain and France, Germany had no tradition of freedom of speech or political freedom. So freedom of enquiry had to be formally incorporated in the constitution of the new universities and this was done by making independent research a necessity for professors and other staff.

Great emphasis was put on ensuring the high quality of those appointed to permanent positions. These arrangements, together with competition for outstanding professors between the established and new universities, greatly stimulated scientific research in Germany. Although this had not been the original intention, many German universities became major centres for scientific research in the 19th century. We have already mentioned the chemists, Liebig and Bunsen, and the physiologist, Müller (see Chapter 4.11). But there were many other comparably distinguished professors who became recognised world authorities and whose research often created new scientific specialities. Examples include the organic chemists Wöhler at Göttingen and Kekulé at Bonn, the physical chemist Ostwald at Leipzig and the physicists Clausius at Bonn and Helmholtz in Berlin.

Helmholtz also did valuable work in biology, as did many others. Schleiden at Jena was one of the originators of the cell theory of living things while Carl Ludwig at Leipzig was a major influence in physiology. Other world-famous names were the pathologist Virchow at Berlin, the embryologist Haeckel at Jena, and Weissman at Freiburg who contributed significantly to the post-Darwinian debates on evolution. In the social sciences there were the psychologist Wundt at Leipzig and the sociologist Weber (see Chapter 4.10) who during his career was a professor of various subjects—political economy at Freiburg, economics at Heidelberg and sociology at Vienna and Munich.[2]

Many of these men created new academic subjects and trained a whole generation of experts, many of whom, in turn, became world-famous. For example the behavioural psychologist Pavlov was a student of Wundt's and the physicist Max Planck studied under Helmholtz. The *Technische Hochschule*, which were advanced institutes of technology, also flourished and were given university status in 1899. Overall, the number of students in universities and advanced institutes of technology increased tenfold between 1840 and 1931—a much more rapid rate of growth than that of the population as a whole which doubled between 1840 and 1910 and then stayed almost constant, partly due to the loss of territory after 1918 (see Table 7.2.1).

Table 7.2.1 Growth of population and university students in Germany, 1840–1931

Year	Population	University students
1840	33 000 000	13 000
1880	46 000 000	26 000
1900	57 000 000	50 000
1913	65 000 000	80 000
1922	62 000 000	121 000
1931	64 000 000	138 000

Another significant event was the establishment of separate research institutes such as the state-funded Imperial Institute of Physics and Technology in 1887 and the Kaiser Wilhelm Gesellschaft which was founded in 1911 and financed jointly by the state and industry.

By the 1920s, the Kaiser Wilhelm Gesellschaft was running an ever-growing network of research institutes throughout the country.

The result of all these developments was that, by the early decades of this century, Germany was probably the foremost scientific nation in the world. Many of the founders of modern physics (see Chapters 6.2, 6.3 and 6.6)—Planck, Einstein, Born, Schrödinger and Heisenberg—were professors at German universities such as the universities of Berlin, Munich and Göttingen during the 1920s. Fig. 7.2.1 shows these men and many others at an international conference in 1927.

In other sciences, too, Germany was pre-eminent. One of the founders of X-ray crystallography, von Laue, was a professor at the University of Berlin. The

Fig. 7.2.1   Participants at the fifth international Solvay conference, Brussels, 1927. *Left to right, first row:* I. Langmuir, M. Planck, Mme Curie, H. A. Lorentz, A. Einstein, P. Langevin, C. E. Guye, C. T. R. Wilson, O. W. Richardson; *second row:* P. Debye, M. Knudsen, W. L. Bragg, H. A. Kramers, P. A. M. Dirac, A. H. Compton, L. de Broglie, M. Born, N. Bohr; *third row:* A. Piccard, E. Henriot, P. Ehrenfest, E. Herzen, T. De Donder, E. Schrödinger, E. Verschaffelt, W. Pauli, W. Heisenberg, R. H. Fowler, L. Brillouin.

chemist Haber, famous for his work on the fixation of atmospheric nitrogen, and the radiochemists Hahn and Meitner, who later helped to discover fission in uranium, were directing work at two of the Kaiser Wilhelm Gesellschaft's research institutes. Fundamental work was also being carried out in what was later to be called biochemistry. At the University of Berlin, Fischer was working on the chemistry of carbohydrates and the structure and synthesis of proteins while Warburg later established the details of some of the key chemical reactions involved in respiration. And in many other sciences German research workers were also making major contributions. (See Table 7.2.2 and Chapter 7.3, Table 7.3.1, page 380.)

THE GERMAN UNIVERSITIES AND NATIONAL SOCIALISM

When Adolf Hitler's National Socialists came to power in January 1933, the independence of the universities and of their scientific communities was threatened almost immediately. On 23rd March 1933, Hitler forced an Enabling Law through the German parliament which in effect allowed him to rule by decree.[3]

Within two weeks, the National Socialists used this law to establish one of their basic aims—to make anti-Semitism legal. They did this by means of a law (7th April 1933) which provided for the dismissal of any public official who was of non-Aryan descent; a non-Aryan was defined as anyone with a Jewish parent or grandparent. And, since the German universities were public institutions, all their staff were affected by this law. Einstein had already resigned, on 28th March, but now many thousands of people, both well-known academics and minor officials, were affected. Bronowski has summarised the impact of National Socialism like this:

> When Hitler arrived in 1933, the tradition of scholarship in Germany was destroyed, almost overnight. . . . Europe was no longer hospitable to the imagination—and not just the scientific imagination. A whole concept of culture was in retreat: the conception that human knowledge is personal and responsible, an unending adventure at the edge of uncertainty. Silence fell, as after the trial of Galileo. The great men went out into a threatened world. Max Born, Erwin Schrödinger, Albert Einstein, Sigmund Freud, Thomas Mann, Bertolt Brecht, Arturo Toscanini, Bruno Walter, Marc Chagall, Enrico Fermi, Leo Szilard. . . .[4]

Perhaps the most extraordinary case was that of the chemist Fritz Haber. Unlike Einstein, who had always been pacifist and internationalist, Haber had been strongly patriotic during World War 1. Haber's research institute had then geared itself completely to the war effort and had been instrumental in developing both war gases and the Haber process for fixing atmospheric nitrogen, which was vital in the manufacture of explosives and synthetic fertilisers. Haber was Jewish but, because of his war service, he himself was exempted from the new law. However, he lost so many of his staff that he resigned in protest on 30th April 1933. This is how he ended his letter of resignation:

> In a scientific capacity, my tradition requires me to take into account only the professional and personal qualifications of applicants when I choose my collaborators—without concerning myself with their racial condition. You will not expect a man in his sixty-fifth year to change a manner of thinking which has guided him for the past thirty-nine years of his life in higher education, and you will understand that pride with which he has served his German homeland his whole life long now dictates this request for retirement.[5]

Planck, who was the President of the Kaiser Wilhelm Gesellschaft, tried to intervene with Hitler on Haber's behalf, but Hitler flew into a rage and was reported to have said:

> Our national policies will not be revoked or modified, even for scientists. If the dismissal of Jewish scientists means the annihilation of contemporary German science, then we shall do without science for a few years.[6]

The effects of this purge of Jewish scientists and academics were enormous. No fewer than 20 present or future winners of the Nobel Prize left their positions in German universities or research institutes in the 1930s as a direct result of the new law. (See Table 7.2.2.)

Table 7.2.2   Nobel prizewinners forced to leave positions in National Socialist Germany,[7] 1933–45

Name	Nobel award	Year of departure	Country of birth	Institution departed
Physics				
Albert Einstein	1921	1933	Germany	Prussian Academy
James Franck	1925	1933	Germany	Göttingen
Gustav Hertz	1925	1935	Germany	TH Berlin
Erwin Schrödinger	1933	1933	Austria	Berlin
		1938		Graz
Viktor Hess	1936	1938	Austria	Graz
Otto Stern	1943	1933	Germany	Hamburg
Felix Bloch	1952	1933	Switzerland	Leipzig
Max Born	1954	1933	Germany	Göttingen
Eugene Wigner	1963	1933	Hungary	TH Berlin
Hans Bethe	1967	1933	Germany	Tübingen
Dennis Gabor	1971	1933	Hungary	Siemens Co., Berlin
Chemistry				
Fritz Haber	1918	1933	Germany	KW Institute for Physical Chemistry, Berlin
Peter Debye	1936	1940	Netherlands	KW Institute for Physics, Berlin
George de Hevesy	1943	1934	Hungary	Freiburg
Gerhard Hertzberg	1971	1935	Germany	TH Darmstadt
Medicine				
Otto Meyerhof	1922	1938	Germany	KW Institute for Medicine, Heidelberg
Otto Loewi	1936	1938	Germany	Graz
Boris Chain	1945	1933	Germany	Charité Hospital, Berlin
Hans A. Krebs	1953	1933	Germany	Thannhauser Clinic, Freiburg
Max Delbrück	1969	1937	Germany	KW Institute for Chemistry, Berlin

Many of these men went abroad, mainly to Britain and the United States. Most were Jewish, some had Jewish wives and a few went because of the treatment of their colleagues or because of the National Socialist attitude to science and scientific objectivity. Hitler himself regarded scientific objectivity as merely a slogan invented by professors to protect their interests. He summed up his own views like this:

That which is called the crisis of science is nothing more than that the gentlemen are beginning to see on their own how they have gotten onto the wrong track with their objectivity and autonomy. The simple question that precedes every scientific enterprise is: who is it who wants to know something, who is it who wants to orient himself in the world around him? It follows necessarily that there can only be the science of a particular type of humanity and of a particular age. There is very likely a Nordic science, and a National Socialist science, which are bound to be opposed to the Liberal-Jewish science, which, indeed, is no longer fulfilling its function anywhere, but is in the process of nullifying itself.[8]

This attitude was symbolised on 10th May 1933, in a public burning of books which manifested the 'un-German spirit'; many of Einstein's books were included.

Moreover, it was not only well-known academics who were purged. Nearly 2000 university professors, lecturers and other scholars lost their jobs between 1933 and 1935—that is nearly 20 per cent of the staff of German institutions of higher education. And it was not only medicine and the physical sciences which were affected. Many dismissals also took place in the faculties of the social sciences, the humanities, law and theology. Consequently, there was hardly a department or subject which was not affected. Even in places where there were no dismissals, the work of the academies was disrupted.[9]

For what was being attempted by the National Socialists was much more than the dismissal of Jews. They were attempting to create a totalitarian state in which everything, including the content of education, was controlled from the centre.

THE IDEOLOGY OF NATIONAL SOCIALISM

Although Adolf Hitler's *Mein Kampf*, partly written while he was in prison in 1923–4 and published in 1925, sets out many of the ideas and future policies of National Socialism, the Italian, Benito Mussolini, was also one of the founders of the ideology of National Socialism. Fig. 7.2.2 shows Hitler and Mussolini in 1938 at the height of their power. Mussolini was originally an extreme socialist but when he came to power in Italy in 1922 he had reshaped his ideology which he now called Fascism. This is how he described it in the *Enciclopedia Italiana* of 1936:

> ... the idea of Fascism is for the state, and for the individual in so far as he coincides with the state. ... It is against classical Liberalism ... which has reached the end of its historical function ... Liberalism negates the state in the interest of the single individual, Fascism affirms the state as the true reality of the individual ... for the Fascist, everything is in the state, and outside of the state nothing legal or spiritual can exist, or still less be of value. In this sense Fascism is totalitarian. ...[10]

The three key principles of Fascism are hierarchy, or government by appointment from the top downwards; authority or decision by command; and discipline which means 'the complete voluntary adjustment of an individual to a

Fig. 7.2.2   Hitler and Mussolini in 1938.

group, the change from an independent human mind to an intellect only capable of working in a group, and through a group. . . .'[11] Fascism does not and cannot tolerate any opposition. '. . . every single critic is a danger. . . . Every mental and physical activity must be brought under the dominating influence of the state. That is why there can be no freedom of speech, of the Press, of peaceful assembly, why science, arts, trades, professions, sports, everything must show the Fascist colouring. . . .'[12] Fig. 7.2.3 graphically illustrates the total commitment required by Fascism.

What the German National Socialist Party added to the ideology was enormous thoroughness in putting into practice its main principle of organisation—the 'cell system'. The party established 'cells' of trusted followers in every building, in the workers' organisations in every factory and in every branch of business, the arts, the sciences and the professions. After Hitler came to power, this system was

Fig. 7.2.3 The simple message of Fascism—the effigy is Mussolini.

greatly extended and very soon there was no office, no workshop, no profession, no artistic or literary occupation, no sport or leisure activity that did not require membership in some organisation set up for the sole purpose of keeping its members under the immediate influence of the National Socialist movement.[13]

National Socialism was thus not just negatively anti-Semitic, it was positively totalitarian in its attempt to control all aspects of life. Joseph Goebbels saw his powerful Ministry of Propaganda as:

> . . . a Ministry for popular education, in which Cinema, Radio, new educational institutions, Art, Culture and Propaganda would be brought together under one administration . . . to serve the purpose of building the intellectual-spiritual foundation of our power and of capturing not only the apparatus of the state but the people as a whole.[14]

Particular attention was given to education—the loyalty of the teachers was exclusively demanded for National Socialism and the curriculum in both schools and universities was to be permeated by the ideology of National Socialism.[15]

The result was to create a society in which there were virtually no checks and balances of the sort which are taken for granted in liberal societies. And it was the absence of any check on centralised power that made possible the transition from the anti-Jewish law of 1933 to the extermination camps of the early 1940s in which it is estimated that about six million Jews were systematically killed.[16]

Put beside an appalling fact like that, the effect of National Socialism on science may seem of little moment. Yet it was important because Hitler was eventually removed from power after a war in which science played a major part.

THE END OF NATIONAL SOCIALIST POWER

World War 2 started in 1939 with Hitler in alliance with the Soviet Union and at war with Britain and France. It ended six years later in 1945 when, as well as Britain, Germany was at war with both the Soviet Union and the United States. Many factors were important in the defeat of Hitler but one which did play a major role was the application of science and technology. Radar, operational research, jet engines for aircraft, vastly improved radio communications and nuclear weapons were only some of the scientific and technological developments which followed from the large-scale mobilisation of science and scientists during World War 2. It is arguable that the damage Hitler had done to Germany's scientific community was one very important factor in his ultimate defeat. Certainly many of the scientists who had to flee from Germany during the 1930s were prominent in the massive war effort of Britain and America. We shall see later (Chapter 7.3) how this happened in the development of nuclear weapons, but it was true in other fields too.

The forced emigration of all these scientists also played a great part in the way the United States took over from Germany as the world's leading scientific nation after World War 2. But since this did not happen until after 1945, it lies outside the scope of this chapter.

So let us end by repeating Hitler's words to Max Planck:

If the dismissal of Jewish scientists means the annihilation of contemporary German science, then we shall do without science for a few years.

There you have what Bronowski calls 'the despots' belief that they have absolute certainty'—the very opposite of the tolerance and humility which come from 'the realisation that all knowledge is limited'.[17] (See Chapter 6.7.)

REFERENCES

1   See Chapter 6.3, references 3, 5 and 6 (page 275), for the sources of these quotations.
2   Ben-David, J., *The Scientist's Role in Society*, (New Jersey: Prentice Hall, 1971), Chapter 7.
3   Hartshorne, E. J., *The German Universities and National Socialism*, (London: Allen and Unwin, 1937), p. 77.

4   Bronowski, J., *The Ascent of Man*, (London: BBC Publications, 1973), p. 367.

5   Beyerchen, A. D., *Scientists under Hitler: Politics and the Physics Community in the Third Reich*, (New Haven: Yale University Press, 1977), p. 42. Copyright © Yale University Press, 1977.

6   Ibid., p. 43.

7   Ibid., p. 48.

8   Ibid., p. 134.

9   See Hartshorne, E. J., op. cit., pp. 87–105 and Beyerchen, op. cit., Ch. 3 for detailed accounts of the dismissal policy and its effects.

10 Mussolini, B., *Enciclopedia Italiana*, 1936; quoted in Ashton, E. B., *The Fascist: His State and Mind*, (London: Putnam, 1937), p. 33.

11 Ashton, E. B., op. cit., p. 35.

12 Ibid., pp. 39–40.

13 Ibid., Appendix—Notes on the Development of Fascist Organisation, pp. 281–92.

14 Goebbels, J., *Vom Kaiserhof zur Reichskanzlei*, 1943, p. 28; quoted in Hartshorne, E. J., op. cit., p. 29.

15 Mann, E., *School for Barbarians: Education under the Nazis*, (London: Lindsay Drummond, 1939) gives a graphic and detailed account.

16 Dawidowicz, L., *The War Against the Jews*, 1933–45 (Harmondsworth, Penguin, 1977) gives an authoritative and chilling account; see in particular Appendix A—The Fate of Jews in Hitler's Europe: By Country, pp. 427 78, and Appendix B—The Final Solution in Figures, pp. 479–80, which estimates that the Germans killed about two-thirds of all the Jews in the countries they occupied during World War 2.

17 Bronowski, J., op. cit., p. 182.

SUMMARY

1   By the early decades of the 20th century, the high quality of scientific research in the German universities had made Germany the foremost scientific nation in the world.

2   When the National Socialists came to power in 1933, the independence of the German universities was immediately threatened and many Jewish scientists and academics were dismissed.

3   During the 1930s, 20 present or future winners of the Nobel Prize—including Einstein, Haber and Schrödinger—were forced to leave positions in German universities or research institutes because of the National Socialists' anti-semitic laws.

4   The National Socialists aimed to create a totalitarian state in which everything—including the content of education—was controlled from the centre. One result of their policies was the death of nearly six million Jews in the extermination camps of the early 1940s.

5   Hitler is reported as having told Max Planck that 'Our national policies will not be revoked or modified, even for scientists. If the dismissal of Jewish scientists means the annihilation of contemporary German science, then we will do without science for a few years.'

7.3 Science, technology and government in liberal capitalist societies

As we saw in earlier chapters, there was an historical association between the rise of science in the 17th and 18th centuries and the evolution of liberal political institutions in England, France and America. Newton's contemporary, John Locke, was the philosopher of the English liberal revolution of 1688 and the liberal ideas of the French *philosophes* were influential in the American and French revolutions of 1776 and 1789 (see Chapters 4.1, 4.2 and 4.3).

In this chapter, however, we are not going to look at the historical roots of liberalism but at the growth of science in modern liberal societies—or at least at some aspects of this enormous subject. We will start by making some general remarks about science, technology and the role of government and then, because the subject is so large, we will look at three case-studies in more detail: the development of a basic science, genetics and molecular biology; of an applied science, nuclear energy; and the career of an individual scientist, J. R. Oppenheimer.

SCIENCE IN LIBERAL SOCIETIES

Let us first look again at Fig. 3.5.1 (page 78) which shows the rapid growth of scientific journals since 1700 and of abstract journals since about 1850. Both have doubled in number roughly every 15 years and growth at this rate has continued well into this century. These are total figures for the whole world. Now look at Fig. 7.3.1 which shows the percentages of papers on chemistry published in different countries over the period 1910–60—the actual numbers of papers were of course increasing exponentially over this period. We can clearly see that these figures reflect the overall growth of science in the Soviet Union and Japan (see Chapters 7.4 and 7.5). You can also see how Germany's share of the total fell dramatically under the Nazis—from a third to quarter earlier in the century to less than 10 per cent by the mid-1960s. But it is also clear that, over the whole period, the majority of these papers have been published in countries which can be regarded as liberal, democratic societies—the British Commonwealth, America, France, Germany for much of the period and Japan since World War 2.

These measures of scientific output—numbers of journals or papers—indicate roughly how much science there is, but give no indication of the quality of the work being done in any particular country. It is not easy to estimate scientific quality but one measure which has been used is the number of Nobel prizes awarded to scientists from different countries. Table 7.3.1 shows the numbers of Nobel prize-winning scientists in the major scientific nations from 1901, when the prizes were first established, up to 1980. Once again the pre-eminence of Germany before 1940 and its subsequent decline are clearly shown, as is the rise

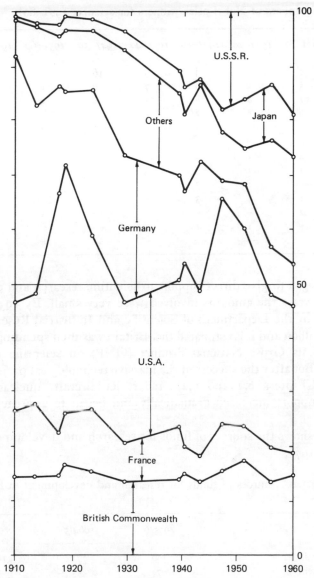

Fig. 7.3.1 Percentages of papers in *Chemical Abstracts* by country, 1910–60. *Redrawn from de Solla Price, D. J., Little Science, Big Science (Columbia University Press, 1963).*

of the United States since that time. And once again the clear majority of the prizes have been awarded to scientists from countries with liberal, democratic governments.

Now let us turn to another very significant development in the 20th century— the involvement of governments in the financing of science and technology. Germany led the way here, having given some direct support to science even in the 19th century. But in Britain it was only in 1916, during World War 1, that the

Table 7.3.1 Numbers of Nobel prize-winners for science, 1901–80[1]

Country	1901–10	1911–20	1921–30	1931–40	1941–50	1951–60	1961–80	Total
USA	I	I	3	9	16	27	64	121
Britain	5	3	8	7	6	9	23	61
Germany	12	8	8	10	4	4	3	49
France	6	5	3	2	—	—	6	22
Sweden	I	2	2	—	I	I	4	11
USSR	2	—	—	—	—	4	4	10
Netherlands	4	I	2	—	—	I	I	9
Switzerland	I	I	—	2	3	—	2	9
Austria	—	I	3	2	I	—	I	8
Denmark	I	I	3	—	—	—	2	7
Italy	2	—	—	I	—	I	I	5
Belgium	—	I	I	I	—	—	2	5
Japan	—	—	—	—	I	—	2	3

government began to give direct support to scientific research and development, and for many years the amounts involved were very small. By 1939 the annual amount spent by the Department of Scientific and Industrial Research was still less than £1 million and it is estimated that Britain was then spending only 0·1 per cent or so of its Gross National Product (GNP) on scientific research and development. But after the success of the massive technological projects of World War 2 (see Chapters 6.1 and 7.2)—radar, jet aircraft, nuclear weapons— government support increased dramatically and began to approach its present levels.

Table 7.3.2 shows the sources of funds for research and development in Britain from 1955 to 1975.

Table 7.3.2 Sources of funds for research and development in Britain, 1955–75[2]

	1955/6	1964/5	1975
Total £ million	300	771	2151
Government %	74	55	52
Industry %	23	37	35

These figures represent about 2 per cent of GNP—rather less in 1955/6 and rather more in the later years—which is more than 20 times as large a proportion of GNP as in 1939. And since GNP itself has increased to a considerable extent in real terms, the actual resources devoted to research and development in Britain are probably now 50 or 100 times greater than in 1939. And most of this money comes from the government.

Now let us look at where money is spent. Table 7.3.3 gives this information for Britain for the same years.

Table 7.3.3   Users of funds for research and development in Britain, 1955–75[2] (Figures in brackets show sources of funds)

	1955/6	*1964/5*	*1975*
Total £ million	300	771	2151
Government %	29 (74)	25 (55)	26 (52)
Industry %	62 (23)	60 (37)	55 (35)
Higher Education %	5 (0)	7 (0)	8 (1)

So although the government finds most of the money, industry and the universities do most of the research and development. Similar trends are visible in other liberal democratic societies as can be seen from Table 7.3.4, which shows the sources of funds and where they were used for nine OECD (Organisation for Economic Cooperation and Development) countries in 1969.

Table 7.3.4   Main sources and users of funds for research and development in 1969 (percentage of total expenditure in each country)[3]

Country	Business		Government		Higher education	
	S	*P*	*S*	*P*	*S*	*P*
France	33	56	62	29	1	14
West Germany	60	68	39	5	0	18
Italy	50	55	41	25	7	20
Netherlands	59	62	38	11	0	18
UK	44	65	51	25	1	8
USA	38	70	58	14	3	13
Canada	30	37	54	35	13	29
Japan	68	67	14	12	18	19
Sweden	57	66	40	15	4	19

S = Source of funds for R & D;   P = Performance of R & D.

In all these countries the government provides more money for scientific research and development than it uses itself—often much more. And all these countries spend enormous sums of money on science—between two and three per cent of their GNP in most cases.

This large-scale involvement by government and, to a lesser extent, industry has thus completely changed the way science is funded and the scale of the support it receives. But has it changed the way the scientific community operates? Are logic and evidence still the main criteria for judging the worth of a piece of scientific research? And is open public criticism by the scientific community at conferences and in journals still of central importance?

Roughly speaking, the answer is that although the scientific community has changed with increasing affluence and patronage, it has not changed anything like as much as might have been expected. This is because both government and industry have operated through the scientific community rather than tried to take it over. Government and industry *have* determined overall priorities and broad areas of research interest. But despite being the main paymasters, they have not, on the whole, tried to interfere in purely scientific matters. The key decisions— who does the work, how it should be tackled, who is to assess its scientific value— have mainly been left to the scientists themselves. Logic, evidence and public criticism are still vital and only the latter has been significantly restricted. This has occurred in two main ways—where national security is involved, as in defence or armament work, and where commercial secrecy is enforced. The first restriction is the more serious since it is difficult to see how some degree of secrecy can be avoided. However, it is recognised to be a source of inefficiency as perhaps I can illustrate from my own experience. Some years ago when I was working for the scientific civil service, the group of which I was a member was given a new research project to develop. We worked on it for about nine months, making a lot of mistakes but, I hope, some progress. Then we discovered that at least three other groups on the same site were all working on the same project—all in complete ignorance of what the others were doing. For security reasons we had not been told of each other's existence and so had probably all been making the same mistakes!

Commercial secrecy, while real enough, is probably of less long-term importance since once a product is on sale it is not long before its main technical features can be discovered. It is generally recognised that even in very high technology fields such as whole-body scanners for medical research, once a product is commercially available, it is only possible to hold on to a technical lead for about 18 months or two years. And to hasten this process, there now exist 'reverse engineering' companies which do nothing else except buy companies' products, analyse how they work and sell the results to other companies.

Now let us ask another question. Why is it that governments and industry in liberal societies have shown such respect for the scientific community and the traditional way it operates? There are probably two main reasons. First, it is widely realised that, if they don't do this, the 'science' they will get for their money will not be the real thing. The scientific community—its values, structures and methods of operation—are a necessary feature of science. You can't do without them.

The second reason is perhaps more fundamental. It is because the values of the scientific community—pluralism, tolerance, individual freedoms and the free-flow of information—are very similar to the underlying values of liberal societies. Ideally, liberal societies are fundamentally decentralised. All the main spheres of human life—economic, political, educational, cultural, religious—are partially separated. And within each sphere, pluralism is, to some extent, both encouraged and realised—diverse centres of economic power, many political parties, diversity of provision in education and other cultural spheres, many co-existing religions. The whole complex society is underpinned by the values of tolerance, pluralism and individual freedoms—and, in particular, by freedom of expression, communication and information. This compatibility of fundamental values makes it easier for liberal societies both to maintain the partial autonomy needed by scientists and the scientific community and to make large-scale use of science and technology in a creative way in the development of society.

These are large claims which are not easy to appreciate when they are made in a general and abstract way. So now let us discuss three case-studies which will illustrate some of these general points and also have an intrinsic interest of their own.

GENETICS AND MOLECULAR BIOLOGY—A FUNDAMENTAL SCIENCE

We have already described in Chapter 6.4 the enormous growth of knowledge in these subjects which has taken place in this century. We identified three main phases in these developments—the early fruit fly experiments of the period 1910–20; the statistical and experimental population genetics of the 1920s and 1930s; and the understanding of the structure and biochemical functioning of important biological molecules like DNA, haemoglobin and other proteins.

Now let us look briefly at where all this work was done and how it was financed. T. H. Morgan's early work on *Drosophila* was done in the biology department at Columbia University in New York under fairly primitive physical conditions—at least by modern standards. Chetverikov's revolutionary work on population genetics was done at the Institute of Experimental Biology attached to Moscow University. Dobzhansky first worked there too and then moved to Columbia in 1927.

The three founders of the statistical theory of population genetics—Fisher, Haldane and Sewall Wright—worked at a number of different institutions. Fisher did most of his best work while he was head of the Statistics division at the Experimental Station for Agricultural Research at Rothamsted in Hertfordshire in England (1918–30); this station had been set up in 1843 by J. B. Lawes and J. H. Gilbert who did much fundamental work there on artificial fertilisers and plant nutrition. Initially Rothamsted was financed from the profits of Lawes' factory for manufacturing superphosphate fertiliser. However, by Fisher's time, the British government were providing some financial support and since 1931 Rothamsted has been run by the government-financed Agricultural Research Council. Later Fisher became Professor of Eugenics at University College, London (1933–43) at the same time as Haldane was Professor of Genetics and Biometry there. Sewall Wright first worked at Harvard University and was then, for ten years, senior animal breeder for the US Department of Agriculture (1915–25). Later he taught at the Universities of Chicago and Wisconsin. Government or state funded agricultural research institutes were also closely involved in the setting up of the two main centres for the study of statistics in the United States—at the Universities of Iowa and North Carolina. Thus the whole subject of statistics and population genetics was developed by a mixture of funding from individual universities and publicly funded practical research institutes.

As we saw in Chapter 6.4 the development of molecular biology and the discovery of the structure of DNA and the genetic code were not the product of an isolated spark of genius but were the outcome of decades of work by many different people in many different university laboratories, both in Europe and America. This work was largely financed by these individual universities and also by government financed bodies like the Medical Research Council in Britain.

However, an important role was also played by the Rockefeller Foundation, which gave grants to nearly all the main workers in the fields I have mentioned and was particularly influential in providing the resources to buy the increasingly expensive physical and chemical apparatus—ultra-centrifuges, apparatus for radioactive measurements and X-ray crystallography, electron microscopes— which the new biology needed. An indication of the importance of the Foundation's support is that it was the director of its natural sciences division, Warren Weaver, who first used the term 'molecular biology' as long ago as 1938.[4]

Conclusion

All this work in genetics and molecular biology clearly represents an enormous increase in basic biological knowledge—an increase which has developed in no definite planned or organised way and which was financed in a variety of different ways: by governments, directly and indirectly, by state and private universities and by private foundations and endowments.

This knowledge has also had great practical importance—as, for example, in the development of new varieties of high-yielding hybrid corn in America in the 1930s. These are estimated to have been worth at least 200 million dollars a year since about 1935 resulting from a total research investment of about 3 or 4 million dollars from 1910 to 1933. Other examples include the new varieties involved in the so-called 'Green Revolution' of the 1950s and 1960s. Many people are now claiming that we are on the verge of large-scale applications of the knowledge gained since 1953 on how to modify the genetic constitution of bacteria (see Chapter 8.4).

NUCLEAR ENERGY—AN APPLIED SCIENCE

In Chapter 6.3 we described the developments in nuclear physics which turned Einstein's famous equation—$E = mc^2$—from a laboratory curiosity into a large-scale practical reality. In the 1930s nuclear physics was an open and public academic science—little known or understood by those outside the subject. During World War 2, it went underground. The first nuclear weapons were produced at Los Alamos in the USA in a top secret project—called the Manhattan Project—in which many European and American scientists collaborated. After World War 2 had been brought to an end by the dropping of two atomic bombs on Japan in 1945, nuclear energy suddenly became headline news.

After 1945, separate programmes to develop nuclear energy were started in Britain, France and China while the USSR accelerated its existing work. The American programme also continued but it now ceased to be an international project. All these early programmes were aimed at the production of nuclear weapons and were conducted in almost total secrecy. The first Russian atomic bomb was exploded in 1949, the first British one in 1952, and similar weapons were tested by France in 1960 and China in 1964. After the first Russian bomb in 1949, an intense debate developed in the United States about whether to try to develop a super-bomb—now called a hydrogen bomb. Much of this debate was not public at the time but its main outlines have since been published. The scientific advisory committee of the United States Atomic Energy Commission, chaired by Robert Oppenheimer—the war-time director of Los Alamos—opposed the development of hydrogen bombs but was overruled by President Truman. The research went ahead and a hydrogen device (not a bomb) was exploded in the Pacific in November 1952. This was closely followed by the first Russian hydrogen bomb in August 1953, and the first American operational bomb, which was tested in March 1954. Hydrogen bombs were later tested by Britain (1956), China (1967) and France (1968). Since the early 1950s many nuclear weapons have been tested by all these countries although, since the partial test-ban treaty in 1963, the tests have mostly been conducted underground.

The peaceful application of large-scale nuclear energy started in Britain as a by-product of the weapons programme. Dual-purpose reactors were designed in 1952 to produce both plutonium for bombs and electricity for the national grid, and in 1956 the first industrial scale nuclear power station opened at Calder Hall in Cumbria.

Since 1956 the installed world capacity of nuclear power stations has increased dramatically—reaching 5 gigawatts (GW)* by 1964, doubling to 10 GW by 1968, and doubling again every few years until it reached 80 GW in 1976, which is around 1 per cent of total world energy consumption. These figures exclude the nuclear power programmes of communist countries (USSR, Eastern Europe, China) since information about these is far from complete. Nuclear power stations now exist in more than 20 countries. The USA has most, nearly a hundred, but six countries have ten or more working stations—Britain, France, Japan, West Germany, Canada and the USSR.

As these nuclear power programmes developed, nuclear power emerged from the secrecy which had surrounded, and still surrounds, the development of nuclear weapons. Information became available about both its technical and economic aspects, and nuclear power became one of the many issues which in liberal societies are on the agenda for public political debate. The safety of nuclear power is obviously a key issue in these debates, and this aspect was highlighted as early as 1957 when a nuclear accident occurred in Britain—at Windscale in Cumbria. The reactor involved was not one of the power reactors but an older experimental one used in the weapons programme. A fuel element overheated, fire broke out and some radioactive material escaped into the atmosphere. Some local milk supplies were contaminated and milk deliveries were banned for a few days over an area of 200 square miles. This accident and the inquiry which followed led to a considerable tightening of safety measures in nuclear power stations. Since then there have been many reports published on safety and other aspects of nuclear power, especially in the United States and Britain. Some of the most

* 1 GW = 1000 MW = 1 000 000 kW.

recent of these are, in Britain, the *Report on Nuclear Power and the Environment* published by the Royal Commission on Environmental Pollution in 1976 and the *Report of the Windscale Enquiry* in 1978—and, in the USA, the *Rasmussen Report* to the US Nuclear Regulatory Commission in 1976 and the *Kemeny Report* in 1979 following the accident at Three Mile Island. And, so far, the safety record of the nuclear industry in liberal societies is very good, comparing favourably on most counts with other industries concerned with large-scale sources of energy (see Chapter 8.2).

THE CASE OF ROBERT OPPENHEIMER

Our third case-study is of an individual scientist, Robert Oppenheimer, whose career as a scientist was severely affected by forces external to science—in this case, the powers of the United States government. The Oppenheimer case is the clearest example in liberal societies in the 20th century of such external interference. Robert Oppenheimer was a brilliant American nuclear physicist who had studied at the University of Göttingen in the 1920s. We described earlier how he took a leading part in the original Manhattan project, being Director of the Los Alamos laboratory from 1943 onwards. After the war, when many scientists returned to research posts in universities or industry, Oppenheimer stayed on. He was one of the two or three leading individuals in the development of the whole United States nuclear weapons programme, right through to the successful testing of a hydrogen bomb in 1953. There had been some doubts about his security clearance in 1943—because he had been associated with some left-wing organisations in the 1930s. However, these doubts had been overruled and his years of invaluable service to the US government since then seemed to have removed all doubts about his loyalty. Then, late in 1953, the whole question of Oppenheimer's loyalty was re-opened, largely on the initiative of the FBI. His security clearance was temporarily withdrawn, and he had to face a secret commission of enquiry. This enquiry met in April, 1954, and for three weeks questioned Oppenheimer and many other prominent scientists about Oppenheimer's left-wing past, his opposition to the dropping of atomic bombs on Japan, his reluctance to develop hydrogen bombs and many other matters. The transcript of these procedings has since been published and it provides a fascinating account of the tensions and dilemmas faced by the academic nuclear scientists when their exciting new science of the 1930s developed into the nuclear power politics of the 1940s and 1950s. The final decision of the enquiry was that Oppenheimer was a loyal American citizen but that sufficient doubts existed for them not to recommend restoring his security clearance. This meant that Oppenheimer could no longer hold any position with the United States Atomic Energy Commission and had to leave all the government committees of which he was a member. He returned to Princeton University and became Director of the Institute for Advanced Studies there, a post he held until his retirement in 1966. Thus Oppenheimer ended his career with 12 years as Director of the same Institute where Einstein had spent the last 22 years (1933–55) of his life.

Oppenheimer's career as a scientist was thus greatly affected by political factors—both in the way he rose to his government office and in the way he lost that position. But the power of the government never queried Oppenheimer's scientific expertise. It was his political reliability and judgement that was criticised. And even after his 'trial' Oppenheimer was able, without hindrance, to carry on his scientific work at one of the most prestigious research institutions in the world.

CONCLUSION

This chapter started with some general comments about the relationship between liberal societies, their scientific communities and the development of science. Then we outlined three case-studies: the development of genetics and molecular biology; the rise of nuclear power; and the Oppenheimer case. In all these cases there is much we could comment on and criticise in the way that liberal governments and societies have dealt with the rise of science and technology. But before attempting to do this we will first look at the relationship between science, technology and government in some other societies. So in the next chapter we will attempt for the Soviet Union the same kind of general survey illustrated by specific case-studies which this chapter has given for liberal societies.

REFERENCES

1   Data collected from Mitsutomo, Y., 'The Shifting Centre of Scientific Activity in the West' in Shigeru, N., Swain, D. L. and Eri, Y., *Science and Society in Modern Japan*, (Cambridge, Mass: MIT Press, 1974), Table 6, p. 90 and from *Science and Technology in Modern Japan* Vol. 1, No. 1, 1982, Fig. 8, p. 9.
2   Compiled from data published by the Department of Education and Science.
3   From *Science and Public Policy*, Sept/Oct 1972, Butterworth Scientific Ltd.
4   See Kohler, R. E., 'The Management of Science: the Experience of Warren Weaver and the Rockefeller Foundation Programme in Molecular Biology', in *Minerva*, Vol. XIV, No. 3, Autumn 1976, pp. 279–306.

SUMMARY

1   Since World War 2, government financial support for science and technology has increased dramatically; most liberal democratic societies now spend between two and three per cent of their gross national products on scientific research and development.
2   Although governments, directly or indirectly, provide most of the money, much of the scientific work is done by industry, research institutes and universities.
3   Massive government support for science has not significantly changed the way the scientific community operates; governments in liberal societies realise that the values, structures and methods of operation of the scientific community are a *necessary* feature of science—without them the 'science' they get for their money will not be the real thing.

4   Knowledge of genetics and molecular biology has grown rapidly in the 20th century and is of great practical importance; but this growth was not planned or organised and was financed in a variety of different ways, both public and private.

5   Governments in liberal societies have taken a more direct role in the development of nuclear energy—both for weapons and for power generation—but they have never, even in the Oppenheimer case, queried the expertise of scientists in *scientific* matters.

7.4   Science and the Russian Revolution

Much has been written about the Russian Revolution of 1917, some of it idealistically optimistic—in the style of the May Day poster shown in Fig. 7.4.1—and some of it grimly realistic like Solzhenitsyn's eye-witness account of the forced labour camps in the Gulag Archipelago (see Fig. 7.4.2).[1] But, until recently, relatively little has been written about the whole development of science in the Soviet Union since the Revolution.

In this chapter we will first distinguish three phases in the development of Soviet science—the immediate post-revolutionary period (1917–29), the Stalin era (1929–53) and the post-Stalin period (1953–present). In doing so we will draw heavily on the writings of the Russian biologist Zhores Medvedev,[2] which are based on his own first-hand experience of working as a scientist in the Soviet Union from 1948–73. Then we will look at three case-studies in a little more detail and use these to illustrate some more general points—just as we did when discussing science in liberal societies. For interest, and to help in making comparisons, we will consider two of the same topics as we did for liberal societies: the development of a basic science, genetics and molecular biology, and of an applied science, nuclear energy. Then we will look at the careers of two Soviet scientists, the geneticist N. I. Vavilov (1887–1943) and the nuclear physicist A. Sakharov (1921–   ). Finally we will consider the present state of Soviet science and technology.

FROM THE REVOLUTION TO THE RISE OF STALIN (1917–29)

Before the Russian Revolution in 1917, the scientific community in Russia was fairly small and, like elsewhere, there was relatively little organised research. The imperial Academy of Sciences had been founded by Peter the Great in 1725 as part of his drive to open up Russia to European influences and St Petersburg (now Leningrad) became the main scientific centre. The most famous 19th-century Russian scientist was Mendeleev (see Chapter 6.6) but others working in Russian universities included the chemists Butlerov, Beilstein and Hess, the physicist Lenz, the astronomer Struve, the mathematician Lobachevski and the physiologist Pavlov.

The communist Revolution marked a turning point in the development of science in Russia. The new government regarded itself as being based on the principles of scientific socialism and it expected the natural sciences to play a key role in the new society which it intended to create. Many scientists were initially hostile to the Revolution but the new government made more resources available for science than ever before and established many new research institutes which were directly supported by the government in a way which was unique in the world at that time.

During the civil war which followed soon after the Revolution, the government

Fig. 7.4.1　May Day poster sending greetings to all the workers of Russia.

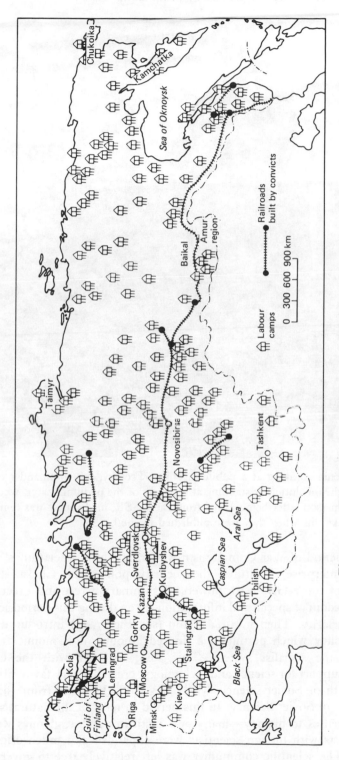

Fig. 7.4.2 The destructive labour camps of the Gulag Archipelago.

Labour camps

Railroads built by convicts

0   300   600   900 km

Chukoika
Kamchatka
Sea of Oknoysk
Baikal
Amur region
Taimyr
Novosibirsk
Tashkent
Aral Sea
Caspian Sea
Tbilisi
Black Sea
Stalingrad
Kiev
Kuibyshev
Kazan
Gorky
Moscow
Sverdlovsk
Minsk
Riga
Leningrad
Gulf of Finland
Kola

Fig. 7.4.3 Lenin speaking at a public meeting; Trotsky, the commander of the Red Army during the revolution, is standing at the right of the podium. Later, when Trotsky became an 'unperson' after being exiled from the USSR, his likeness was removed from photographs like this when they were published in the USSR.

instituted a period of War Communism which, amongst other things, abolished all private enterprise, instituted the security police, the Cheka—the direct predecessor of the KGB—with powers of summary arrest and execution, and abolished freedom of speech and information. Even during this period science was given high priority. Then in 1921 Lenin (see Fig. 7.4.3) introduced his New Economic Policy which permitted a limited amount of economic freedom to complement the socialist system. This step, together with the continuing strong state support for science and technology, created a very favourable climate for the growth of Soviet science. Medvedev calls the years from 1922–29 the Golden Years of Soviet Science. In this period in most of the natural sciences—from mathematics to biology—many new institutes and academies were established, contacts with foreign scientists were expanded and much original work was begun. The scientific community was left relatively free to govern its own affairs—provided that it gave a vague general assent to Marxism or at least refrained from direct criticism of the government. However, this was not true in the humanities and the social sciences. Here there was much more direct ideological control, both of individuals and of the content of what was written and taught.

THE STALIN YEARS—1929–53

The whole situation changed dramatically around 1929 with the rise of Stalin (see Fig. 7.4.4) to virtually supreme power. The extraordinary events of Stalin's rule—the forced collectivisation of agriculture, the rapid expansion of forced labour camps and the creation of the Gulag Archipelago, the Great Terror of the 1930s and 1940s, the Soviet pact with Hitler from 1939–41 and the life and death battles with National Socialist Germany from 1941–45—all tended to overshadow what was happening in the world of science. But here too there were drastic changes as Stalin attempted to tighten his control over the whole of Soviet life. Communist Party policy and the principles of dialectical materialism, the official Marxist philosophy, became the touchstones of what was acceptable in much of science as well as in the social sciences, history, literature and the arts. And it was not only decisions about *who* was to be allowed to work as a scientist which came under Stalin's power. Even the very *content* of science was subjected to control and censorship. Biology and agriculture were most affected, as we shall see, but parts of chemistry, Einstein's theory of relativity, and the whole development of cybernetics and computing were also severely distorted. Not only was valid work suppressed and banned during this period but many dubious ideas were given widespread credence and publicity even when there was little or no evidence to support them. This situation lasted until the death of Stalin in 1953.

Fig. 7.4.4   Early photograph of Stalin (*circa* 1905) from police files; at this time Stalin was organising bank raids in order to finance Bolshevik revolutionaries.

THE POST-STALIN PERIOD (1953–PRESENT)

After Stalin's death, many of the terrible events of his rule were admitted by his successor Khrushchev in a secret speech to the Soviet Communist Party congress in 1956. After this the tight control exercised by Stalin was relaxed a little but the fundamental nature of the Soviet State did not change.

Soviet society, like all other existing Marxist states, is centralised and monolithic. The government attempts to control all aspects of life—economic, political, educational, cultural and religious. The economy is centrally planned and only one political party, the Communist Party, is allowed to exist. This party controls all educational and cultural activities. Religious freedom is written into the constitution but, in practice, it is severely limited. Freedom of expression and of access to information are expressly prevented, and the values which are encouraged are those of collective commitment to the centralised rule of the party. Anything which the party does is to be approved and anything which threatens its central role is likely to be condemned and suppressed. The only major daily newspaper permitted in the Soviet Union is published by the Central Committee of the Communist Party and its name *Pravda* means 'Truth'.

Most of these features of the Soviet state have existed ever since the Revolution—in Lenin's time, in Stalin's and under Khrushchev, Brezhnev and Andropov. So the values of Soviet society are very different from the values of the scientific community as I have described them (see Chapter 7.1). Therefore we might expect conflicts to arise between the government and the scientific community and also for there to be problems in the development of science-based industries in the Soviet Union. And such conflicts and problems are inherently likely to arise even though the government claims that it bases itself on scientific principles and wishes to foster the development of science and technology.

But before discussing these general points any further let us first turn to the three case-studies—genetics and molecular biology, nuclear energy, and the careers of Vavilov and Sakharov.

GENETICS AND MOLECULAR BIOLOGY

As we saw in Chapter 6.4, in the early post-revolutionary period the Soviet Union was in the forefront of world research into genetics with Chetverikov and his pupil Dobzhansky working on the application of Mendelian genetics to the evolution of natural populations of the fruit fly *Drosophila*.

In the 1920s the Soviet Union was also a world leader in another field of genetics—the collection of a wide variety of naturally occurring and cultivated plants to act as a general genetic pool for selection and hybridisation. This work was carried out by N. I. Vavilov (1887–1943) in the Laboratory of Applied Botany which was set up in Petrograd, the former St Petersburg (renamed Leningrad in 1924 after Lenin's death) in 1919. In the mid-1920s more than 200 expeditions were organised to 65 countries and over 150,000 plant varieties, forms and species were brought to the Soviet Union with the specific aim of improving agriculture and increasing crop yields and disease-resistant strains.

Thus by the end of the 1920s, the Soviet Union was well established as a world centre for both basic and applied research in genetics.

All this changed dramatically in 1929 as the Communist Party began to interfere drastically in scientific matters. The first all-Union Conference of Agrarian Marxists was held in December 1929:

> ... the climax of the conference came on the seventh day when Stalin appeared to address the participants. In a speech which began rather blandly, he startled his audience with the announcement of the impending liquidation of the kulaks.* Stalin's declaration heralded a radical shift in official policy towards the rural sector. Since the early months of 1928, force had been used almost without respite to compel the peasant to deliver his grain, but the use of force had never been publicly acknowledged or officially justified. Nor had there ever before been any proposal to use force on a large scale in the countryside. The effect of Stalin's announcement upon rural studies was that delegates understood that the debates they had been having for six days had been superfluous. Henceforth, the criterion of validity would not even be derived from Marxian theory as they themselves understood it; conformity with the rural policy of the Communist Party would be the criterion of truth. 'Theory' would now have to prove the correctness of this policy and research would now have to confirm the existence of the 'facts' which were postulated by this theory and thus the correctness of theory. The research workers' findings would have to demonstrate that what the politicians said was true.[3]

From now on '... party policy created strong pressure on social scientists to use political criteria in the assessment of scholarly work. Towards the end of 1929, total correspondence between the findings of research and the factual postulates and assertions of rural policy became a prerequisite of all investigators in this field. With these developments, the field of rural studies became politicised and the scholarly study of the countryside came to an end.'[4]

The results of this shift in policy soon became evident. In the practical field Stalin's policy of collectivisation was rigidly enforced over the period 1929-32. Millions of peasants had their land forcibly incorporated in collective farms. There was widespread resistance and agricultural output was greatly reduced. Yet, even so, grain continued to be requisitioned and much even continued to be exported. The result was famine—according to Robert Conquest writing in 1971: '... perhaps the only case in history of a purely man-made famine',[5] although since then there has been a similar man-made famine in Cambodia which has affected a much greater proportion of the population. Returning to Russia in the 1930s, Conquest estimates that there were 'over 5 million deaths from hunger and from the diseases of hunger. ...' In addition 'mass execution also played its part. Stalin later told Churchill that 10 million kulaks had to be dealt with, and that "the great bulk" were "wiped out", others being transferred to Siberia. Over 3 million seem to have ended up in the newly expanding labour-camp system'.[6]

* The kulaks were the more prosperous peasants.

In science, too, drastic changes took place. T. D. Lysenko (1898–1976) and his followers rapidly acquired positions of power and influence in Soviet agricultural and biological science. Their largely incorrect ideas on the inheritance of acquired characteristics seemed to promise much more rapid practical results than the more cautious genetic approach of scientists like Vavilov. There was no reliable evidence to support Lysenko, and his rise to power and influence can be attributed to three things: his proletarian origins which enhanced his credentials as a pioneer of socialist science as opposed to the middle-class origins of Vavilov leading to what was claimed to be an essentially bourgeois science; the claim that the gene was an idealist concept incompatible with the Marxist philosophy of dialectical materialism—hence genetics could not possibly be correct; and Lysenko's backing by the Communist Party and Stalin—a corollary of his willingness to accept Party policy as the criterion of truth. The harassment of genetics and geneticists escalated during the 1930s in the wake of the Great Terror. Chetverikov was exiled to Siberia in 1931 and for the rest of his life—he died in 1959—he was never able to do any significant work in genetics again. Vavilov continued to argue against Lysenko and his ideas but it was a losing battle. Then, in 1940, Vavilov was arrested while on a scientific expedition. Later he was convicted of spurious offences and died in prison in 1943. The height of Lysenko's power came in 1948 when another personal intervention by Stalin at a scientific conference led to the dismissal, or worse in many cases, of all who taught genetics or who opposed Lysenko.[7] Decrees were issued banning the teaching of genetics in any school or university, withdrawing books on genetics from libraries, and deleting all mention of Vavilov or other geneticists from more general books. It was even decreed that all stocks of *Drosophila*, the fruit-fly used in so many classic genetic experiments, were to be destroyed.[8] Lysenko and his supporters brought off what was in effect a political coup.

After Stalin's death in 1953, Lysenko began to lose some of his influence. However, when Khrushchev's agricultural policies began to fail, he too, like Stalin before him, turned to Lysenko with his promise of quick results. So, in the late 1950s, Lysenko's power and influence again increased. But, the weight of scientific opinion was beginning to make itself felt—and this increased after a disastrous nuclear accident in 1958 which affected tens of thousands of people and in which the lack of any kind of widespread genetic knowledge greatly increased the suffering and loss of life (see below).

In the 1960s the campaign for the recognition of genetics grew, with Zhores Medvedev playing a leading part. But it was not until 1966, two years after Khrushchev's fall from power, that Lysenko was removed from his principal positions, and many of his followers are still in post. Even today the story cannot be told in the Soviet Union, and Medvedev was confined to a mental hospital soon after his book *The Rise and Fall of T. D. Lysenko*[8] appeared in the West. The story of his confinement and release is told in the book *A Question of Madness*.[9] In 1973 Medvedev had his passport withdrawn while on a trip to Britain and he now works for the National Institute for Medical Research in London.

Thus the ideological perversion of science went on for more than 40 years, during which the science of genetics in the West made enormous strides, not least in the development of new genetic hybrid cereal strains with greatly increased crop yields. However, these improvements were not available in the USSR because the essential genetic technique of polyploidy was 'not a socialist concept'.

Instead, an ideologically predetermined straightjacket was imposed on the development of natural science. Medvedev has recently summed up the effects of the Lysenko affair like this:

> Many important fields of research, like genetics, cybernetics, quantum-resonance theories in chemistry and even Einstein's theory of relativity, were under attack as anti-marxist and anti-materialist. 'Pseudoscientists' became very influential in many fields of science.
>
> The Lysenko affair is well known in the West. What is less well known is the disruptive influence which it had in many other practical fields. Lysenko's pseudoscientific ideas dominated Soviet biology from 1937–64. It was the only form of biology taught in all schools, universities and agricultural colleges. Even medical or human genetics was not taught in the USSR from 1938–65.
>
> The research institute of medical genetics was shut down in 1937 and it was only reopened in 1970. Theoretical aspects of human genetics (inheritance of psychological or behavioural characteristics) were not recognized officially until 1981. At a special general meeting of the Academy of Sciences of the USSR in that year, it was agreed after heated debate that a genetic approach could be used to study human psychology and special talents like musical, poetic, literary etc. Lysenko's influence was especially harmful in agriculture where methods of selection and hybridization were delayed for at least two decades.
>
> In other theoretical fields like physiology, cytology, organic chemistry and soil sciences similar pseudosciences dominated at various times. Research work in cybernetics was considered reactionary and 'idealistic' until ten years after the war. This caused a ten-year delay in computer design, which is responsible for the backwardness of Soviet computer technology.[10]

That such appalling distortions should have been possible, even in natural science, for more than 40 years raises several significant questions. What ideological distortion can be perpetuated in other, more vulnerable, areas of knowledge? And how long would it have taken to overthrow Lysenkoism in the USSR if 'bourgeois' science had not existed in the West as a basis for correction? Would it ever have been overthrown?

Finally, why did it happen? Medvedev gives five main reasons:

the arguments about the existence of a 'bourgeois' as opposed to a proletarian or socialist science;
gross mistakes in agricultural policy;
the existence of a centralised press—boosting the official view;
the isolation of Soviet science and scientists from science in the rest of the world;
the total centralisation of life in the USSR.

MOLECULAR BIOLOGY

The history of work on molecular biology before 1977 in the Soviet Union can be described very briefly. There wasn't any.

NUCLEAR ENERGY

The Soviet nuclear weapons programme started in a small way during World War 2, but became a top priority project immediately after the dropping of the Hiroshima and Nagasaki bombs. The project was under the control of Kurchatov, the distinguished atomic physicist and of Beria, the head of Stalin's secret police, the NKVD—which has since been renamed the KGB.

Beria and the NKVD were in charge of the millions of slave-labourers in a vast number of labour camps (see Fig. 7.4.2). Almost all the construction work in the Soviet nuclear programme was carried out by slave-labourers from these labour camps, many of them living and working under appalling conditions.[11] There was nothing unusual in this—most large construction projects in the Soviet Union, including the rocket and space programmes, have been built by slave-labour, and the earliest such Soviet labour camps had been set up in 1918, within a year of the Revolution. Estimates vary as to the number of those who have died at the hands of the NKVD, either by execution in the Great Terror or from hunger and brutality in the labour camps. The estimates range from 15 million to as high as the 66 million given by Solzhenitsyn in *The Gulag Archipelago*. Nobody really knows since, unlike the case of the Nazi concentration camps, nobody has access to the records of the NKVD. What does seem certain is that the real figure is many times greater than the 6 million killed in Hitler's camps.

Much of the research work on nuclear power was also done by prisoners—not in the labour–camps but in special prison research institutes like the one described by Solzhenitsyn in his book *The First Circle*.[12] These prison research institutes had first been set up in the late 1930s and they developed rapidly during World War 2. Engineers and scientists, who had been imprisoned during the Great Terror, were collected from many different labour camps in order to work on military projects. They included S. P. Korolev, who designed the first Sputnik, and A. N. Tupolev, the well-known aircraft designer.

Many imprisoned scientists worked on the Soviet crash programme to develop nuclear weapons. They worked side by side with scientists like Andrei Sakharov at a site near Chelyabinsk in the Urals mountains. Sakharov has described this programme thus:

In 1950 our research group became part of a special institute. For the next eighteen years I found myself caught up in the routine of a special world of military designers and inventors, special institutes, committees and learned councils, pilot plants and proving grounds. Every day I saw the huge material, intellectual, and nervous resources of thousands of people being poured into the creation of a means of total destruction, something potentially capable of annihilating all human civilisation. I noticed that the control levers were in the hands of people who, though talented in their own way, were cynical. Until the summer of 1953 the chief of the atomic project was Beria, who ruled over millions of slave-prisoners. Almost all the construction was done with their labour. Beginning in the late fifties, one got an increasingly clearer picture of the collective might of the military-industrial complex and of its vigorous, unprincipled leaders, blind to everything except their 'job'.[13]

As a result of the crash programme, the first large Soviet reactor started up in 1947 and the first atomic bomb was exploded in 1949—much quicker than had been expected in America and three years ahead of Britain whose programme had started at roughly the same time. And when it came to the development of a hydrogen bomb, the Soviet Union was even more successful—exploding their first bomb in 1953, six months before America and four years before Britain (see Chapter 7.3). This Soviet success was largely due to a completely new approach—a three-stage or fission-fusion-fission bomb—which was suggested and largely developed by a team headed by Sakharov.

Since 1953 the Soviet Union has developed and tested many nuclear weapons and has also started a nuclear power programme. However, it is not easy to find out very much about this programme as very little information is made public on nuclear power or on many other important issues in the Soviet Union. To illustrate just how tight the censorship is in the Soviet Union we will now discuss a nuclear accident which took place in 1957 or 1958 at the original Soviet nuclear weapons centre near Chelyabinsk in the Urals; 1957 was also the year in which the Windscale accident occurred in Britain (see Chapter 7.3).

In 1976, Zhores Medvedev wrote an article called *Two Decades of Dissidence* in the popular British scientific weekly *New Scientist*. The article mainly described the role of Soviet scientists in the dissident movement but it contained this section:

A tragic catastrophe occurred in 1958, which made nuclear physicists extremely sensitive to the radiobiological and genetics issue. The catastrophe itself could have been foreseen. For many years nuclear reactor waste had been buried in a deserted area not more than a few dozen miles from the Urals town of Blagoveshensk. The waste was not buried very deep. Nuclear scientists had often warned about the dangers involved in this primitive method of waste disposal, but nobody took their views seriously. The alternative of drowning the containers in the very deep waters of the Pacific or Indian Oceans had been rejected as too expensive and protracted. Dispersing the highly radioactive materials over other parts of the country was also considered unnecessary. The large nuclear industry, concentrated in the Urals, just continued to bury its waste in the same way it had done since the beginning of the atomic race. Suddenly there was an enormous explosion, like a violent volcano. The explosion poured radioactive dust and materials high up into the sky. It was just the wrong weather for such a tragedy.

Strong winds blew the radioactive clouds hundreds of miles away. It was difficult to gauge the extent of the disaster immediately, and no evacuation plan was put into operation right away. Many villages and towns were only ordered to evacuate when the symptoms of radiation sickness were already quite apparent. Tens of thousands of people were affected, hundreds dying, though the real figures have never been made public. The large area, where the accident happened, is still considered dangerous and is closed to the public. A number of biological stations have been built on the edge of this—the largest gamma field in the world—in order to study the radioactive damage done to plants and animals.

The irradiated population was distributed over many clinics. But no one really knew how to treat the different stages of radiation sickness, how to measure the radiation dose received by the patient, how to predict what the effects would be both for the patients and their offspring. Radiation genetics and radiology could have provided the answer but neither of them was available. There was no laboratory in the whole of the country which could make a routine investigation of chromosome aberrations—the most evident result of radiation exposure; marrow stocks did not exist; there was no chemical protection against radiation exposure available for immediate distribution.

Many more towns and villages, where the radioactive level was moderate or high, but not lethal, were not evacuated. The observation medical teams established in them were not well prepared for serious tests.

All this greatly shocked the nuclear scientists, and their opposition to Khrushchev's anti-genetic stand became too strong to resist. The government was forced to legalise classical genetics, at least for radiology, radiobiology and medicine. Lysenko's only remaining power base was agriculture.[14]

In 1976 virtually nothing was known in the West about this disastrous accident and Medvedev's article caused a sensation. Sir John Hill, the chairman of the United Kingdom Atomic Energy Authority, dismissed the story as 'pure science fiction' and 'a figment of the imagination'.[15] Stung by this, Medvedev set about justifying his story. But this was not easy to do, since the accident has never been mentioned in any Soviet publication. However, Medvedev knew that scientific studies had been made of this radioactive pollution—he had even been asked to take part in them himself. So he searched the Soviet literature and found that:

More than 100 works on the effect of strontium-90 and caesium-137 in natural plant and animal populations have been published since 1958. In most of these publications, neither the cause nor the geographical location of the contaminated area are indicated. This is the unavoidable price of censorship. However, the specific composition of the plants and animals, the climate, soil types and many other indicators leads to the inevitable conclusion that it lies in the south Urals. (In one publication, the Cheliabinsk region is actually mentioned—a censorship slip.) The terms of observation—10 years in 1968, 11 in 1969, 14 in 1971, and so on—reveal the approximate date of the original accident. Finally the scale of the research (especially with mammals, birds and fish) indicates clearly that rather heavy radioactive contamination covered hundreds of square miles of an area containing several large lakes.[16]

Medvedev has since published further evidence in a book *Nuclear Disaster in the Urals*[16] and his version of events has been largely confirmed in an independent report published by the Oak Ridge National Laboratory in the United States.[17] The precise cause of the explosion is not known—and is unlikely to become known unless the Soviet Union releases more information. However it seems likely that it was a chemical and not a nuclear explosion which released wastes which had been carelessly dumped during the crash programme to develop nuclear weapons.

To the rulers of the Soviet Union this accident—the worst ever in the history of nuclear power—is an event which never happened. But the obliteration of real events is not rare in Soviet 'history'. The *Great Soviet Encyclopedia* does not contain the names of five of the eight men who have been prime minister there since the Revolution. And when Beria, the NKVD chief who ruled the Gulag Archipelago and controlled the early nuclear weapons programme, fell from power, he too was removed from the *Encyclopedia* and became what George Orwell called an 'unperson'. Subscribers to the *Encyclopedia* were solemnly sent an article on the Bering Straits and told to paste it over the one referring to Beria.

Such an extraordinary disregard for the truth is especially alarming in a country in which 'it is expected that already in the near future nuclear power generating facilities will be developed on a tremendous scale.'[18]

INDIVIDUAL SCIENTISTS

For our third case-study we will outline briefly the careers of two of the greatest Russian scientists of this century, N. I. Vavilov, and A. D. Sakharov.

N. I. Vavilov (1887–1943)

Vavilov (see Fig. 7.4.5) grew up in Tsarist Russia and was already a professor at Saratov University before the Revolution. In the 1920s he directed and organised the new Institute of Experimental Agronomy in Leningrad and Moscow and carried out the great botanical collecting expeditions mentioned above. For this work he was later elected a Foreign Member of the Royal Society of London. He became the first president of the Soviet Academy of Agricultural Sciences when it was established in 1929.

After 1932, when foreign contacts virtually ceased for Soviet scientists, Vavilov's expeditions were severely curtailed. He continued with his work as best he could and, throughout the 1930s, he battled against Lysenko and his ideas both in private and in public. He was demoted to vice-president of the Academy of Agricultural Sciences in 1935 and was eventually arrested while on a scientific expedition. He was interrogated for 11 months, tried in 1941 on spurious charges and sentenced to death. The sentence was not carried out but Vavilov was held in the condemned cell in Saratov prison for almost a year until he died on January 26 1943.[19]

Fig. 7.4.5 N. I. Vavilov (1887–1943) (above) and Andrei Sakharov (1921–   ) (below).

In 1948 he became an 'unperson' and his books, articles and ideas were banned. Then in 1955, after Stalin's death, he was partly rehabilitated and his name was restored to the list of deceased members of the Soviet Academy of Sciences. But the story of his life and of his battle with Lysenko can still not officially be told in the Soviet Union.

A. D. Sakharov (1921–   )

Sakharov (see Fig. 7.4.5) was born in Moscow in 1921, after the Revolution. He graduated from Moscow State University in 1942 and obtained the equivalent of a Ph.D. in 1948. He worked in the Soviet nuclear weapons programme for 18 years from 1950 to 1968. He is widely regarded as the man who made the major contribution to the development of the Soviet hydrogen bomb. He has also published significant work on the possibility of controlled fusion reactors. He was elected a full member of the Academy of Sciences in 1953 at the age of 32, the youngest man ever to be elected. For his secret research work he received—also in secret—the Stalin Prize and three Orders of Socialist Labour. And he became one of the privileged élite of the Soviet Union receiving—along with top Communist Party members, and other key individuals—special housing, chauffeurs, access to Western consumer goods, and a very high salary.

In the late 1950s Sakharov began to protest, in private, about Soviet nuclear tests and he also, in the early 1960s, took part in the campaign against Lysenko. He started to mix more freely with other Soviet intellectuals, such as Solzhenitsyn, Zhores Medvedev and his twin brother, the historian Roy Medvedev.

In 1968 Sakharov circulated a Manifesto called *Progress, Coexistence and Intellectual Freedom*[20] which was eventually published in nearly every country in the world—except the Soviet Union and other communist states. This led to his security clearance being withdrawn and he was dismissed from his job in the nuclear weapons programme.

Since 1968 Sakharov has taken a prominent part in the Soviet dissident movement and has issued many public statements defending those who were imprisoned for 'crimes' such as exercising the right to free speech. He helped to get Zhores Medvedev released when he was illegally confined in a psychiatric hospital. As a result Sakharov's position has become more and more uncertain, many of his family have lost their jobs and attempts have even been made to expel him from the Academy of Sciences. He has often been attacked in the Soviet press and early in 1980 was banished from Moscow to the provincial town of Gorky.

Sakharov quite clearly is a man who speaks with exceptional authority on both Soviet society and on scientific matters. So let us end this third case-study by quoting from his book *My Country and the World*:

> The consequences of the Party-State monopoly are especially destructive in the sphere of culture and ideology. The complete unification of ideology at all times and places—from the school desk to the professorial chair—demands that people become hypocrites, timeservers, mediocre, and stupidly self-deceiving.

The tragi-comic, ritualistic farce of the loyalty oath is played over and over, relegating to the background all considerations of practicality, common sense and human dignity.

Writers, artists, actors, teachers and scholars are under such monstrous ideological pressure that one wonders why art and the humanities have not altogether vanished in our country. The influence of those same anti-intellectual factors on the exact sciences and the applied sciences is more indirect but no less destructive. A comparison of scientific, technological and economic achievements in the USSR and abroad makes this perfectly plain. It is no accident that for many years in our country new and promising scientific trends in biology and cybernetics could not develop normally, while on the surface out-and-out demagogy, ignorance and charlatanism bloomed like gorgeous flowers. It is no accident that all the great scientific and technological discoveries of recent times—quantum mechanics, new elementary particles, uranium fission, antibiotics and most of the new, highly effective drugs, transistors, electronic computers, the development of highly productive strains in agriculture, the discovery of other components of the 'Green Revolution' and the creation of new technologies in agriculture, industry and construction—all of them happened outside our country.

The significant achievements in the first decade of the space age, which were due to the personal qualities of the late academician S. P. Korolev and to certain fortuitous features of our programs for building military rockets, which made possible their direct use in space, constitute an exception which does not disprove the rule. And certain successes in military technology are the result of an enormous concentration of resources in that sphere.[21]

SOVIET SCIENCE AND TECHNOLOGY TODAY

Sakharov's statement raises important questions about the nature of Soviet technology. From the time of the Revolution up to 1965, there is considerable evidence that the vast majority of technological processes used in Soviet industry originated abroad—mainly in the USA but also in Britain, Germany and other European countries.[22]

After Stalin's death, Khrushchev tried to raise the level of Soviet science and technology by developing large, new research institutes and by explicitly copying the most advanced designs and machinery from abroad. According to Medvedev '"Creative assimilation" of Western achievements became the slogan of the day. . . . The research establishments became the largest in the world. But in spite of generous grants, copying Western science could not give scientists a desirable lead. . . . No sooner had experts mastered and produced a piece of technology or equipment than it was already obsolete.'[23] So, in 1966, the Soviet Union accepted the international conventions on licences and patents and started to make much equipment under licence—sometimes, as with the Fiat car plant in Togliattigrad, importing whole industrial complexes.

But this new policy has not necessarily benefited the scientific community. Again according to Medvedev:

Unfortunately the new era of active scientific cooperation has not made the scientific community freer from rather strict forms of political control. The scientific community in the USSR has become much more exposed to Western influence. This has resulted in a much more vigorous control of the ideological and political maturity of scientists. The party no longer interferes with advice on how to solve a particular problem and it does not try to decide which methods of scientific research are materialistic and which are not. But the party and state apparatus is more than ever involved in deciding who should take part in particular scientific and research missions.[24]

Many thousands of scientists, in addition to Sakharov and Medvedev, were involved in the growing dissident movement in the 1960s and this eventually led to a change in official policy.

The Soviet intervention in Czechoslovakia in August 1968 marked the end of tolerance towards the liberalism of the Soviet scientific community. Thousands of those who had openly protested were reprimanded, demoted, expelled from the party or dismissed, particularly if they refused to 'repent'. Those who continued to protest were arrested.

It took several years to remove vocal political dissent from the scientific community. The measures did not only include legal or administrative repression. Structural changes introduced political screening at every stage of scientific graduation, award and academic promotion. Party bureaux and committees now have far more power in the running of institutes and universities.[25]

For senior appointments

... the appointment is subject to confirmation by the local district or regional party committee or the Central Party Committee of the National Soviet Republic. The same procedure is followed for election of members of the academies, the most prestigious scientific positions in the USSR. Nominees for all important prizes (Lenin prizes, annual state prizes) are also subjected to intensive preliminary political screening.[26]

Perhaps in this way it is hoped to prevent the emergence in future of eminent dissident scientists like Sakharov—for there is no doubt that Sakharov's prestigious position as a leading member of the Soviet Academy of Sciences has been an important factor in enabling him to withstand political pressures.

Other even more restrictive measures are also now in force:

In 1976 the rules of awarding academic degrees and titles became even more restrictive. A new government decree stipulated that degrees can be awarded only in cases 'where high scientific standards are combined with a good mastery of Marxist-Leninist theory, a broad cultural level, active participation in political life, and following the principles of communist morality in all actions.'

The most comprehensive review of political loyalty is, however, reserved for those who apply to make scientific trips abroad. In the USSR only diplomats have passports valid for foreign travel. Scientists receive travel passports before their journey, and they are valid for one trip only. Obtaining permission for a foreign trip takes several months and it involves the approval of the institute, the academy, local and central party and government officials, the KGB and specialized departments of the Ministry of Foreign Affairs.[27]

It remains true that the Soviet Union devotes enormous resources to science and technology. About 4·5 per cent of its GNP goes on research and development and there are nearly one and a half million research scientists working in about 6000 research institutes and 850 universities and other institutions of higher education. But the real question is how effective will this investment prove to be? Overt political control of the content of research may be a thing of the past, at least in the natural sciences. But increasing political control over senior appointments could prove just as restrictive to genuine scientific creativity. And one final question. Is it possible for the Soviet Union to apply science and technology creatively without drastically changing its system of centralised economic planning and social control in ways that would ultimately destroy the Communist social order itself?

REFERENCES

1   Solzhenitsyn, A., *The Gulag Archipelago*, vols. 1–3, (London: Collins/Harvill, 1974, 1975, 1978).
2   In particular see—Medvedev, Zh., *Soviet Science*, (Oxford: Oxford University Press, 1978).
3   Solomon, Susan Gross, 'Controversy in Social Science: Soviet Rural Studies in the 1920s' in *Minerva*, vol. XIII, No. 4, Winter 1975, p. 580.
4   Ibid., p. 582.
5   Conquest, R., *The Great Terror*, (Harmondsworth: Penguin, 1971), p. 45. Quotation: © Robert Conquest 1968, Curtis Brown on behalf of Robert Conquest.
6   Ibid., pp. 46–7.
7   *See* Joravsky, D., *The Lysenko Affair*, (Cambridge, Mass: Harvard University Press, 1970), for a list of more than 100 scientists who were repressed at this time.
8   Medvedev, Zh., *The Rise and Fall of T. D. Lysenko*, (New York: Columbia University Press, 1969), p. 125.
9   Medvedev, Zh. and R., *A Question of Madness*, (London: Macmillan, 1971).
10 Medvedev, Zh., *From Lysenko to Sakharov* in *Times Higher Educational Supplement* (*THES*) March 26 1982, p. 12.
11 Medvedev, Zh., *Soviet Science*, op. cit., pp. 50–2.
12 Solzhenitsyn, A., *The First Circle*, (London: Collins/Harvill, 1968).
13 Sakharov, A., *Sakharov Speaks*, (London: Collins/Harvill, 1974), pp. 30–1.
14 Medvedev, Zh., *Two Decades of Dissidence* in *New Scientist* November 4 1976, pp. 265–6. This first appeared in New Scientist, London, the weekly review of science and technology.
15 Letter to *The Times*, November 11 1976.

16   Medvedev, Zh., *Nuclear Disaster in the Urals*, (London: Angus and Robertson, 1979).

17   Trabelka, J. R., Eyman, L. N. and Auerbach, *Analysis of the 1957–58 Soviet Nuclear Accident*. Oak Ridge National Laboratory, December 1979.

18   Dollezhal, N. and Koryakin, Y., *The Soviet Nuclear Energy Programme*, 'Kommunist', No. 14; reprinted in *Marxism Today*, December 1979, pp. 23–5.

19   Conquest, R., op. cit., pp. 435–6. Quotation: © Robert Conquest, 1968, Curtis Brown on behalf of Robert Conquest.

20   Sakharov, A., op. cit., pp. 53–97.

21   Sakharov, A., *My Country and the World*, (London: Collins/Harvill, New York: Alfred, A. Knopf, Inc., 1975), pp. 30–1. Copyright © Andrei D. Sakharov.

22   Sutton, A., *Western Technology and Soviet Economic Development 1917–1965*, 3 vols., (Stanford: Stanford University Press, 1968, 1971, 1973). Sutton's work is also summarised in van der Elst, P., *Capitalist Technology for Socialist Survival*, (London: Institute of Economic Affairs, 1981).

23   Medvedev, *THES*, op. cit., p. 13.

24   Ibid.

25   Ibid.

26   Ibid.

27   Ibid.

SUMMARY

1   After the Russian Revolution science was strongly encouraged, many new research institutes were established and much original work was done; in particular by the end of the 1920s the Soviet Union, through men like Chetverikov and Vavilov, was in the forefront of world research into genetics.

2   Under Stalin, the Communist Party extended its control even as far as the *content* of science; valid work in genetics, chemistry, relativity and cybernetics was suppressed and many dubious ideas were publicised even when there was little or no evidence to support them.

3   Stalin backed the ideas of the charlatan T. D. Lysenko who became extremely influential in Soviet biology and agriculture; Chetverikov and Vavilov were dismissed, genetics was banned and it was even decreed that all stocks of *Drosophila* be destroyed.

4   Much of the highly successful Soviet nuclear weapons programme was carried out by slave labour from the labour camps of the Gulag Archipelago and by imprisoned scientists working in special prison research institutes. During this programme careless storage of highly radioactive nuclear waste led to a disastrous nuclear accident in 1958, the occurrence of which has never been admitted by the Soviet government.

5   Massive financial support for science and technology continues but new forms of political control over scientific degrees and appointments now exist which, together with the centralised Soviet system of economic planning, are likely to limit the fruitful development of science and technology in the Soviet Union.

7.5   Science and technology in Japan since the Meiji restoration

In 1869 Mutsushito, an emperor of the Meiji dynasty, regained supreme power in Japan after centuries of rule by the feudal Shoguns. This could be seen as a retreat into the past but in fact it led to the rapid growth of Japan as a technological power which has continued, almost unchecked, ever since.

For nearly 250 years before 1869 Japan had been a closed society—almost isolated from the rest of the world. This had been a deliberate policy designed to exclude European influence, particularly Christianity. A few Dutch trading posts were all that were permitted although some foreign books were imported after 1720. Then in the 19th century the growth of European power and influence (see Chapter 4.11) across the world began to affect Japan. In the 1850s Japan was virtually forced to conclude trading treaties first with America and then with Britain, Holland, Russia and France. These treaties led to much more trade with Europe and America which gradually undermined the traditional feudal structure of Japanese society.

When the emperor regained power in 1869 he represented those who wanted to reverse the isolation policy and open up Japan to Western influence. The emperor proclaimed that 'Knowledge shall be sought throughout the world so as to strengthen the foundations of imperial rule'.[1]

In this chapter we will describe how that precept was put into practice in the 19th and early 20th centuries and how, in very different circumstances, it is still important in Japan today.

THE MEIJI RESTORATION AND SCIENCE AND TECHNOLOGY—1869–1900

Once the decision had been taken to import Western science and technology, the Japanese government set about the task with characteristic thoroughness. They made detailed surveys of the engineering industries in Europe and America. Then they acted on a broad front. For the short-term they imported foreign engineers and scientists; for the medium-term they sent students abroad and set up colleges in Japan staffed by foreign lecturers; and for the long-term they set up universities and numerous research institutes. Throughout, the emphasis was on the practical application of existing knowledge. In the words of the Prime Minister Prince Ito in 1886:

> The only way to maintain the nation's strength and to guarantee the welfare of our people in perpetuity is through the results of science. . . . Nations will only prosper by applying science. . . . If we wish to place our own country on a secure foundation, insure its future prosperity, and to make it the equal of the advanced nations, the best way to do it is to increase our knowledge and to waste no time in developing scientific research.[2]

The importance of new techniques

The foundations for the industrial revolution in Japan were laid by the Engineering Ministry established in 1870. Hundreds of foreign engineers were employed to build railways and establish a telegraph network. Modern technology was imported to develop the mining industry and to establish factories for cotton spinning. Most of these foreign engineers were British, but some were from France and the other European countries. Many of these engineers were paid salaries which were four or five times greater than those paid to Japanese government ministers.

Technical education

Great emphasis was placed on technical education and many foreign lecturers were employed in Japanese schools and colleges. They primarily taught practical subjects like engineering, agriculture, medicine and geology together with supporting basic subjects like mathematics, physics, chemistry and biology. They came primarily from Germany, Britain, France and America (see Fig. 7.5.1) and taught in their native languages. Again they were often paid much more than native Japanese.

One particularly important development was the College of Engineering in Tokyo which was staffed mainly by British engineers and which began teaching in 1873. The aim of the College was 'to train men who would be able to design and

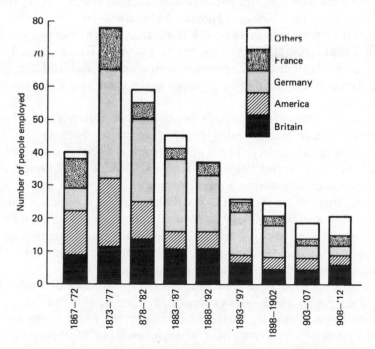

Fig. 7.5.1 Numbers of foreign lecturers in Japan, 1867–1912. *Redrawn from Nakayama, S., 'A Century's Progress in Japan's Science and Technology', in Technical Japan, vol. 1, part 1, 1968.*

superintend the works which were necessary for Japan to carry on if she adopted Western methods.'[3] The Prime Minister Prince Ito later said '. . . that Japan can boast today of being able to undertake such industrial works as the construction of railways, telegraphs, telephones, shipbuilding, working of mines, and other manufacturing works entirely by the hands of Japanese engineers is mainly attributable to the College . . . .'[4]

Many students were also sent abroad and later came back to teach the next generation of Japanese students.

Universities, research institutes and scientific societies

In the period from 1875 to 1900 the Japanese government established many of the same kinds of institutions which make up the scientific community in Europe and North America. The Imperial University of Tokyo was set up in 1877 and, a few years later, it absorbed the Tokyo College of Engineering. Similar universities were established at Kyoto in 1897 and Tohoku in 1911. Again the emphasis was on practical knowledge as can be seen from this extract from the charter of Tokyo University:

> The aim of the Imperial University shall be to teach and study such sciences and practical arts as meet the demands of the State.[5]

The government also set up a number of research establishments during the years following the Meiji restoration. Examples include the Naval Hydrographic Division in 1871, the Tokyo Hygenic Laboratory in 1874, the Central Meteorological Observatory in 1875, the Geological Survey Bureau in 1878, the Electro-Technical Laboratory in 1891, the Institute for Research on Infectious Diseases and the Agricultural Experimental Station in 1892 and the Chemical Industrial Research Institute in 1900. Once again there is a clear emphasis on practical research.

Many scientific societies were also established in this period. The Tokyo Mathematical Society was founded in 1877 and later became the Japanese Mathematico-Physical Society. The Tokyo Chemical Society originated in 1878 and in the following year the Tokyo Academy of Sciences was founded, although initially natural scientists were in a minority on this body which was renamed the Imperial Academy of Sciences in 1906. Other societies were established for medicine in 1875, physical geology in 1879, pharmacology in 1881, meteorology and botany in 1882 and zoology in 1888. Societies for heavy engineering tended to be set up a little later—for mining in 1889, construction in 1886, electrical engineering in 1888 and mechanical engineering only in 1897.

Many of these societies have grown extremely rapidly since that time. See, for example, Fig. 7.5.2 which shows how the membership of the Japanese Mathematico-Physical Society increased from 1877 to 1945. From about 1888 onwards the growth was exponential with the number of members doubling roughly every ten years, except for a brief period during World War 1. This growth is much more rapid than the growth in the population of Japan (see Fig. 7.5.3) which has increased from about 36 million in 1875 to about 110 million

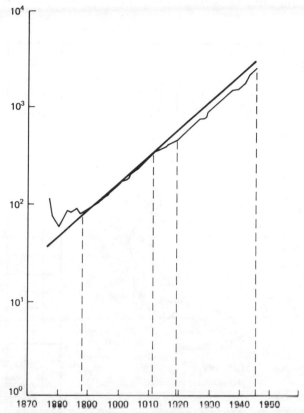

Fig. 7.5.2 Exponential growth in the membership of the Japanese Mathematico–Physics Society, 1877–1945 (logarithmic scale); the membership approximately doubled every ten years. *Redrawn from Yagi, E., 'The Statistical Analysis of the Growth of Physics in Japan' in Nakayama, S., Swain, D. L., and Yagi, E. (eds), Science and Society in Modern Japan, Cambridge, Mass: MIT Press, 1974.*

today—a doubling time of about 60 years. And it is even faster than estimates of the growth of Western science where the doubling time is usually estimated as approximately 15 years (see Fig. 3.5.1, page 78 and Chapter 6.1).

This extremely rapid growth in the numbers of Japanese scientists took place in a number of stages as can be seen from Fig. 7.5.4 which shows the number of physicists active in Japan from 1860 to 1960. At first these were mainly foreign physicists together with a few foreign-trained Japanese (Group I); by about 1910 they had either left Japan, died or retired. These men trained a group of Japanese physicists (Group II) who from the 1890s onwards became the teachers of the first generation of Japanese physicists who were both trained in Japan and taught in Japanese. After this the physics community in Japan entered on a period of self-sustained growth (Groups III and IIIb) during which both the number of physics graduates (IIIa) and of postgraduate students (IIIb) doubled roughly every seven years.

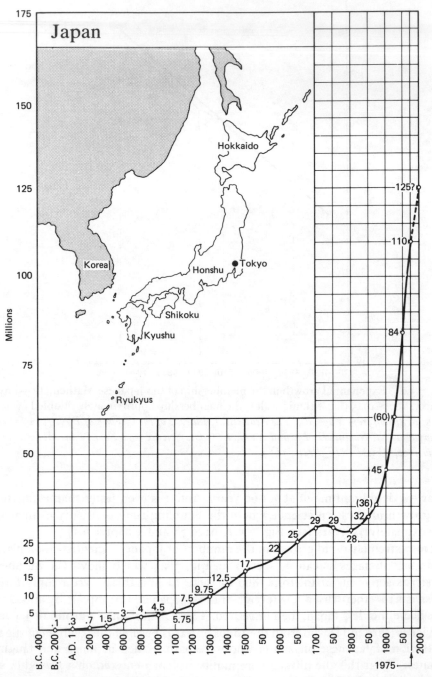

Fig. 7.5.3 Growth in population of Japan. *Redrawn from McEvedy, C. and Jones, R., Atlas of World Population History, page 181, copyright © Colin McEvedy and Richard Jones, 1978, reprinted by permission of Penguin Books Ltd.*

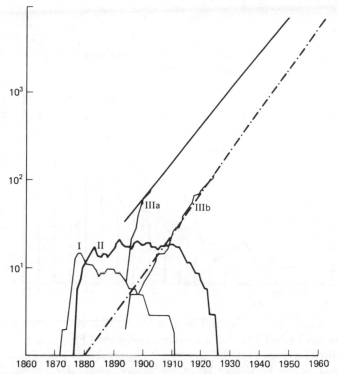

Fig. 7.5.4   Numbers of physicists in Japan, 1860–1960; I—Foreign or foreign-trained physicists, II—Japanese students of group I; III—Japanese physicists taught in Japanese, (a) graduates, (b) post-graduates continuing for D.Sc. *Redrawn from Nakayama, S., Swain, D. L., and Yagi, E. (eds), Science and Society in Modern Japan, Cambridge, Mass: MIT Press, 1974.*

THE RISE OF JAPAN AS A MILITARY POWER—1890–1945

The development of science and technology in Japan was one important factor in the growth of Japan as an important military power in the Far East. In 1894–5 Japan defeated China in war and nine years later, in 1904–5, also defeated Russia—the first time in modern times that a European country had been defeated in war by a country outside Europe.

From the 1890s onwards Japan began to develop its heavy industries. Iron and steel production expanded rapidly and much of it was concentrated in the Yawata Iron Works (see Fig. 7.5.5) which was financed by the government and operated under military control. The output from this foundry was used mainly in the construction of warships for the Japanese navy. Other heavy industries were also developed at this time such as coal-mining, and the manufacture of rolling stock for the railways and of equipment for the electrical supply industry. From 1897 onwards steam power replaced watermills and by 1914 all heavy industry had switched to electrical power supplies.

During World War 1 Japan also began to pay closer attention to the development of scientific research aimed specifically at the growth of science-

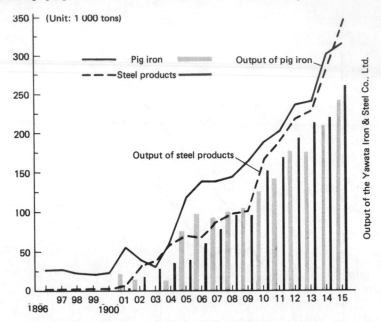

Fig. 7.5.5 Production of pig-iron and steel in Japan, 1896–1915; the output from the government-owned Yawata Iron Company was mainly used for warships; compare with Figs. 4.11.3 and 4.11.4 (page 215). *Redrawn from Nakayama, S., Swain, D. L., and Yagi, E. (eds), Science and Society in Modern Japan, Cambridge, Mass: MIT Press, 1974.*

based industries. Probably the most important initiative was the founding of the Institute of Physical and Chemical Research in 1917 which was partly modelled on Germany's Imperial Institute for Physics established in 1887. However, other research institutes were also founded at this time, both by the government and by private industry. Japan was now beginning to develop its own scientific research community, but in contrast with research in Europe, its attention was primarily directed towards Japanese industry and to military technology in particular.

During the 1930s the emphasis on military technology increased and Japanese military power grew rapidly. In 1932 she effectively annexed Manchuria. Then in 1937, taking advantage of the civil war in China, Japan declared war on China and rapidly occupied much of the country, including the capital Peking and the major cities of Shanghai and Nanking. At this time science and technology in Japan became almost completely devoted to military purposes. A National Mobilisation Law was passed in 1938 and a Science Mobilisation Council established in 1940. However, it was also in the 1930s that the theoretical physicist Yukawa did his famous work on the theory of mesons for which he was awarded Japan's first ever Nobel prize in 1949. However, this kind of pure scientific research was very much the exception rather than the rule.

Then in 1941 Japan made a surprise attack on the American fleet at Pearl Harbor. This attack brought America into World War 2 and it marked the beginning of the war in the Pacific which finally ended in August 1945 with the dropping of the first atomic bombs on Hiroshima and Nagasaki.

SCIENCE AND TECHNOLOGY IN JAPAN SINCE 1945

In 1945 Japan was a defeated nation—her productive capacity had fallen to only 10 per cent of previous levels and there was a threat of food shortages and epidemics. Since then Japan has become one of the most prosperous nations in the world. Science and technology have clearly been important in this transformation but it is much less clear precisely how science and technology have influenced Japanese prosperity and what role the Japanese government has played in the rise of Japan as a major technological power. In this section we will try to illuminate these questions by describing some of the changes which have taken place since 1945.

Post-war reconstruction, 1945–55
In the early years the clear priority was to avert food shortages by the improvement of agriculture. Better strains of rice, more fertilisers and pesticides and improved agricultural machinery all led to greater output. Productivity also increased substantially which meant that more people were available to work in the growing industries of the 1960s.

As in the past, industries like mining and manufacturing were revived by the import of foreign technology. But by contrast with the 1930s, there was virtually no military or defence expenditure on science and technology. One result has been that, since 1945, private companies have provided by far the largest share of the resources devoted to research and development (see Table 7.3.4, page 381). But the government did control licences for the import of foreign technology and set limits on the foreign ownership of Japanese firms. One important development was the import of quality control technology from the United States.

Economic growth, 1955–73
In this period Japan's industries expanded rapidly. The production of household electrical goods such as TVs, radios and refrigerators grew very fast and major developments also took place in the transport industries—railways, shipbuilding and car manufacture—and in the production of artificial fibres. These changes also led to rapid growth in the production of iron and steel and in the output of the chemical industry. Towards the end of the 1960s the electronics industry also grew very rapidly and a considerable increase took place in expenditure on both research and development and on investment in equipment for the manufacture of semiconductors and integrated circuits.

In all these industries, great emphasis was put on the application of new techniques and many industrial research laboratories were established. The government also became more directly involved in research and development with the establishment of the Science and Technology Agency in 1956 and the Council for Science and Technology in 1959. These agencies have set up a number of research organisations and laboratories such as the Atomic Energy Research Institute in 1956 and the National Centre for Space Development in 1964. In addition, they have produced a series of reports on the state of science and technology in Japan which have considerably influenced government policies.

Foreign or Japanese technology? 1973 onwards

Since the mid-1970s there has been a significant change in the major aims of science policy in Japan both in industry and in the government agencies. The emphasis is now much more on trying to develop specifically Japanese technology rather than on the efficient application of technology imported from abroad.

The annual reviews published by the Council for Science and Technology identify some weaknesses in Japanese science but also clearly show that Japan is now one of the top six countries involved in large-scale scientific research and development—the other five are the Soviet Union, the USA, France, Britain and West Germany.

Fig. 7.5.6 shows, for the mid-1970s, the share of these six countries in the world's total GNP and total expenditure on research. The top six countries share about 65 per cent of world GNP but spend nearly 85 per cent of the total expenditure on research. Japan is third, after the USA and the Soviet Union, in both categories with just under 10 per cent of each of the totals. But Japan, like France, only spends about 2 per cent of its GNP on research compared with about 4·5 per cent in the Soviet Union and about 2·5 per cent for the USA, Britain and West Germany. Fig 7.5.7 shows, for the same six countries, their share of the world's total number of research workers and total population. Together the six

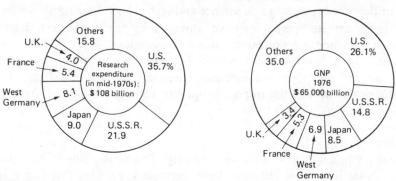

Fig. 7.5.6 Proportions of total world Gross National Product (GNP) and research expenditure in the mid-1970s. *From Science and Technology in Japan, vol. 1, no 1, January, 1982, Japanese Government Publication.*

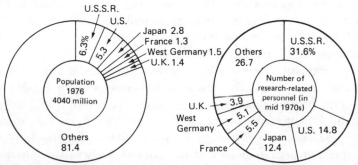

Fig. 7.5.7 Proportions of total world population and research workers in the mid-1970s. *From Science and Technology in Japan, vol. 1, no. 1, January 1982, Japanese Government Publication.*

countries have about 75 per cent of the research workers but less than 20 per cent of the total population, while Japan is again third after the Soviet Union and the USA in numbers of research workers. More detailed analysis of the available statistics shows that, between the 1960s and the 1970s, Japan has roughly doubled its share of the current world technological capability and of the world potential for technological development. However, Japan still originates very little of the technology it uses.

In order to try to correct this imbalance, Japan is now attempting to foster genuine innovation in science and technology both by bringing together outstanding researchers and by investing heavily in specific projects such as the development of nuclear power—both fission and fusion; satellites for meteorology and communications; short take-off and landing aircraft; biotechnology; and an ambitious programme to develop the resources of the oceans.

Only time will tell whether these projects and policies really will make Japan, for the first time, a net exporter of science and technology.

Another and perhaps more important question remains about the remarkable growth in Japanese technological strength ever since the Meiji restoration and since 1945 in particular. Does Japan provide a model of technological development which other countries could follow? Or is the rise of Japan, in just over a hundred years, from a relatively poor agricultural and feudal country to one of the world's richest technological powers, due primarily to specific features of the Japanese character and social structure?

REFERENCES

1 Fox, G., *Britain and Japan, 1858–1883*, (Oxford: Oxford University Press, 1969), p. 261, quoted in Brock, W. H., *The Japanese Connection*, British Journal for the History of Science, Vol. 14, No. 48, 1981, p. 229.
2 Brock, W. H., op. cit., p. 229.
3 Ibid., pp. 232–3.
4 Ibid., p. 239.
5 Tuge, H. (ed.), *Historical Development of Science and Technology in Japan*, (Tokyo: Kokusai Bunka Shinkokai, 1968), p. 101.

SUMMARY

1 In 1869, after centuries of isolation, Japan adopted a policy of rapid industrialisation.
2 Expert foreign engineers were recruited to introduce new techniques and to staff new colleges and universities; many research organisations and scientific societies were founded.
3 Heavy industries were developed from 1900 onwards and Japanese science and technology became increasingly devoted to military purposes culminating in war with China in 1937 and World War 2 in 1941–5.
4 Since 1945, the efficient application of imported science and technology has played a major part in Japan's emergence as one of the most prosperous nations in the world.
5 The majority of Japan's investment in scientific research and development is made by private industry but the government is now attempting to foster genuine innovation in science and to develop specifically Japanese technology.

7.6 Science and technology in China before and after 1949

Why 1949? It was in 1949 that the Chinese Communist revolution took place. Led by Mao Tse-tung, the Chinese Communist Party overthrew the existing government and it has ruled China ever since. So 1949 forms a natural division in 20th-century Chinese history in the same way as 1917 does in the history of Russia. In Russia science was an import from 18th-century Europe and the flourishing science that did exist before the revolution was clearly in the European tradition. For China the situation was very different. As we saw in Chapter 5, in traditional China science and technology had developed independently and in many ways were more advanced than in Europe. This situation changed after the extremely rapid growth of European science and technology which followed the 17th-century scientific revolution. This growth of European technology and economic power later led to a world-wide expansion of European power and influence (see Chapter 4.11). And it was this expansion which, in the 19th century, began to undermine the centuries-old stability of the Chinese empire.

EUROPEAN INFLUENCE ON CHINA IN THE 19TH CENTURY

In the early part of the 19th century, European influence in China was confined to a few trading ports. But then in 1840–2 came the Opium War, in which Britain defeated China because of the superiority of European weapons and tactics. It led to the conclusion of what came to be called the unequal treaties with, first, Britain and then, in later years, with many other countries—America, France, Belgium, Sweden, Norway, Portugal, and, later still, with Russia and Germany. These unequal treaties gave vast concessions to the Europeans: foreigners in China lived in special settlements and were not subject to Chinese law; very low customs duties were levied on imported goods; and free movement was allowed for foreign ships around China's coasts and along its inland rivers and canals. Gradually the European powers extended their trade and influence. More wars followed—and again the military weakness of China compared with the European countries meant that the Europeans virtually took control of much of China. In 1860, British and French troops advanced to Peking and British troops sacked and destroyed the great imperial palace outside Peking—an act which confirmed the Chinese in their traditional view that all foreigners were uncivilised barbarians. This domination by foreign powers was deeply resented in China. And this resentment grew stronger when Japan joined in too and defeated China in war in 1894–5. As we saw in Chapter 7.5, Japan had developed its science and technology on the European model—particularly after the Meiji restoration in 1869—and this led to pressure for China to do the same.

However, there were disputes amongst the Chinese about how to do this. Some wanted to use the technical and military devices on which European power depended while leaving the traditional structure of Chinese government intact— that is, rule by the emperor and by the mandarins, the experts in classical literature and in the writings of Confucius (see Chapter 5). Others wanted to introduce science, technical subjects and knowledge of European societies into the examinations through which the mandarins were selected. Such a reform would effectively change the whole system of government in China, and it aroused great opposition. These disputes went on for many years. Some changes were made— for example, the founding of the Peking National University in 1898 after the pattern of Japan's Imperial University. But it was not until the early years of this century that any major changes took place.

THE END OF TRADITIONAL CHINA

The two changes which, more than anything else, opened Chinese society to the influence of Europe were the abolition of the traditional examination system in 1905 and the overthrow of the Emperor in 1911 which was followed by the establishment of a republic in China. The abolition of the traditional examinations opened up the whole of China's education system to European ideas. And the new republic was founded by Sun Yat-sen and his supporters, most of whom had spent many years abroad as students and were familiar with European ideas and institutions. However, overthrowing the Emperor was relatively easy compared with the problems of establishing a stable system of government in China. The republic lasted nearly forty years—from 1911 to 1949—but its history over all that period was stormy and turbulent, with civil war and war with other countries going on most of the time. We can't describe this period in any detail— all we can do is indicate some major events.

In 1921, after years of civil war, two major political groups were established in China. The nationalist Kuomintang was reformed under the leadership of Sun Yat-sen and Chiang Kai-shek. And, partly inspired by the Russian revolution, the Chinese Communist Party was formed in Shanghai; Mao Tse-tung and Chou En-lai were founder members. During the early 1920s the two groups co-operated but it was always an uneasy alliance. Then in 1927 Chaing Kai-shek attacked the Communists and drove them from Shanghai. After that the Communists' main power base was the countryside and it was at this time that Mao Tse-tung first emerged as the leader of a peasant revolt. Civil war between the Kuomintang and the Communists intensified and in 1934 Mao Tse-tung led the Red Army in its famous Long March of more than 6000 miles from its old base in southern China to a new base in the northwest. After the war with Japan started in 1937, the Kuomintang and the Communists united against the invader but, when Japan was defeated in 1945, the civil war started up again. Then in 1949 the Red Army finally triumphed, the Communists came to power and Sun Yat-sen's republic collapsed.

SCIENCE AND TECHNOLOGY IN REPUBLICAN CHINA—1911–49

Despite all this turmoil in republican China, there was a considerable growth of science and technology. How was this growth brought about? The method used was to found and develop the same kind of institutions that, as we have seen, make up the scientific community in Europe and America—universities, research institutes, scientific societies and the journals they publish. In short, we can say that republican China followed the Western model of scientific development—in much the same way as Japan had done 40 years earlier but without the firm Japanese emphasis on the practical application of knowledge.

Students sent abroad
First of all many students were encouraged to go abroad to study—to the United States, to Europe and, most of all, to Japan. It is estimated that about 40,000 Chinese students graduated abroad during the first half of the 20th century, about 40 per cent of these in the sciences and engineering subjects. About 10 per cent of these students obtained Ph.D.s, mostly from North American or European universities.

Universities in China
There was also a considerable expansion in the numbers graduating from Chinese universities. Probably about ten times as many students graduated in China as did so abroad, although, as with the overseas students, the majority of these were studying the social sciences and the humanities. Following the Confucian tradition, the study of social sciences such as law, political science and economics still had greater prestige than the natural sciences and engineering. However, the numbers of graduates were only a very small proportion of the vast Chinese population which was around 500 million at this time.

Scientific societies and research institutes
A considerable number of research institutes and scientific societies were set up under the republican government; most of the societies began publishing their own journals as well as translating Western books and journals and building up specialist libraries.

One of the first research institutes was the National Geological Survey, founded in 1912, and the Geological Society of China dates from 1922. In 1928 the Academia Sinica, or Chinese Academy of Sciences, was founded to advise the government and to supervise research institutes in major areas. In its first year, 1928, it set up research institutes in physics, chemistry, engineering, meteorology and geology, and in 1929 it established the Museum of Science and Natural History which became the Research Institute of Zoology and Botany in 1934.

This period also saw the establishment of many scientific societies—amongst them are the Chinese Meteorological Society in 1924, the Chinese Institute of Mining and Metallurgy in 1927, the Institute of Oceanography in 1930, the Chinese Institute of Engineers in 1931, the Chinese Physical Society and the

Chinese Medical Association in 1932, the Chinese Chemical Society in 1934 and the Chinese Institute of Mathematics in 1935.

There was a genuine upsurge in science and technology in the 1920s and 1930s which was temporarily halted during the war with Japan from 1937 onwards but which gave a solid foundation for the later developments which took place after the war with Japan and the civil war were over.

CHINA SINCE 1949—THE COMMUNIST REVOLUTION

The People's Republic of China was set up on 1st October 1949, after the People's Liberation Army, or Red Army, under Mao Tse-tung had finally captured Peking. Mao Tse-tung became the Chairman of the Governing Council. Ever since 1949 the Chinese Communist Party and the Red Army have been the only important political forces in China and Mao Tse-tung remained the most powerful man in China until his death in 1976.

Like communist governments all over the world, the Chinese Communist Party set out to control and dominate all aspects of life in China. So it is impossible to discuss the subsequent development of science and technology in China without outlining the major political developments since then.

There have been five main periods in the development of communist China:

1949-57   Revolution on the Soviet model
1957-59   The 'Hundred Flowers' movement and 'The Great Leap Forward'
1960-66   The break with Russia
1966-76   The Great Proletarian Cultural Revolution
1976-   Post-Mao China—the 'Four Modernisations' and the end of the 'Gang of Four'.

There have thus been at least four major reversals of policy by the Chinese communists since they came to power in 1949 and innumerable minor reversals. In all these five main periods of communist rule, the aim has been to turn China into a great power as fast as possible—a great industrial power, a great military power and a great political power. This was to be done under the control and leadership of the Chinese Communist Party and by making the best possible use of science and technology. The major question was—how to do this? This question was answered very differently in each of these main periods.

SCIENCE AND TECHNOLOGY IN CHINA SINCE 1949

The Soviet model—1949-57

In its early years, the Chinese communist revolution was strongly influenced by the Soviet Union. China not only modelled its institutions and political structures on those established in the Soviet Union—it also received considerable direct Russian aid both in men and materials.

If we remember that this was the period in which Stalin's rule was at its most despotic in the Soviet Union, it will not surprise us to learn that the Chinese revolution adopted many of the same totalitarian features which were, by this time, well-established in Russia. As in the Soviet Union the aim was a centralised state in which the governing party, the Communist Party, controlled all aspects of life—economic, political, educational, cultural and religious. The economy was to be planned centrally: the first Five-year plan on the Russian model was adopted in 1953. Central control by the Communist Party was maintained using the same methods as in the Soviet Union: control of ideas by controlling and censoring the media and the educational system; and control of individuals by means of the secret police who carried out purges of dissidents and set up forced labour camps similar to those in the Gulag Archipelago. However, there were some distinctively Chinese features to the Chinese revolution. One of these was the emphasis on 'thought-reform'. 'Thought-reform' is a process of re-education which aims to change completely the consciousness of individuals, using any means ranging from condemnations at public meetings through physical imprisonment and torture to every kind of psychological technique. The importance placed on 'thought-reform' is shown by the fact that the Chinese communist criminal code has a sanction which is unique in the world—the death sentence suspended for two years pending satisfactory 'thought-reform'. This was the sentence passed on Chiang Ching, Mao Tse-tung's widow, following the show trial of the Gang of Four in 1981.

'Thought-reform' was much used against intellectuals in the early days of the revolution, and in particular against those trained in the social sciences. Subjects such as sociology and political science were abolished in 1952 and courses in economics, law, philosophy and history were completely revised. All those social scientists who had previously taught these subjects, or who were of bourgeois origins, or had been educated in a Western country, or who had co-operated with the previous government were subjected to the full rigours of the 'thought-reform' process.

However, this policy was not adopted to anything like the same extent towards those who had studied the natural sciences or engineering. Like the Soviet Union, the Chinese communists put great emphasis on science—both because they thought they were reconstructing their society on the principles of scientific socialism and because they saw that science and technology were vital to industrialisation, without which China would never become a great power.

So in the early years science and engineering were strongly promoted and scientific subjects were closely integrated in the Five-year Plan. In addition there was a specific long-term science plan for the period 1953–67 which was drawn up with the help of the Academia Sinica. Also there was a great influx of trained scientists and engineers from abroad. Many Western-educated Chinese scientists returned home and took up key positions in Chinese universities and research institutes, and many Russian scientists and engineers came to work in China on major engineering projects.

In this period many aspects of applied science and technology were given special priority and, while scientists had to conduct their work under the guidance of the Communist Party, they were not subject to the same pressures as the social scientists. For these key people the party line at this time was that it was more important for them to be expert than to be red—but if they were red and expert, so much the better.

The 'Hundred Flowers' movement and the 'Great Leap Forward'—1957–60

By 1957 the first Five-year Plan was well on the way to being successfully completed and Mao Tse-tung felt confident enough to relax the control of the Communist Party a little. In February 1957, he called on the people of China to strengthen Chinese Communism by public criticism and counter-criticism. After eight years of censorship and 'thought-reform' for dissidents, people were at first reluctant to speak out. But early in May 1957, Mao made his famous speech in which he said 'Let a hundred flowers bloom; let a hundred schools (of thought) contend'. This speech marked the beginning of a short period of growing public criticism of communist rule in China. It only lasted about six weeks—until about the middle of June—but in that time the criticism grew and became so widespread and far-reaching that the Communist Party had to call a halt or risk losing all control. After the clamp-down, many of those who had spoken out were purged, and some have seen the whole episode as a deliberate trick by Mao Tse-tung to expose and root out those he saw as 'poisonous weeds'. But a more likely explanation is that he really believed that his rule was in the interests of the great mass of the Chinese people and that by involving the masses in this way in the process of government he would be able to govern China better.

The same belief in the role of the masses is seen in the Great Leap Forward of 1958, which was also forced through by Mao. This was a major change in the centralised economic policy of the first Five-year Plan. For example, as well as investing in large new steel mills, Mao called on workers and peasants all over the country to smelt their own steel in small furnaces and workshops. At the same time agriculture was to be collectivised even further. For a short time, output increased but then the dislocation caused by the rapid changes led to drastic falls in both industrial output and the output of food. By 1960 the situation was desperate and was made worse by another dramatic event—the break with the Soviet Union.

The break with the Soviet Union—1960

Despite close co-operation in the early years of the Chinese revolution, from 1957 onwards relations between China and the Soviet Union became more and more strained as Mao Tse-tung tried to take a more independent line. Then in 1960 there came a complete break. In July of that year the Soviet Union suddenly decided to withdraw at once—within a month—all the Soviet students and engineers in China, about 1400 men altogether, and to cancel all agreements on technical co-operation. According to one source, when they went the Soviet experts took their blueprints with them together with everything moveable. This, of course, was a staggering setback for the Chinese programme of industrialisation. What it meant was that China had now to be self-sufficient in science and technology as well as everything else. And this was the policy which was adopted from 1960–66. It was once again more important to be expert than to be red.

One of the main points of difference between China and the Soviet Union concerned nuclear power and nuclear weapons. The Soviet Union exploded its first hydrogen bomb in 1953 and then went on to develop a combined nuclear power and weapons programme.

China expected to share in this knowledge but Khrushchev—perhaps fearful of what China might later do with this knowledge—said no. So China had to go it alone and start its own crash programme of nuclear research which was given top priority. Many Western-trained Chinese nuclear physicists played leading parts in this work, which was very successful. The first Chinese atomic bomb was exploded in 1964 and their first hydrogen bomb only three years later, in 1967 (see Chapter 7.3). China, too, has had a programme to develop high-powered rockets to carry these bombs.

Although these top-priority projects have been successful, there is no doubt that the sudden withdrawal of Soviet aid had an extremely damaging effect on China's economy. This may be one reason for the fierce hostility between China and the Soviet Union during the 1960s and 1970s.

The Great Proletarian Cultural Revolution 1966–76

The next major phase in the history of communist China was the Great Proletarian Cultural Revolution—or Cultural Revolution, for short—which Mao Tse-tung initiated in 1966. Once again Mao appealed to the masses to criticise the Communist Party leadership but this time it was because that leadership was now opposing Mao himself. The spearhead of the Cultural Revolution were bodies of students, the Red Guards, who were encouraged by Mao to criticise publicly all those in authority—including their teachers and lecturers. The Red Guard movement grew rapidly and they soon effectively controlled most schools and colleges. 'The Chinese newspapers were filled day after day with pictures of intellectuals being paraded through the streets in dunce caps and being humiliated verbally and physically at mass meetings.'[1]

It was now more important to be red than to be expert—in fact to be expert was to run the risk of severe criticism, dismissal and public disgrace, not to say physical injury or even death.

The Cultural Revolution did enormous damage to Chinese education. Most universities and colleges were closed *for four years* and even when they reopened in 1970 the education they gave was not good, for many teachers had been appointed because of their political reliability rather than their knowledge and qualifications. According to M. Goldman, an expert on China's intellectuals:

> . . . A whole generation of westernized intellectuals were decimated. Those who did not die of natural causes were purged or killed or committed suicide only because they were intellectuals. The institutions they helped to create—the universities, research institutes, journals, libraries, and other creative

enterprises—were also decimated. The destruction of China's cultural life may take generations to repair.[2]

At first Mao had 'decreed a Cultural Revolution from below against the party bureaucracy, but one that did not interfere with industrial and agricultural production or with science and technology. Scientists were exempted from the Cultural Revolution.'[3] and 'the Red Guards were initially restrained from attacking scientists and scientific institutions.'[4] But later:

> ... the linkage of the scientists with the party bureaucracy spurred the Red Guards to disrupt high-level ministries concerned with science and technology. Not even the ministries involved in defense were unscathed. The Seventh Ministry of Machine Building, which directed aircraft and missile production, came to a standstill because of Red Guards' rampages through the ministry. There were some exceptions, but most scientists at research institutes and universities were sent with other intellectuals to factories and farms to learn through practice. In the face of an accelerating, runaway movement, even Chou En-lai, who sought to curb the excesses of the Cultural Revolution and restrain the attacks on intellectuals and officials, was unable to protect the scientists from Red Guard abuse.[5]

Universities and colleges became centres for indoctrination in the thoughts of Mao rather than of academic study. An editorial in the Shanghai *Liberation Daily* 'prophesied that, when Mao's thought was used to direct science and technology, the result would be the "creation of a host of miracles". Whereas before the Cultural Revolution Mao's thought had been a guide, now it was to be the major source for scientific research. Workers who rejected "reliance on experts" and "discarded all foreign conventions and rules" in favor of Mao's thought were reported to produce thousands of technological innovations.'[6]

The result was that:

> ... In most areas, scientific research stagnated and even retrogressed because of disruptions, isolation from world scientific developments, and the closure of universities in order to politicize their curricula. A generation of scientists was lost. Even China's nuclear weapons program, which Mao tried strenuously to insulate, was upset by Red Guard interference. Mao authorized Chou En-lai in April 1968 to stop the disruption caused by the Red Guards in the defense agencies, but to no avail. In the following month, Red Guards again attacked leading scientists and ransacked their offices. ... Virtually all academic and scientific journals stopped publication and virtually all universities, libraries and museums were closed.[7]

So the education of a whole generation has been almost irretrievably damaged and departments of science and technology have been affected along with the rest—even those working on top-priority defence projects like the development of nuclear weapons. In these areas Mao considered that it was still permissible to be expert but the Red Guards thought otherwise.

Post-Mao China—The 'Four Modernisations' and the end of the 'Gang of Four':
1976–present

After Mao's death in 1976, the so-called 'Gang of Four' attempted to seize power and carry on the policies of the Cultural Revolution. However, they were defeated and the new Communist Party leaders, as well as bringing the Gang of Four to trial, have reversed their policies. They now strongly emphasise the importance of education in science and technology and of high academic standards in these subjects. This is one of the famous 'Four Modernisations' by which China is to be transformed into a major industrial power—the other three are agriculture, industry and defence.

A leading part in shaping these new policies has been played by the vice-premier, Teng Hsiao-ping, who had been disgraced during the Cultural Revolution but rehabilitated in 1973. Teng was appalled by a report he commissioned from the Chinese Academy of Sciences on the damage done to Chinese science and technology during the Cultural Revolution. So Teng has emphasised China's need for experts even if they are not ideologically sound. As he said in 1975 'What does it matter if one is a little white and expert?'[8]

Early in 1978, in a major speech, Teng stated that 'The crux of the four modernisations is the mastery of modern science and technology. . . . Modern science and technology are undergoing a great revolution. . . . How [world wide] have the social productive forces made such tremendous advances and how has labour productivity increased by such a big margin? Mainly through the power of science, the power of technology.'[9] So the import of Western technology is now being strongly encouraged and Chinese students are once again going abroad to study technical subjects and languages.

And Teng, in the same speech, redefined the class status of Chinese intellectuals and experts. 'The overwhelming majority of them are part of the proletariat. . . . Those who labour, whether by hand or brain, are all working people in a socialist society.'[10] To be expert is once again more important than to be red.

CONCLUSION

In China since 1949 we have seen, probably even more clearly than in the Soviet Union, the tension which exists in a society which tries to base itself on science and technology and yet which, at the same time, tries to organise its whole culture in a centralised and monolithic way. In Chapter 7.1 we put forward two main hypotheses about the interaction of science and society: that for science to flourish in a society, it is necessary for there to be some compatibility between the values of that society and those of the scientific community; and that such compatibility is even more vital if a society wants to make fruitful use of science and technology.

In a society like China, where there is such close central control of all sources of information, these conditions do not exist. The sort of apparently inexplicable oscillation between redness and expertness which we have observed in China becomes understandable if our hypotheses are valid. And this could well mean that the present phase of enthusiasm in China for Western science and technology is unlikely to last *unless* China evolves towards a more liberal type of society. Yet there is no sign of this happening for if it did it would surely mean that the Chinese Communist Party would cease to be the sole centre of power in China. The Democracy Wall in Peking in 1978–9, which carried criticisms of the government, didn't last much longer than Mao's 'Hundred Flowers' period.

> Although party leaders repeatedly urge a hundred flowers to bloom and a hundred schools to contend, there is still no institutional or legal way for intellectuals to separate themselves from the party's overall determination of what flowers will bloom and which schools will contend.[11]

and the question remains as to

> whether a Communist Party leadership, no matter how flexible, can accommodate a group of experts whose work at times might diverge from the party's political demands.[12]

REFERENCES

1   Goldman, M., *The Persecution of China's Intellectuals* in *Radcliffe Quarterly*, September 1981, vol. 67, no. 3, p. 12. Copyright 1983, Radcliffe College, reprinted with permission of Radcliffe Quarterly.
2   Ibid.
3   Goldman, M., *China's Intellectuals: Advise and Dissent*, (Cambridge, Mass: Harvard University Press, 1981), p. 135.
4   Ibid., p. 136.
5   Ibid., p. 137.
6   Ibid., pp. 137–8.
7   Ibid., p. 138–40.
8   Ibid., p. 219.
9   *Peking Review*, No. 12, March 24 1978; quoted in Barnett, A. D., *China's Economy in Global Perspective*, (Washington: The Brookings Institution, 1981), p. 61.
10   Ibid., p. 61.
11   Goldman, M., op. cit., p. 243.
12   Ibid., p. 231.

SUMMARY

1   In the 19th century the world-wide expansion of European power and influence—backed by science and technology—eventually undermined the centuries-old stability of the Chinese empire, and a republic was set up in 1911.
2   In republican China (1911–49), despite years of war and civil war, there was a considerable growth of science and technology; many universities, research institutes and scientific societies were established including the Chinese Academy of Sciences in 1928.

3   Since the communist revolution in 1949, the Chinese Communist Party have aimed to turn China into a great industrial, military and political power; this was to be done *under* the control and leadership of the Communist Party *and* by making the best possible use of science and technology.

4   These two aims—redness and expertness—have frequently been in conflict since 1949; then, during Mao's Cultural Revolution, redness won and enormous damage was done to China's science and technology.

5   Since the death of Mao, China has designated excellence in science and technology as one of the 'Four Modernisations'; for the moment it is once again more important to be expert than red but the potential for future conflict remains.

PART EIGHT

Case-studies of specific issues

In this section a number of topical issues are considered—economic development in the third world, energy resources, the spread of computers, biological and population problems and the differences between the natural and social sciences. Each of these topics is discussed in the light of the material dealt with in earlier sections of the book. In particular, some relevant historical developments are described and an attempt is made to consider these issues and problems as they affect a number of different societies—rather than focusing exclusively on one society or type of society. Throughout, the indispensable part played by reliable scientific knowledge in attempts to solve these problems is emphasised.

These brief chapters clearly cannot be comprehensive. However, it is hoped that they will indicate the kinds of arguments and considerations which need to form part of any realistic assessment of these complex topics.

8.1   Science, technology and economic development in the third world

Science and technology are sometimes seen as the key to economic development—both in the third world and elsewhere. As we saw in Chapters 4.4, 4.9 and 4.11, science and technology were important in the Industrial Revolution but they were by no means the only causes of the growing power and prosperity of Europe during the 18th and 19th centuries. And today the application of science and technology is only one factor amongst many which influence the complex development of modern industrialised societies. This is also true for the many, very different, societies which make up the third world.

In this chapter we clearly cannot deal with all the complex interactions between science, technology and economic development in the third world today. However, we can review some of the ways in which science and technology have influenced third world countries in the past and hence try to establish some guidelines for the future. We will therefore first discuss some of the prerequisites for economic development. Then we will outline the role of technology in a number of different routes to economic development. Finally, we will consider the vital question of choice between different technologies. But first a question.

WHAT IS THE THIRD WORLD?

The very term 'third world' implies the existence of first and second worlds. Yet very few people could say what these other worlds are. Roughly speaking, the first world is the group of liberal capitalist countries which we discussed in Chapter 7.3—the USA, Canada, the countries of Western Europe, Australia and New Zealand. By contrast, the second world consists of the Marxist socialist countries—the USSR and the countries of Eastern Europe with their centrally planned economies. The third world is all the rest, which means that it comprises a very diverse collection of countries. It includes countries as different as India and Saudi Arabia, Egypt and the Philippines, Brazil and the Central African Republic. All these countries differ enormously in terrain, climate, natural resources, population, culture and religion. Probably their only common attributes are that they are less prosperous than the countries of the industrialised first and second worlds, and that they wish to industrialise in order to increase their prosperity but without losing their national identities.

It is worth emphasising this enormous diversity as it indicates how difficult it is to generalise about the problems of the third world.

Finally, there are some countries such as China and Japan which are difficult to classify. Should China be placed with the Marxist socialist countries of the second world rather than as the biggest third world country? Should Japan be regarded as a liberal capitalist country or as the most prosperous third world country? And should the very different approaches to science and technology in China and Japan (see Chapters 7.5 and 7.6) be taken as models for other third world countries to emulate?

PREREQUISITES FOR ECONOMIC DEVELOPMENT

Economic activity involves combining factors of production—land, labour, materials and capital—to satisfy expressed demand for goods and services. Economic development occurs when a community does not use all its resources to meet consumption needs but devotes a surplus, in money or labour, and usually referred to as savings, to some profitable or productive investment in agriculture, manufacturing, transport or community services. A high rate of savings alone will not cause economic development, contrary to the belief of many development economists in the 1950s and early 1960s; there must also be profitable opportunities. The developmental process also requires certain pre-conditions— law and order, an efficient and flexible form or forms of economic organisation, people with necessary skills, a community with attitudes sympathetic, or at least not opposed, to change and an adequate infrastructure of transport, distribution and financial services.

The Industrial Revolution in Britain did involve technological developments— for example, when Abraham Darby successfully combined the use of coke as a furnace fuel and techniques of casting iron to overcome the growing scarcity and high price of timber for charcoal (see Chapter 4.4). But it was not just new

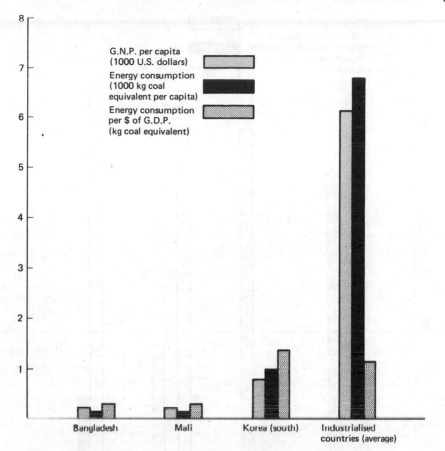

Fig. 8.1.1   Energy consumption and Gross Domestic Product.

technology that prompted Britain's development. The other pre-conditions existed too—law and order, the joint stock company as a form of economic organisation, craftsmen with skills, a generally favourable attitude within society, an ability and willingness to invest in transport improvements, as well as manufacturing ventures, and potential customers from increasing farm incomes—made possible by the agrarian revolution—and from profitable overseas ventures.

   The process of economic development results in scarce resources, usually labour or land, being replaced by capital goods—tools and machines dependent on power other than the human muscle. This inevitably leads to an increase in energy and capital per unit of labour as one of the main characteristics of development—the familiar 'power at the elbow' of the typical worker. Fig. 8.1.1 shows some of the variations in energy consumption which exist. These vast differences conjure up images of transport by head load in Mali, or bullock cart in Bangladesh compared with a forty-tonne lorry in Europe; of one person at a hand loom compared with one person supervising 25 immensely faster automatic looms; of a woman cooking a meal on a dung fire compared with the electric

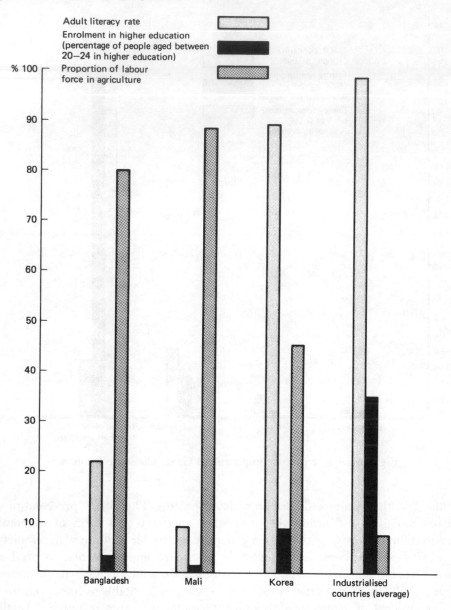

Fig. 8.1.2   Education and structure of employment, 1976.

cooker supplied with electricity from a network of computer-controlled power stations. Note also the variations in energy consumption per unit of gross domestic product (GDP).

The manpower needs of industrialised societies compared with some third world countries are illustrated in Fig. 8.1.2. Note that there are far fewer people tilling the land, but many more are literate and have higher education.

TECHNOLOGY IN DIFFERENT ROUTES TO DEVELOPMENT

A number of distinct routes to development have been or are being followed by third world countries. We will now describe some of these routes in order to investigate further the role played by technology and other factors in the economic development of different countries.

1 *Exports from peasant farms*

The development and expansion of exports from the farms of peasants has been one of the main origins of development for many countries in Asia and Africa. We can think of rice exports from Burma and Thailand, cocoa from Ghana, palm oil and other oil seeds from Nigeria and jute from India. The technological pre-conditions for these exports have been transport innovations, principally the steamship and the motor vehicle; and communications facilities, principally postal and telegraph services, to enable the local agent of the trading firm to communicate speedily and efficiently with his principals in the importing country—frequently the colonial power. For some exports innovation in the processing of the commodity was also a pre-condition of widespread demand. Jute is one example. For centuries Indian peasants had made domestic use of the fibre from fast-growing corchorus plants for coarse cloth. Victorian entrepreneurs were aware that jute could become a cheap replacement for flax to meet the growing demand for sacks for corn and other agricultural crops. But jute would not spin or weave well in mill conditions because of continual thread breakage until it was discovered by chance in Dundee, an important whaling port, that, when some whale oil was added to the jute fibre, it would respond to machine spinning and weaving as well as cotton or flax.

Beyond these generally available transport and communications technologies and a few processing technologies, science played a very minor direct part in the early stages of development along this route. Initially all the expansion in output usually came from the peasants using spare land and spare family labour to grow or collect the additional output. At first peasants tended to continue to produce all or most of their own food, but later some peasants became near exclusive producers of the cash crop and purchased their food and other consumer needs with the income from sale of the export crop. This stimulated the growth of a cash economy and development in other sectors of the economy. In some cases—cocoa in Ghana or oil seeds in Nigeria—where Marketing Boards were established to handle peasant exports and help iron out price fluctuations, pricing policies were used, particularly after independence, to tax the peasants and channel the resultant surpluses or savings into industrial development and also into some rather grandiose prestige projects.

A further stage in development along this route has frequently occurred involving the use of technology—the so-called 'Green Revolution'—to improve crop yields and sometimes product quality. High yielding or improved varieties of seeds have been produced by research stations and their use popularised among the peasants. The use of fertiliser and pesticides, and sometimes of simple farm machinery and better cultivation practices, has also contributed to increases in output per unit of land and per unit of labour. Additional income that has resulted has contributed to development in the new manufacturing and service sectors of the growing economy.

2 *Mines and plantations*

In some countries development started with mines and plantations set up to provide raw materials for industrial countries. Once again technological pre-conditions were developments in sea transport and the technical discoveries which had led to the new industries in Europe and North America. Although, generally speaking, mines and plantations were large-scale enterprises, they tended to be very labour-intensive and made little use of technical discoveries in their own operations. Some 'low-level' technology was used in mining and also in agronomy and pest control, but that is all. Even where the mines or plantations were in sparsely populated areas this did not stimulate the introduction of labour-saving technology. Rather, enterprise was devoted to increasing the supply of labour through labour-contracting from nearby countries, as with the South African mines, or organising migration of labour—usually from India or China—as with plantations in East Africa, the West Indies or Assam.

There were a number of reasons for this limited use of technology:

- Indigenous labour was cheap, low in productivity and not suited to operating machines; migrant labour was also relatively cheap.
- Usually indigenous labourers were only willing to work for limited periods as they saw wages as a supplement to family wealth under a subsistence economy. There was thus little or no incentive to train labour.
- Cyclical demand for the products of mines and plantations, resulting in wide price fluctuations, inhibited the high capital investment needed either to develop a stable community or a more highly productive work force. Since independence, political risks of nationalisation or expropriation have added to the economic risks.
- Attitude of mind among foreign managements, often with racial overtones, which resulted in local labour being regarded as unsuitable for skills training.

Mines and plantations provided jobs—although not always for the indigenous people, and they made major contributions to export earnings—much of which, however, went to pay for imports required by the mines and plantations themselves. They also contributed to government tax revenues—but these were often used disproportionately in urban areas. Their contribution to overall economic development has not been great and their linkage with the rest of the economy has frequently been weak, often being restricted to demand for food crops from peasants.

3 *Domestic industrialisation*

This has taken various forms. Import substitution was popular in the first five-year development plans in the 1950s and early 1960s. Generally speaking, it has had a poor success rate, although the level of industrial technology used has been low. Invariably machinery, and often most materials, had to be imported and, in the early stages, foreign management and some technical skills too. Often the machinery would be under-utilised. Frequently components would be imported for final assembly and so the value added would be small. This is particularly the case where the transfer price for the component includes a payment, often excessive, for technology, as in the case of some pharmaceuticals. In addition the more expensive imported article, sometimes higher in price only because of customs duties, is often regarded as preferable to the cheaper local product.

Then there is what may crudely be described as 'export substitution' or the processing of commodities within the country where previously they were exported unprocessed: instant coffee instead of coffee beans, cotton piece goods or garments instead of raw cotton; jute yarn instead of raw jute. This route to industrialisation has many virtues, provided the level of technology used is fairly labour-intensive, and fullest possible use is made of local supplies for other materials and services. If this is done, what economists call the 'multiplier effect'—economic activity stimulated elsewhere in the economy—can be quite substantial.

Building a base of heavy industry is yet another approach, favoured principally by centrally-planned economies such as Russia and China in the early stages of their revolutions. The process involves savings, usually forced (see Chapter 7.4), from the agricultural sector being invested in heavy industry, such as iron and steel manufacture, the products of which are then used to make machinery for new factories or large-scale state farms. It is a route which is only feasible where the country has a wide range of mineral resources and a system of government which can direct labour, control prices and overcome people's natural reluctance to defer consumption for long periods. A wide range of technology is involved which is developed indigenously or brought from outside.

Finally, there is the development of a modern consumer goods industry as in Hong Kong and Singapore. This is an approach particularly suited to highly urbanised countries with abundant but naturally skilled labour. It involves the use of sophisticated technology and combines the capital and knowledge of industrialised countries with the cheap labour of the under-developed. Some of the main institutions involved in such development have been the multi-national companies. Much of the opposition to their activities is attributable to the covert political motives of a few and to the opposition from trade union leaders in developed countries whose members are suffering from so-called 'unfair' competition. Amongst the work force of Hong Kong's electronic factories there has been little complaint since earnings in other jobs are low and such jobs are few and far between.

4   Balanced growth

This rather blandly named route now features in most development plans. It aims to combine development in the agricultural sector with developments in the industrial and other sectors. In agriculture the aim is increased supply by using more inputs and better techniques to produce higher yields; by opening up new land or increasing the cropping intensity of existing land, for example, through irrigation; and by introducing new crops. In industry, priority is given to activities which use locally produced materials from farms, forests or mines. In transport, distribution, financial and community services, priority is given to those activities which will meet the needs of the expanding productive sectors. Development in agriculture should generate additional demand for the products of local industry, whether as agricultural inputs such as fertiliser, or as bicycles, radios and other consumer goods for better-off peasants. It should also supply the growing food needs of the urban population and some of the materials used by the new factories. This route was the route followed, with little planning, by the USA and with more deliberation by Japan, Taiwan and South Korea. It has also been the basis of China's recent economic policy.

Many of what the World Bank classifies as middle-income countries are experiencing rapid economic growth following one or other of the routes to domestic industrialisation or balanced growth. In development along any of these routes, technology plays a key role at many points. Nevertheless, there is frequently considerable scope for choice between technologies at different levels of sophistication and hence with different labour and capital requirements. It is to these problems of choice that we will now turn.

CHOICE BETWEEN ALTERNATIVE TECHNOLOGIES

Probably the most important characteristics of alternative technologies are their different capital and labour requirements. Table 8.1.1 illustrates this point by comparing the capital and labour requirements of technologies available for a number of industrial operations; the differences speak for themselves.

Table 8.1.1   Differing labour and capital requirements for alternative manufacturing technologies (ratios of highest to lowest)

	Numbers employed	Capital cost	Capital cost per employee
Maize milling	1·5	2·2	3·2
Sugar milling	3·0	2·5	10·0
Brewery (bottling and packing stage)	3·5	—	—
Nut and bolt manufacture	10·5	3·0	3·5
Iron founding (melting stage)	—	16·3	—
Urea manufacture	1·4	1·5	2·1
Farm implement manufacture	27·4	5·7	156·6

Note: for each technology compared, same level of output assumed.

We can make the same point by comparing the average investment per work-place in different countries. In the UK, to create one new work-place a *private* investment of about £12,000 is required. By contrast, the current five-year plan for India implies a *total* investment per work-place of only £1000 and this includes all the *public* investment in roads, education, health services, etc. Or, to take an extreme case in goods transport, Table 8.1.2 shows the labour and capital requirements for headload, rickshaw, country boat and five-tonne lorry in Bangladesh. Country boat, rickshaw and lorry are practical alternatives in many situations, but headload is only appropriate where the quantities to be moved are relatively small.

Table 8.1.2 Resource requirements for different modes of goods transport in Bangladesh

	Ratios of labour c.f. lorry	Capital cost c.f. lorry	Imported energy
Headload	1800	negligible	negligible
Rickshaw	55	about same	negligible
Boat	3	about same	negligible
Lorry	1	—	substantial

Note: Assuming daily mileage of headload 15, rickshaw 45, country boat 30, lorry 200.

Given these large differences in resource requirements, it is abundantly clear that choice of technology should be influenced by the relative abundance of labour and capital in an economy. In fact, the choice of technologies is one of the most important decisions facing a developing country. It is a choice which affects the whole pattern of income distribution and the economic and social fabric of society. It determines *who* works and who does not; *where* work is done and therefore the balance of economic activity and prosperity between urban and rural areas; what goods and services are produced and to what extent they are relevant to meeting basic human needs. All this adds up to determining for *whose* benefit the resources of a country are used. To what extent these choices should be made by governments or left to market forces is a complex and difficult question. So it may be helpful to consider briefly a few examples of technologies which, having been adopted, have turned out to be inappropriate for various reasons:

– *the establishment of efficient diesel-powered deep tube wells in Bangladesh* resulted in: unreliable supplies of irrigation water; increase in power and land-holdings of larger landowners; move away from optimum distribution of land for maximum food output and rural employment; environmental and social side effects of lowering water table.
– *the continued use of traditional bullock carts in Southern Asia* despite increasing costs of materials (wood) and growing scarcity of land for grazing; the potential for improving design in response to need was not exploited.
– *a hospital in Turkey equipped with sophisticated equipment*, such as cobalt radiation and electronic body-scanning equipment, but lacking sufficient nursing staff for even the minimum of post-operative care; benefits of high technology very limited; cost of medical services higher and effectiveness lower than would be the case with a better balance between labour and capital.

– *the use of chemical pesticides to control pests in cotton* such as bollworm, boll weevil and leaf worm; initial increase in output only sustainable by use of larger quantities and different types of pesticide as pests gradually become tolerant; costs increase and, in some areas, pest infestation ultimately worse than before; undesirable side-effects on all forms of wild life, sometimes upsetting ecological balance and eroding fishing and hunting resources.

– *the use of Sultzer automatic looms for jute cloth manufacture in Bangladesh*; because of technical sophistication, inadequate maintenance and lack of spare parts, about two-thirds of looms out of action; remaining looms working at well below design speed; ratio of employees to machines two or three times the recommended level; capital cost per employee and operating costs much higher than with simpler alternatives.

– *the establishment of 'turnkey' breweries in many developing countries*; costs of 'Rolls Royce' design are significantly higher than least-cost alternative involving similar technological processes; work-place cost and capital costs per unit of output higher than necessary; beer more expensive or less profitable than need be.

– *match manufacture in India in sophisticated modern factory*; 30 per cent of Indian match output produced by workforce of 15,000, while remaining 70 per cent produced by about 5 million people in numerous small simple establishments.

These examples are by no means isolated instances. Innumerable other cases can be quoted and may well be familiar from experience both in developed and developing countries. Why then is inappropriate technology adopted? From these examples we can identify five types of defect or inappropriateness.

First, there is the straightforward misuse of resources; the spending of more than is necessary on machinery; the use of expensive imported materials, skills or organisational capacities when cheaper, locally available, resources, would have been adequate; the misjudgement of the scale of operations and hence the type of machinery needed.

Second, there is the use of technology which is too capital-intensive, and so uses less labour than is appropriate to a country's economic and social objectives.

Third, there is the use of technology which results in unexpected environmental damage.

Fourth, there is neglect in applying technological inventiveness to a problem area.

Fifth, there is bias in government spending, government policy or in laws reinforcing the choice of a technology which is inappropriate for other reasons.

For example, the Bangladesh government subsidises either the purchase price or rental cost of deep tube wells *and* both diesel fuel and electricity for their operation. In many developing countries investment by foreign companies in modern manufacturing plant has attracted financial benefits, such as tax holidays or remission of import duty, greater than those available to local entrepreneurs investing in simpler and more appropriate operations. More often than not, clinching a turnkey operation involves high-level bribery.

CAUSES OF INAPPROPRIATE CHOICE OF TECHNOLOGY

Amongst the complex of factors which give rise to inappropriate choice of technology, we will identify and discuss briefly four specific causes.

First, businessmen and planners in developing countries are frequently unaware of the range of choices open to them. This imperfect knowledge of alternatives reflects both the absence of a comprehensive international reference system about technologies—due both to the high cost of collecting and evaluating information on technologies and the rapidity of technical change—and the frequent lack of skills in interpreting comparative data about technologies. Local businessmen may know about the potential domestic market, skills available within the local workforce and sources of finance. However, they often lack the technical knowledge to evaluate alternative options. Foreign experts may be able to undertake technical evaluations but lack sufficient appreciation of the local economic situation, and it is not easy to marry these two contributions. In addition, important knowledge about the direct and indirect environmental consequences of using particular techniques is frequently lacking.

Then there is the pressure to use capital-intensive methods. The bulk of the world's research and development effort is within or under the control of organisations in industrialised countries and has, as its main objectives, increasing productivity within developed countries and producing new products for developed country markets. The result is an orientation towards techniques which save labour or energy, or improve product quality, or lessen pollution. There is a great deal of information available, openly or under licence, about such technologies. Considerable effort, both commercial and political, is put into selling new technologies to developing countries, both to increase the direct returns from the original research and development expenditure and also to stimulate future exports of machinery and technical manpower. These are activities which frequently cause political and economic commentators to make accusations of 'economic imperialism'. However, such activities are generally self-interested responses to business opportunities rather than evidence of any concerted design.

A third factor underlying many of the decisions about choice of technology has been the world-wide cult of 'bigness'. This has led to a tendency to give more weight to theoretical technical economies of scale provided by large plants rather than to the disadvantages of their market inflexibility, their propensity to provide monotonous jobs, the hidden costs of poor industrial relations and the probability of plant stoppages.

Finally, there are various kinds of divergence between private and public interests. A large landowner in a developing country is less likely to be interested in the intensive cultivation of every square foot of land than a marginal farmer whose survival depends upon maximum output. Again, the large landowner can afford and may prefer to adopt mechanised farming, particularly where oil is subsidised, in preference to employing an army of labourers, although the 'national interest' may require the latter. Unless forced to do so, few industrialists will spend money on treating waste products from their factories and mines, but it is in the public interest and indeed in the interest of the industrialists as private citizens that pollution of the environment should be minimised.

CONCLUSIONS

The problems of choosing an appropriate level of technology—which has been defined as 'that level of technology which best achieves the economic and social aims of a society, given its resource endowment'—are clearly very great. And they are inevitably bound up with the kind of political and economic debate which is exemplified by the recommendations of the Brandt Report [1] and the contrasting views of development economists like Professor Bauer in his book *Dissent on Development*. [2]

Whatever the outcomes of these debates, it would surely be helpful for international organisations to try to make more comprehensive information about levels of appropriate technology available to third world countries. Also, these countries should try to make the technical education available to their citizens more relevant to the real life of the country rather than following the pattern of education in first world countries. In this context, the example of Japan (see Chapter 7.5), both after the Meiji restoration and since 1945, might be the most appropriate model for third world countries to seek to emulate.

REFERENCES

1   *North–South: A Programme for Survival*, (London: Pan, 1980); *see also What's Wrong with the Brandt Report* in *Encounter*, December 1980.
2   Bauer, P., *Dissent on Development*, (London: Weidenfeld and Nicolson, 1976).

SUMMARY

1   Science and technology are two important factors in the economic development of the many diverse third world countries, just as they were in the industrialisation of the liberal capitalist countries of the first world and the Marxist socialist societies of the second world.

2   Other important prerequisites for economic development are law and order, efficient and flexible forms of economic organisation, people with the necessary skills, a community with attitudes sympathetic to change and an adequate infrastructure of transport, distribution and financial services.

3   Third world countries have followed a number of distinct routes to development such as exports from peasant farms or from mines and plantations; domestic industrialisation; and development plans emphasising 'balanced growth'. Science and technology have been of most importance in the last two of these routes.

4   Since different technologies require very different amounts of capital, labour and other resources, the choice of the most appropriate technology is one of the most important decisions facing a third world country; it can determine who works and who does not, where work is done and for whose benefit the resources of a country are used.

5   Many factors leading to the numerous existing examples of inappropriate technological choices are not easy to modify; however, some progress could be made by following the Japanese example of trying to make the technical education available in third world countries more directly relevant to the real life of those countries.

8.2 Nuclear power and energy resources

Since the Industrial Revolution the number of people in the world has increased dramatically. In the hundred years from 1870 to 1970, the world's population trebled from approximately 1300 million to nearly 4000 million (see Fig. 6.1.8, page 249). But over roughly the same period the world's consumption of energy increased more than six times as fast as the population. Between 1870 and 1960 the world's energy output increased by a factor of 20—from about 1700 million to 34,000 million megawatt hours (see Fig. 8.2.1). These figures show just how much we now depend on sources of power and energy in addition to those which were traditionally available in pre-industrial societies—wind, water, fire, and muscle power, both animal and human. We also depend on our store of knowledge about the different forms of energy and about how energy can be converted from one form into another. And we depend for the maintenance of our way of life on the continuing availability of large amounts of energy.

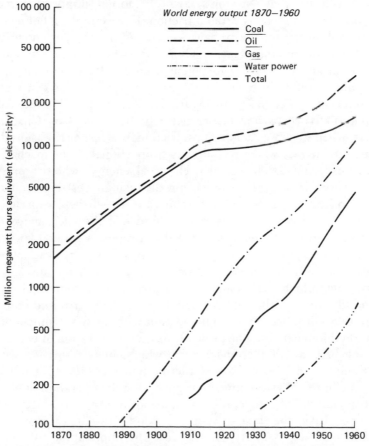

Fig. 8.2.1 World energy output, 1870–1960; note that the energy scale is logarithmic.

Therefore in this chapter we will first outline the growth of our current knowledge about sources of energy and their utilisation. Then we will briefly consider some data concerning various energy sources and rates of energy consumption. Finally we will outline some of the main arguments involved in that most contentious area of energy policy—the present and future use of nuclear power.

THE DIFFERENT FORMS OF ENERGY AND THEIR INTERCONVERSION

The principle of the conservation of energy states that energy cannot be created or destroyed but only converted from one form into another. This principle is one of the most securely based and fundamental principles in science and the history of how it evolved is a fascinating illustration of the increasing generality and cumulativeness of much scientific knowledge.

The conservation of energy has its roots in the age-old fruitless search for a mechanical perpetual motion machine. Nobody has ever been able to make such a machine and this empirical fact was given theoretical support when it was realised in the 19th century that Newton's laws of motion imply that mechanical energy is conserved in systems without friction. For example, a falling stone or a swinging pendulum continually exchanges energy of position, or potential energy, for energy of motion, or kinetic energy. Or a water-wheel operates by water falling from a reservoir and exchanging its potential energy for kinetic energy of both wheel and falling water. In each case the total energy remains constant.

In the 19th century systems involving friction were included in the analysis too. Careful quantitative experiments—particularly by Joule (see Chapter 4.7)—showed that, when work was done against frictional forces, heat was produced at a definite measurable rate which was directly proportional to the mechanical work done. Other experiments showed that electrical energy, whether produced by mechanical work via a dynamo or chemically using a battery, could also be converted into heat at a definite measurable rate. In all these experiments with different kinds of energy it was invariably found that the total energy remained constant. So a number of scientists were led to propose a general law, called the principle of the conservation of energy, which states that energy cannot be created or destroyed but only converted from one form to another. This law was also found to be true in biological systems and seemed to be of universal application.

In the middle years of the 19th century the study of energy and how it can be used was put on a firm mathematical foundation with the development of the laws of classical thermodynamics. The principle of the conservation of energy came to be called the first law of thermodynamics which, in combination with a new general principle—the second law of thermodynamics, led to a much deeper understanding of the physical limitations governing the operation of heat engines of all kinds including, of course, the workhorse of the Industrial Revolution, the steam engine. Like the first law, the second law of thermodynamics is concerned with an impossibility—in this case the impossibility of continuously taking heat from a reservoir and turning it completely into work or indeed into any other form of energy. Whereas Joule and others had shown that mechanical work or electrical energy can be interconverted or converted into heat with 100 per cent efficiency, the second law showed that the reverse process—the conversion of heat into work—could only be carried out with much lower efficiencies. This doesn't mean that the first law is broken. Rather it means that even ideal heat engines, like the ones Carnot studied (see Chapter 4.6), must operate with the equivalent of a hot boiler and a cold condenser and that some heat must be given up to the condenser rather than turned into useful work. The efficiency improves if the temperature difference between boiler and condenser increases but, in practice, it can never be more than about 40 per cent (see Fig. 4.4.13, page 140). These natural laws about the interconversion of different forms of energy are of central importance in any discussion of the world's energy resources. To paraphrase George Orwell, the first law of thermodynamics states that all forms of energy are equal while the second law states that some are more equal than others!

By the end of the 19th century the laws of thermodynamics were deeply entrenched in all branches of the physical sciences. Together with other very general laws—like the law of conservation of matter which stated that matter cannot be created or destroyed—they had found innumerable applications and had brought order to an enormous amount of experimental data. As we saw in Chapter 6.8, Kelvin had even applied thermodynamic principles to the earth and the sun and had concluded that neither of these bodies could be old enough to satisfy Darwin and the geologists.

In the 20th century the law of conservation of energy has been generalised further. Following Einstein's work on the special theory of relativity (see Chapter 6.2), he predicted that matter and energy were in some senses equivalent and could be converted into one another according to the well known formula $E = mc^2$.

This equivalence was first demonstrated in the 1920s and '30s during experiments on collisions between atomic nuclei which led to nuclear reactions between the colliding particles. These nuclear reactions in some ways resemble chemical reactions between colliding atoms or molecules but with one important difference. In chemical reactions matter and energy are both conserved—the total mass *and* the total energy are each the same before and after the reaction. Whereas in nuclear reactions this is not true. Almost invariably in these reactions some matter is converted into energy, or vice versa, so that energy and mass are both not conserved. But when, using Einstein's conversion formula, we calculate the total mass *and* energy involved, we find that mass and energy together are conserved.

So the two conservation principles—of matter and energy separately—now coalesce into a single principle of total energy conservation in which mass is regarded as one of the possible forms of energy.

In fact mass is in some ways the single most important form of energy. It is the mass lost in the decay of radioactive minerals deep in the earth that provides enough heat to make the earth habitable. And it is the mass lost in nuclear reactions in the sun and other stars which is the source of all other forms of energy on which we depend (see Chapter 6.8). This is true of all the sources of energy used by man—fossil fuels like coal, oil and natural gas, renewable sources like solar radiation or wind, tidal and geothermal power, hydroelectric power and of course nuclear power itself.

So in any discussion of energy resources two basic facts must be recognised. First, all our sources of energy are ultimately nuclear in origin, and second, in utilising sources of energy we need to draw on a vast amount of scientific knowledge. In particular we are inevitably constrained by the laws of thermodynamics.

ENERGY SOURCES AND RATES OF ENERGY USE

Now let us return to Fig. 8.2.1 which shows the rapid increase in the world consumption of energy between 1870 and 1960. As we saw in Chapter 4.11, coal was overwhelmingly the major fuel in the 19th century. By about 1920, oil was providing about 10 per cent of world energy output and by 1950, natural gas had risen to 10 per cent and oil to nearly 30 per cent of the larger world output. By 1960 coal was providing 50 per cent of the world output, oil just under 35 per cent and natural gas under 15 per cent. Hydroelectric power had grown substantially since the initial plant at Niagara in 1896 but was still only providing about 2 per cent of world output in 1960.

So up to 1960 the world's major sources of energy were the fossil fuels—coal, oil and natural gas—which had been produced many millions of years ago by the decay of dead organic matter. The world has been spending this energy capital and spending it increasingly rapidly. Only the very small proportion derived from water power was obtained from a renewable source of energy—energy income rather than energy capital.

Since 1960 these trends have continued. World consumption of oil now exceeds that of coal and consumption of natural gas has risen too. But a new source of power—nuclear energy—has been developed and now provides about 1 per cent of world energy output and more than 10 per cent of total output in some parts of the world. All existing nuclear power stations are based on nuclear fission of heavy elements like uranium or plutonium and are direct descendants of the first atomic pile built by Fermi in Chicago in 1942 (see Fig. 6.3.3, page 271).

Nuclear power has been developed for a variety of reasons. Nuclear power stations are similar in many ways to coal- or oil-fired power stations. All these stations generate electricity by heating steam to high temperatures and pressures in order to drive turbines and dynamos which again are the direct descendants of the steam-driven generators used in the early power stations (see Fig. 4.9.7, page 194). The main difference lies in how the steam is heated. In coal- or oil-fired power stations heat is generated by burning large amounts of fuel whereas in nuclear power stations, the heat is obtained from the fission of relatively small amounts of uranium or plutonium. So nuclear power stations use very small amounts of fuel and generate electricity in large quantities at costs which are comparable with those of conventional power stations.

However, there is another major reason for the development of nuclear power. The rate at which fossil fuels are being used is now so large that known reserves of oil and gas are unlikely to last for more than a few decades or even less if total energy consumption increases. Known reserves of coal may last longer—a few hundred years or so—but again are finite. Reserves of nuclear fuel are much larger and one type of nuclear reactor, the fast breeder, actually produces more fuel than it uses.

So the attractions of nuclear power are considerable, particularly as a replacement for oil and gas as the availability of these fuels drops in the decades ahead. However, nuclear power also has a number of disadvantages, which we will discuss later, and they have stimulated the search for other large-scale sources of energy which, like hydroelectric power, are inherently renewable and so do not use up our stock of natural resources.

The main possibilities here are solar power, power from winds, tides and waves, and geothermal power which taps the heat from radioactive minerals inside the earth. However, there are two main problems with these sources of power. First, unlike fossil fuels or nuclear power, they are very dilute sources of energy which would need considerable concentration before they could be used to replace existing power sources. For example, it is estimated that a collecting area for solar radiation of several hundred square metres would be needed to supply the energy needs of one person at the average rate of consumption in the United States.[1] Secondly, sources of energy like sunlight or wind are intermittent and cannot be turned on and off as required. In this they also differ dramatically from power stations driven by concentrated types of fuel.

There are two other possibilities for large-scale energy generation which may not suffer from the disadvantages of being too dilute and too intermittent. The first is to use solar power in the way that nature did in producing our current stocks of fossil fuels—to use the energy concentrated in vegetable matter by photosynthesis in growing plants as a source of hydrocarbon fuels. This, of course, is what was and is done in traditional economies based on wood as a fuel. Now experiments are well advanced with new forms of biomass, as it is called, in which methane or alcohol will be produced from crops such as sugar cane or cassava.

The second possibility is to use the same nuclear process by which the sun and stars obtain their energy—the fusion of lighter elements like hydrogen which, like nuclear fission, also releases large amounts of energy (see Chapter 6.8). Nuclear fusion has been achieved on earth—on the small scale in experiments with nuclear particle accelerators and on the large scale in hydrogen bombs. However, it has not yet proved possible to control the large scale release of fusion energy and a practical power plant based on nuclear fusion is very far from realisation. However nuclear fusion does have one important attraction—the hydrogen in the world's oceans is a virtually inexhaustible source of fuel.

This brief review of alternatives to fossil fuels shows that substantial energy production from nuclear fission needs to be seriously considered if future world energy supplies are to be maintained even at current levels.

POWER FROM NUCLEAR FISSION

We have outlined above some of the main advantages of obtaining power from nuclear fission reactors. In this section we will first discuss some of the specific risks involved; then we will describe the current and planned location of nuclear power reactors around the world; and finally we will look at energy needs in the wider political and social contexts.

The particular risks involved in using nuclear power reactors are all related to the fact that nuclear fission, by its very nature, produces large amounts of radioactivity. These risks concern three main areas—emission of radiation and other risks during normal operation, risks of accidents, and the problems of storage and disposal of radioactive wastes.

The assessment of risks of any type ought to have two main characteristics. It needs to take into consideration any and all of the available background knowledge. And it needs to be aware of the risks involved both in specific alternative courses of action and in other activities of daily living. From the moment of our birth, and before, we are at risk. It makes little sense to calculate an absolute risk of death from one specific cause as say 1 in 10,000 per year and to declare this as categorically unacceptable if at the same time we unthinkingly accept, as we do, comparable risks from everyday activities such as living in a house or venturing onto the streets.[2]

We will therefore try to consider the risks of nuclear power in comparison both to the risks of other types of power generation and to the risks of other activities. Such a discussion obviously cannot be comprehensive but will hopefully indicate the types of argument which need to be considered in any realistic assessment.

First let us consider risks during normal operation—both from emission of radioactivity and from accidents. The amounts of radioactivity emitted by nuclear power stations are very small both in absolute terms and in comparison with other sources of natural and man-made radiation. And because of the extreme sensitivity of electronic instruments for measuring small amounts of radioactivity, such emissions as do occur can be very easily monitored. These measurements show that such emissions contribute considerably less than 1 per cent of the background radiation to which people are exposed from other sources. Where does this other radiation come from? About a quarter comes from cosmic rays from outer space but more than 50 per cent comes from naturally occurring radioactivity in the air we breathe, the food we eat, the houses we live in and even in our own bodies—mainly from radioactive potassium in the blood. As we saw in Chapter 6.8, even the earth is radioactive. In fact everything is naturally radioactive and always has been. And this exposure to natural radioactivity can vary enormously depending on where and how we live. For example, living in a house built from granite rather than wood can double the natural background while living in the monazite sand area in Kerala, India—as 100,000 people do—can multiply the natural background by ten. Man-made radiation, primarily from medical X-rays and radiotherapy, is also important, making up between 10 per cent and 20 per cent of the total background, while fallout from nuclear weapons tests contributes around 2 per cent.

So the radioactive emissions from nuclear power plants are negligible compared with other radiations to which we are all exposed and also in comparison with the extra radiation we may receive by moving house or having a routine chest X-ray. And it is also likely that the radioactive emissions from coal-fired power stations are considerably greater than those from nuclear power plants.

This brings us to another factor which needs to be considered in assessing risks from the normal operation of power plants—the overall risks of accidents or deaths to workers or to the whole population from such normal operation. A fair basis for comparison is to assess the risks for a given quantity of energy produced, taking into account all the risks involved including such things as mining accidents, accidents during transport of fuel and in producing the materials used in the construction of power stations. When this is done for the whole range of possible technologies, it is found that two sources of power are the safest—natural gas and nuclear power—and that coal, solar and wind power are considerably more dangerous.

The actual figures obtained from such calculations are inevitably subject to error—not least because some of the newer technologies are not yet being used on any appreciable scale. However, there are two main reasons underpinning these somewhat surprising conclusions and these are unlikely to change. First there are the dangers inherent in mining the fuel and in transporting it to power stations. Both coal and nuclear power involve risks but these are much more important for coal because the production of a given amount of energy requires many more tonnes of coal than of uranium. Secondly, the risks for solar and wind power are much greater because these energies are very dilute. These sources of energy therefore require the building of much larger installations, for a given energy output, than would be necessary for power stations using nuclear or fossil fuels.

Now let us consider the risks from major accidents in nuclear power plants. In Britain and the United States, the safety of nuclear power plants has been on the agenda of public debate ever since the Windscale accident in Britain in 1957 which, incidentally, occurred in a military reactor and not one designed for the civil nuclear power programme. This public debate has greatly increased our knowledge about the risks involved and has led to continual improvements in the design of safety measures. Probably the most comprehensive study of nuclear safety is the Rasmussen report of 1975, which attempted to calculate the risks of major accidents of various sizes at nuclear power plants in the United States and to compare these risks with the risks associated with other man-made accidents like air crashes, fires, explosions or dam failures, and with accidents from natural causes such as tornadoes, hurricanes and earthquakes. The basis of comparison was to compute the risk per year of accidents causing a given number of deaths in the United States and to compare these risks with those from accidents at 100 nuclear power stations also in the United States (56 then in existence and 44 planned). The results are shown in Fig. 8.2.2. Once again the actual figures may be subject to error particularly when, as is the case with nuclear power, there have

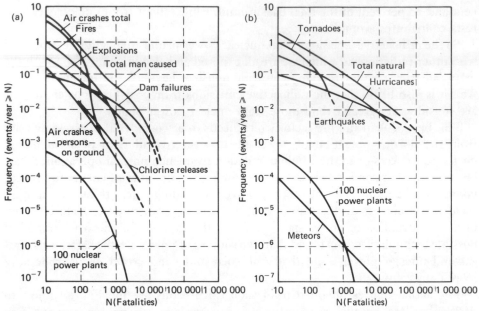

Fig. 8.2.2   Frequencies per year of accidents with a given number of deaths: (a) man-made events; (b) natural events. *From Rasmussen Report; WASH 1400, 'Reactor Safety Study: An Assessment of the Accident Risks in US Commercial Nuclear Power Plants', October 1975.*

been so few accidents that estimates rather than actual accident statistics have to be used. But, even if these estimates of the risks of nuclear power are ten times too low, the conclusion is still that nuclear power is, with one exception, at least a hundred times less dangerous than any of the other accidents considered, man-made or natural, and more than a thousand times less likely than the sum of either the natural or the man-made risks. The exception is the risk of death from being struck by meteorites, which was estimated to be roughly comparable with the risk of death from accidents at nuclear power stations.

These conclusions are not altered by the subsequent accident at the Three Mile Island nuclear power plant in Pennsylvania in 1979. In this accident a number of serious mistakes were committed which made the initial problem much worse. Yet the outer containment vessel was not breached and very little radioactivity was released to the environment. And nobody was killed or injured. In fact, the accident has probably decreased the chances of future accidents, since it will certainly lead to more careful enforcement of routine safety procedures.

Finally, let us consider the problem of disposing of radioactive wastes from nuclear power stations. That this problem needs to be taken seriously is shown by the Soviet nuclear accident in the Urals in 1958 in which carelessness in waste disposal led to highly radioactive materials being scattered over a wide area (see Chapter 7.4). However, there is every sign that, with care and technical skill, the waste products from nuclear reactors can be disposed of safely. There are three main problems with waste disposal from nuclear power plants: reprocessing spent fuel elements; storing the resulting high-level waste until short-lived radioactivity has largely decayed; and long-term storage of wastes with very long half-lives. The first two of these problems can be dealt with by existing procedures, although it is likely that techniques will improve as more experience is gained. However, there is, as yet, no general agreement on the best method for long-term waste disposal. The likeliest possibility is that the wastes will be sealed in vitreous glass and buried either deep underground or in deep ocean trenches. The main objections to these possibilities are that some of the waste has such a long half-life that it will have to be stored for many thousands of years and so over long periods of time the protective materials may not remain effective, and leaks of radioactivity may occur. In considering these objections it is important to bear in mind a number of relevant facts. First, because nuclear energy is a very concentrated form of energy, the actual volume of waste is relatively small, particularly when compared with the enormous volume of waste from coal-fired power stations. Secondly, the wastes from coal-fired power stations contain considerable quantities of radioactive materials with even longer half-lives than wastes from nuclear power stations, together with poisonous materials such as arsenic and mercury compounds which never decay at all. And these wastes are frequently stored in surface tips adjacent to the power stations. Thirdly, it should be remembered that the nuclear fission fuel cycle starts with the mining of some of the enormous amounts of radioactive minerals which are buried deep in the earth and which make the earth warm enough to be habitable (see Chapter 6.8). In due course, any radioactive waste buried in the earth will decay until its activity is less than that of the originally mined uranium or thorium ore.

These considerations do not mean that the risks of nuclear power are insignificant. By the very nature of energy itself, no major source of energy can be provided on a large scale without substantial risks. What we need to do is to try to assess the risks involved in each situation as fairly as possible. Of course it remains true that nuclear reactors do contain enormous amounts of radioactivity which, if released, could cause many injuries and great loss of life. That is the main reason why so much more care is taken over safety precautions in the nuclear power industry than in any other industry. And that is why nuclear reactors in most, but not all, countries are fitted with containment vessels to prevent this radioactivity from escaping even if all the other complex safety measures fail. In fact, the very good overall safety record of the nuclear power industry to date is probably due to the extreme care taken to ensure that the risk of releasing large amounts of radioactivity is negligible. However, in discussing the safety of nuclear power, we also need to ask whether such attention to safety is universal and whether, if nuclear power stations are built all over the world, it will be possible to maintain the high standards of safety achieved so far.

NUCLEAR POWER STATIONS IN DIFFERENT COUNTRIES

As we saw in Chapter 7.3, nuclear power stations are now operating in more than 20 countries and are being built in another dozen or so countries around the world. Most have been built in the United States, where the installed capacity is more than 50,000 megawatts (MW) with another 120,000 MW planned. There are half a dozen countries each with substantial installed capacities of around 5–10,000 MW—Britain, France, West Germany, Canada, Japan and the Soviet Union. Most of these countries also have plans to increase their outputs; for example, France plans to have about another 40,000 MW of installed capacity by 1990. And there is a long list of other countries which have nuclear power stations either working already or under construction—Argentina, Austria, Belgium, Brazil, Bulgaria, Czechoslovakia, Finland, East Germany, Hungary, India, Iran, Italy, Korea, Luxembourg, Mexico, the Netherlands, Norway, Pakistan, the Philippines, Poland, Romania, South Africa, Spain, Sweden, Switzerland, Taiwan and Yugoslavia.[3]

The source of this information, the journal *Nuclear Engineering International*, gave very little detail about the Soviet nuclear programme and no information at all about communist China. So it is interesting to consider the account of the Soviet nuclear power programme given in a recent article in the Soviet journal *Kommunist*.[4] This article claimed that 'Atomic sources of energy are undoubtedly a historical necessity' and that '. . . it is impossible to build up the power supply basis for advanced socialist society without the development of nuclear power generating facilities'. Consequently '. . . it is expected that already in the near future nuclear power generating facilities will be developed on a tremendous scale'.

> . . . in the period from 1954 . . . to 1970, the USSR put into operation several atomic power stations with a total capacity of 1500 MW. . . . By the end of the next decade the USSR will be increasing the capacity by 5–8000 MW a year. It is quite probable that by the end of the century this indicator will rise to 10,000 MW.

The article also makes it clear why this large programme is being carried out.

> The atomic power stations to be commissioned in 1976–80 are to save approximately 50 million tons of conventional fuel. It would have been impossible to extract this quantity of fuel on top of the plan and to deliver it from the eastern part of the country. If these atomic power stations had not been built it would have been impossible to solve a whole range of vital economic problems dependent on increased supplies of electricity. . . . In 1980 atomic power stations will account for 10% of all the electricity generated in the European USSR as compared to 3·2% of all the electricity generated in 1975. Subsequently atomic power stations will account for a larger part in the increase of capacity of the power generating facilities in this region.

The article also states that 'another type of reactor used in the USSR is the pressurised water reactor'—the same type of reactor much used in the United States, including Three Mile Island, and proposed for use in Britain in the 1980s.

The article also stresses the growing importance of protecting the environment in any future plans for power generation and advocates a new policy of building large nuclear power generating complexes containing many nuclear stations and all supporting facilities for fuel reprocessing and waste storage. However, the article does not mention one significant fact about the Soviet nuclear power programme—that, according to an article in the journal *Commentary*, '. . . the vast majority of Soviet reactors have no containments'.[5] This article goes on 'Had a Three Mile Island-type accident happened at one of these uncontained reactors, there would almost certainly have been very serious releases of radioactivity.'

A further worrying feature of the Soviet nuclear programme is the absence of any informed public debate about safety or other aspects of the programme.[6] No doubt the Soviet nuclear industry has learnt some lessons since the disastrous nuclear accident in the Urals in 1958 (see Chapter 7.4). But the Soviet Union has still never publicly admitted that the accident ever took place. And when, at international conferences, Soviet nuclear engineers are asked about the accident, they give evasive replies and imply that it was all a long time ago. Such secrecy does little to inspire confidence in the effectiveness of current safety measures for the greatly expanded Soviet nuclear programme, particularly when, in the United States '. . . A congressional delegation was told that, yes, there had been some accidents in Soviet reactors, but none that harmed the public.'[7]

FUTURE ENERGY SUPPLIES

Of all the benefits of an industrialised technological society, the availability of large amounts of cheap power is the one that is most easily taken for granted. Yet, as we have seen, this cannot be taken for granted indefinitely. In order to maintain energy supplies, hard decisions will need to be taken—often many years in advance of actual needs, since large power stations of any type take many years to build. In taking such decisions, it seems prudent, in a rapidly changing world, for any society to obtain its energy from a diversity of sources rather than to rely on a single source or type of power. And, of course, such decisions must be firmly based on the laws of thermodynamics and on the best available scientific and technical information.

The balance of the facts and arguments we have discussed indicates that, amongst the various possible sources, nuclear fission reactors need to be taken seriously. Indeed, it seems likely that nuclear power may prove to be considerably safer than power from coal. However, this does not mean that coal is now too dangerous to use. In discussing the risks of various types of power generation, it is important to include the risks involved in *not* maintaining adequate supplies of energy. Modern industries, modern agriculture and modern systems of public health all require the input of large amounts of energy. There is nothing more certain than that a return to the levels of energy use which existed in pre-industrial societies would mean wars, famines, the spread of disease and a world-wide loss of life on an enormous scale.

Yet it also seems certain that energy generation in the future will be increasingly dependent on high levels of technology, much of it relatively new and untried. In assessing the acceptability of such new technologies, the importance of continuous and informed public debate should not be underestimated. It may be that any solution to the world's energy problems will depend on the continued existence of social systems like those of the liberal capitalist societies in which such informed public debates can occur.

REFERENCES

1   Hoyle, F. and Hoyle G., *Commonsense in Nuclear Energy*, (London: Heinemann Educational, 1980), pp. 40–2.
2   A risk of 1 in 10,000 per annum implies a death toll of about 5000 each year for the British population of about 50,000,000; the risks of death from road accidents, and from accidents in the home, are of approximately this size (see the Mortality Statistics published by the Government Statistical Service).
3   Information from *Nuclear Engineering International*, July supplement, 1979.
4   Dollezhal, N. and Koryakin, Y., *The Soviet Nuclear Energy Programme*, 'Kommunist' No. 14; reprinted in *Marxism Today*, December 1979, pp. 23–5.
5   McCracken, S., 'The Harrisburg Syndrome' in *Commentary*, Vol. 67, June 1979, p. 34. Quotation reprinted from *Commentary*, June 1979, by permission; all rights reserved.
6   See Lendvai, P., *The Bureaucracy of Truth: How Communist Governments Manage the News*, (London: Burnett, 1981), pp. 65–7 for a discussion of the muted Soviet reporting of the Three Mile Island accident and the absence of any public debate on nuclear policy.
7   McCracken, S., op. cit. Quotation reprinted from *Commentary*, June 1979, by permission; all rights reserved.

SUMMARY

1   In the last hundred years the world's population has roughly trebled but the world's energy consumption has increased more than twenty times.
2   Energy exists in many forms whose interconvertibility is governed by the first and second laws of thermodynamics.
3   Einstein showed that matter can be regarded as a particularly concentrated form of energy, and this has led us to realise that all our sources of energy are ultimately nuclear in origin.
4   The production of energy on a large scale inevitably involves substantial risks, whether the source be nuclear fission, fossil fuels like coal, oil or natural gas, or renewable sources like solar, wind, wave or hydroelectric power.
5   The risks from nuclear power stations, for a given power output, are less than for most other sources of power, and nuclear power stations are now in operation in more than twenty countries.
6   The risks involved in power generation need to be set against the risks involved in *not* maintaining adequate supplies of energy, which could include wars, famines, disease and widespread loss of life.

8.3 Semiconductor physics and the computer revolution

In Chapter 6.7 we touched briefly on the development of computers and on the possibility that silicon chip microprocessors will spark off a second industrial revolution which will change the way people live as much as the first industrial revolution and be as important in spreading knowledge as the invention of printing. In this chapter we will consider some of these topics in a little more detail. We will first discuss the origins of modern semiconductor physics in the 1920s. Then we will outline the rapid technical developments which, since 1948, have made the silicon chip microprocessor into a mass-produced practical reality. Thirdly, we will describe some of the social factors which enabled these rapid technical changes to occur and finally, we will try to establish some guidelines for assessing how the growth of computers will affect all our futures.

THE ORIGINS OF MODERN SEMICONDUCTOR PHYSICS

Modern electronics, like nuclear power, is a clear example of a science-based industry. Nearly all modern electronic devices are made from semiconductors like silicon whose properties were first systematically understood in the 1930s. At that time both semiconductor physics and nuclear physics were laboratory curiosities—of interest only to a few academics and of virtually no practical importance. However, most subsequent practical applications of semiconductors have been based on fundamental knowledge about the conduction of electricity in solids.

Solids probably differ more in their electrical properties than in any other way. Most solids fall into two distinct classes—they are either very good conductors, like most metals, or they are insulators, like glass or porcelain. The differences in electrical conductivity between metals and insulators are enormous—for example, copper is about ten million million million times as good a conductor as glass. Somewhere in between come the semiconductors like silicon and germanium— with silicon conducting electricity about a hundred million times better than glass but a hundred thousand million times less well than copper. These simple facts had long been known but there was no theory to explain the existence of these enormous differences. Then, in the late 1920s, Schrödinger began to apply his wave mechanics, which had been so successful in explaining the behaviour of electrons in isolated atoms (see Chapters 6.2 and 6.6), to the behaviour of electrons in solids. His theory was to prove just as successful in this very different field as it had been in explaining the energy levels of the hydrogen atom. In particular, it proved possible to explain in detail the electrical conductivities of insulators, good conductors and semiconductors and, most important for later developments, to understand how the electrical properties of semiconductors could be significantly altered by the addition of very small amounts of impurities. For our purposes we do not need to understand the details of the theory. All we need to know is that it predicts that electrons in solids, rather like the electron in the hydrogen atom, can only exist with certain definite energies and that the gaps between the allowed energies are such as to explain the very wide range of electrical conductivities described above.

In the 1930s this theoretical understanding of the behaviour of electrons in solids had virtually no effect on practical electronics. All the early developments in electronics—radio and television, amplifiers and counters for scientific instruments, radar during World War 2, the early electronic computers (see Chapter 6.7)—were based primarily on the thermionic valve and in particular on the triode valve invented by De Forest in 1906. A triode consists of an evacuated glass bulb, rather like an electric light bulb, which contains three electrodes—a cathode which emits electrons when heated, an anode which collects the emitted electrons and a grid which controls the flow of electrons from cathode to anode. By the end of World War 2 thermionic valves had been developed to a high degree of precision and were being mass-produced for a wide range of military, scientific and domestic purposes. However, they had several inherent disadvantages—they were inevitably bulky and fragile and they generated a large amount of heat. Consequently there had been a number of attempts to make devices which would perform the same functions as triode valves but which would be more compact, less fragile and which would not generate so much heat. The most promising approach seemed to be to make a three-electrode device from a solid, but none of the early attempts was successful, primarily because the techniques for manufacturing very pure materials were not sufficiently well developed. It was not until 1947 that the first working solid-state amplifier—later christened the transistor—was made.

THE FIRST TRANSISTOR

The transistor was invented by an industrial research organisation, the Bell Telephone Laboratories in New Jersey, which is primarily concerned with research into all matters concerned with communications technology. Bell Laboratories had played a major part in the development of electronics during World War 2. The laboratories were large, well-staffed and well funded, and they encouraged the kind of collaboration between physicists, chemists, metallurgists and engineers which made the invention of the transistor possible.

Three physicists were primarily involved in the invention—William Shockley, Walter Brattain and John Bardeen—and they were later awarded the Nobel prize for their work. All three had many years experience in solid state physics, both theoretical and practical. They were working not with silicon but with the chemically related element, germanium, whose existence had been predicted by Mendeleev in 1871 (see Chapter 6.6). Like many other key developments in science, the discovery was in some ways accidental. The first transistor was a point-contact transistor made by placing two fine wires close together on the

The News of Radio

Two New Shows on CBS Will Replace 'Radio Theatre' During the Summer

Two new shows are announced by CBS to serve as summer replacements for the hour-long "Radio Theatre" on Monday evenings. The first, to be heard at 9:30 P. M., will be "Mr. Tutt," based on the stories of the late Arthur Train. Willard Wright will have the title role and Arnold Perl will do the adaptations. It will make its debut next week.

The second, to open on July 12, will be "Our Miss Brooks," with Eve Arden playing the role of a school teacher who encounters a variety of adventures. It will be written by Al Lewis and Lee Loeb, with Larry Berns serving as director. "Our Miss Brooks" will be offered at 9 P. M. Mondays.

A situation comedy with musical overtones, headlined by Mel Torme, has been selected as the replacement for the Dinah Shore-Harry James program at 8 P. M. Tuesdays over NBC, starting next week. The supporting company will include Janet Waldo and John Brown. Harmon Alexander and Ben Perry will do the script. Frank Danzig will be the producer and Ben Elliott, the musical director.

Station WFUV, Fordham University's frequency modulation outlet, will observe the completion of its first year of operation today. Throughout the day there will be special programs to mark the occasion and a number of personages on local commercial stations will participate, including John McCaffrey, Eileen O'Connell, Pat Barnes, Mary Small, Alma Dettinger and Arlene Francis.

At 8:05 P. M. there will be a critical discussion of present-day radio, the members of the panel including the Rev. Robert I. Gannon, president of Fordham; Morris Novik, John Garrison and F. W. Carlington.

"On Your Mark," a new audience-participation item, will be added to WOR's schedule next Monday. It will be heard at 2:30 P. M. each weekday afternoon thereafter and will include prizes for questions which are correctly answered. Paul Luther will produce and announce the program.

George Shackley's original composition, "Anthem for Brotherhood," will have its first television performance at 5 P. M. Sunday over WPIX. The rendition will be a part of the station's "Television Chapel" program.

Beginning tomorrow, "Waltz Time" will be heard for a full hour for three successive Friday evenings at 9 o'clock over NBC.

A memorial tribute to Col. David Marcus, who died in action while commanding the Israeli forces on the Jerusalem front, will be presented at 10:03 this evening over WMCA. Gov. Thomas E. Dewey and Mayor O'Dwyer will be among the speakers to be heard by means of transcriptions.

Station WNEW will transmit reports on traffic conditions on the major highways in the metropolitan area for those who plan to spend the holiday week-end out of town. On Friday, the reports will be heard on the hour at frequent intervals between 4 and 11 P. M. The same schedule also will be followed on Sunday and Monday afternoons and evenings.

"The Better Half," which is to go under commercial sponsorship on Sept. 16 over the Mutual network, will take to the air as a sustainer on Aug. 19.

A device called a transistor, which has several applications in radio where a vacuum tube ordinarily is employed, was demonstrated for the first time yesterday at Bell Telephone Laboratories, 463 West Street, where it was invented.

The device was demonstrated in a radio receiver, which contained none of the conventional tubes. It also was shown in a telephone system and in a television unit controlled by a receiver on a lower floor. In each case the transistor was employed as an amplifier, although it is claimed that it also can be used as an oscillator in that it will create and send radio waves.

In the shape of a small metal cylinder about a half-inch long, the transistor contains no vacuum, grid, plate or glass envelope to keep the air away. Its action is instantaneous, there being no warm-up delay since no heat is developed as in a vacuum tube.

The working parts of the device consist solely of two fine wires that run down to a pinhead of solid semi-conductive material soldered to a metal base. The substance on the metal base amplifies the current carried to it by one wire and the other wire carries away the amplified current.

MORNIN

4:00—WINS—Recorded Music
5:30—WNBC—Recorded Music;
5:45—WOR—Farm News
        WJZ—Recorded Music
        WCBS—News; Reveille 1
6:00—WNBC—Market Reports
        WJZ—Farm News—Phillip
        WCBS—Arthur Godfrey
        WMCA—News; Recorded
        WHN—Radio Newsreel
        WNEW—Anthing Goes—
6:15—WINS—John Clark
6:30—WOR—Rambling With C
        WJZ—Kiernan's Corner
        WINS—Art Scanlon Show
        WNEW—News; Anything
6:45—WOR—Morning Watch
6:55—WNYC—News; Sunrise S
7:00—WNYC—News; Bob Smit
        WOR—News—Melvin Elli
        WJZ—News—Don Gardin
        WMCA—News; Music
        WINS—News; Art Scanla
        WHN—Todd Lawrence S
        WLIB—News; Recorded
        WQXR—The New York
7:05—WQXR—Breakfast Symp
7:15—WOR—Gambling's Music
        WJZ—Kiernan's Corner
        WMCA—News; Recorded
7:25—WQXR—Weather Report
7:30—WNBC—News; Bob Smit
        WNEW—News; Anything
        WQXR—Breakfast Symph
        WLIB—In Town Tonight
7:45—WCBS—News Reports
        WMCA—Unity Viewpoint
        WNYC—Weather and No
        WLIB—Recorded Music
7:55—WJZ—News; Weather; J
8:00—WNBC—News; Bob Smit
        WOR—Prescott Robinson
        WJZ—News—Martin Agr
        WCBS—News Round-Up
        WMCA—News; Recorded
        WNYC—The Musical Clo
        WINS—Prize Package
        WLIB—Young Peoples C
        WQXR—The New York

12:00—WNBC—Radcliffe Hall—I
        WOR—Kate Smith, Tall
        WJZ—Interviews With
        WCBS—Wendy Warren
        WMCA—News; Mr. and
        WINS—Don Goddard, Ne
        WHN—G. H. Combs, Ne
        WNEW—Let Yourself Go
        WLIB—News; Racing an
        WQXR—The New York
12:05—WQXR—Luncheon Conce
12:15—WNBC—News Round-Up
        WCBS—Aunt Jenny—Sk
        WINS—Peter Donald Sh
        WHN—Program Summar
12:30—WNBC—Norman Broken
        WOR—News—Henry Glad
        WJZ—News; Nancy Crai
        Mrs. Doris Levy
        WCBS—Helen Trent—Sk
        WINS—It's in the Bag—
        WHN—Ella Mason Quiz
        WNEW—News; Show Bu:
12:45—WOR—The Answer Man
        WCBS—Our Gal Sunday
12:55—WNYC—News; Four Stri
1:00—WNBC—Mary Margaret I
        WOR—Luncheon Progran
        WJZ—Baukhage—News
        WCBS—Big Sister—Sketc
        WMCA—News; Mr: and
        WINS—Listen to Lacy
        WHN—Morey Amsterdam

6:00—WNBC—Kenneth Banghs
        WOR—Lyle Van. News
        WJZ—News Reports
        WCBS—Eric Sevareid, N
        WMCA—News; Listen to
        WINS—Midget Auto Rac
        WLIB—News, Racing an
        WQXR—The New York
6:05—WJZ—Joe Hasel—Sports
        WQXR—Music to Remem
6:15—WNBC—Sports—Bill Ster:
        WOR—Bob Elson, Comm

Fig. 8.3.1 Announcement of the discovery of the transistor on page 46 of the *New York Times*, July 1 1948 (the last item in the second column).

surface of a single crystal of germanium. However, Shockley and his co-workers had been trying, without success, to make another kind of transistor—a field-effect transistor*—and it was while they were trying to understand why this device didn't work that they discovered the principle of operation of the germanium point-contact transistor. This discovery was made on 23 December 1947 and the very next month Shockley developed the theory of another kind of transistor—a junction transistor. However, this was not successfully made until 1951 and ultimately it proved easier to mass-produce than the point-contact transistor.

Bell kept the invention secret until July 1948, when the first scientific paper on the transistor was published.[1] Even after the discovery was publicly announced it caused very little stir. Fig. 8.3.1 shows the very brief item which appeared in the *New York Times*; the news was tucked away on page 46, the radio page.

Perhaps this is not surprising because in 1948 the transistor was little more than a laboratory curiosity. Even its inventors did not realise its full potential. As the title of their first paper shows—'The transistor—a semiconductor triode'[1]—they regarded it primarily as a potential replacement for the thermionic valve.

How the laboratory curiosity of 1948 evolved into today's mass-produced microprocessor is a long and complex story. Before we describe how that evolution took place, let us recognise the length and complexity of the evolution of the transistor itself since Schrödinger's wave mechanics first significantly advanced our understanding of how semiconductors work. According to Braun and MacDonald[2]:

> It is . . . unrealistic to see the transistor as the product of three men, or of one laboratory, or of Physics, or even of the forties. Rather its invention required the contributions of hundreds of scientists, working in many different places, in many different fields for many years. For example, one could examine the contribution of scores of chemists and metallurgists to an understanding of semiconductors and the development of their ability to produce them in a form useful to physicists. Without this materials research effort, and particularly the acceleration it received during the war, there could have been no transistor.[2]

FROM POINT-CONTACT TRANSISTOR TO SILICON CHIP MICROPROCESSOR

There are two main stages in the technical evolution of the silicon chip microprocessor—the production of separate transistors of various kinds and the development of integrated circuits which each contain many transistors. We will first discuss these technical developments. Then we shall describe the hectic and unplanned growth of the semiconductor industry. Finally we will consider some possible future applications of semiconductor electronics.

The early germanium point-contact transistors were first produced commercially in 1951. They were difficult to make and very variable and unreliable in

---

* A few years later field-effect transistors were successfully manufactured.

their performance. However, the invention of the transistor stimulated a large amount of research aimed both at understanding the properties of semiconductors and at improving techniques of transistor production. This led to the rapid development of different types of transistor. Shockley's junction transistor was first produced in quantity in 1952. It consisted of a three-layer germanium sandwich with each layer containing definite amounts of impurities which determined the electrical properties of the transistor. The junction transistor was easier to produce and potentially more reliable than the point-contact transistor.

Then, in 1954, a relatively new firm called Texas Instruments succeeded in producing the first transistor to be made of silicon rather than germanium. This success was of great importance for the future development of the electronics industry and, like so much else in that development, was totally unexpected. Again according to Braun and MacDonald:

> The accomplishment was announced by the man most responsible, Gordon Teal, to the Institute of Radio Engineers and directly followed other papers which had declared that a silicon transistor would be impossible for many years.[3]

The next major technical development was the introduction in 1959 of the planar process of transistor manufacture by Fairchild Semiconductors, a new company which had only been founded in 1957. In this process a wafer of pure semiconductor is first masked with a layer of nonconducting oxide. Then microphotographic techniques are used to etch holes in the mask through which suitable impurities can be diffused. Finally, electrical connections are made to the completed transistors by depositing an evaporated metal film at suitable points on the wafer. The planar process works best with silicon and is ideally adapted to batch production in which many identical transistors are made on a single wafer which is then cut up into individual transistors. The process rapidly became the standard technique throughout the industry and led to a great increase in the yield of satisfactory components which might now be as high as 90 per cent compared with figures of 10–30 per cent for many earlier transistor manufacturing techniques. Consequently, demand for transistors and transistor output rose rapidly, many new firms were set up and intense price competition forced prices down so much that many transistor companies were in financial difficulties in the early 1960s.

Then there came the next major technical advance–the move from single transistors to integrated circuits. Up till then each individual transistor had, like thermionic valves before them, been connected individually to other components to make a complete circuit. Now it was realised that, instead of cutting up a wafer into separate transistors, it should be possible to connect two or more transistors together on the wafer in order to make complete integrated circuits which could then be packaged separately.

This technical development—achieved in 1959 by Jack Kilby at Texas Instruments and Robert Noyce at Fairchild—marked a significant watershed for the electronics industry. After this transistors rapidly outgrew their origins as replacements for thermionic valves and semiconductor electronics developed in ways which would have been previously inconceivable.

The rapid development of the planar process meant that, by the mid-1960s, it became feasible to make integrated circuits with tens or hundreds of components. It was only in 1959 that transistors first outsold valves. Yet, by 1965, the total number of components sold in integrated circuits exceeded the sales of separate transistors and, by 1971, 95 per cent of all components sold were in integrated circuits. Since then the number of components per chip has increased so much that chips containing hundreds of thousands of components are now feasible and there is no physical reason why even greater packing densities can't be achieved. Yet, as so often in semiconductor electronics, even the electronic experts have frequently been surprised by the pace of developments. Jack Kilby, one of the pioneers of integrated circuits, is reported to have said 'Certainly at that time (1961) I did not visualise anything comparable to a one-chip calculator or that level of complexity in the foreseeable future.'[4]

What are these extremely complex integrated circuits used for? First of all, they have improved the performance of most conventional electronic applications. Extremely sensitive electronic amplifiers like those used in radioastronomy, counters for detecting low-levels of radioactivity, the radio transmission of pictures of distant planets from spacecraft millions of miles away—all these things can be done more reliably and more cheaply as a result of the development of integrated circuits and microelectronics.

Then there has been the replacement of the magnetic memories used in most early computers by semiconductor memories which are more compact and much cheaper. This replacement is now universal and Fig. 8.3.2 shows how the cost per bit (binary digit) of memory has dropped dramatically over the last few years—a process which shows every sign of continuing.

Another development has been the design of special purpose integrated circuits which contain all the circuits needed for very complex operations on a single silicon chip. The main examples here are calculators and digital watches which have been produced in very large numbers at prices which have dropped sharply year by year in the same way as transistor prices and the prices of computer memories. The market for calculators was totally new whereas the digital watch had to compete with the established watch industry. Yet both were able to command large enough markets to justify the large costs of designing complex special purpose chips and to gain the benefits of the very large cost reductions which come from mass production.

Circuits for many other complex operations could, in principle, also be incorporated on special purpose chips. However, these chips are unlikely to be economic unless they can be sold in huge numbers like calculators and digital watches.

So, just as Charles Babbage switched from special purpose to general purpose digital computers in the 19th century (see Chapter 6.7), much interest has centred on the development of general purpose silicon chips which could have a multiplicity of uses.

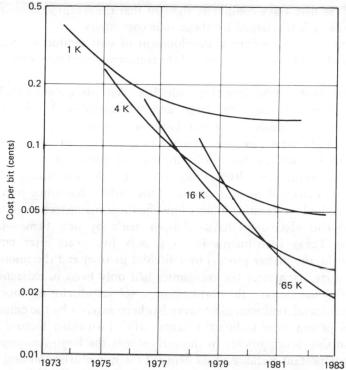

Fig. 8.3.2 Cost per bit of semiconductor memories for computers. *Redrawn from Noyce, R. N., 'Microelectronics'. Copyright © 1977 by Scientific American, Inc. All rights reserved.*

These general purpose silicon chips, or microprocessors, are integrated circuits which can carry out all the functions of the central processor of a computer. They were first developed by Hoff at Intel Corporation in 1971 while he was redesigning the circuits of a calculator. Hoff reduced the design to three chips: the microprocessor and two memory chips, one to read the data in and out and one to hold the program which operated the central processor. He had thus designed not just a calculator but a tiny general purpose computer which, just like any other computer, could be programmed to perform a multiplicity of different tasks.

Since 1971 microprocessors and the other chips needed to operate them have undergone rapid development and have spiralled downwards in price just as transistors and integrated circuits had done. The result is that computing itself now costs virtually nothing. The cost of a microcomputer is now primarily that of the screen, keyboard and other peripherals together with the cost of developing its programs.

The low cost of microprocessors almost certainly means that the number of digital computers in the world will increase dramatically in the years ahead. The total number of digital computers in the world could be measured in thousands in the mid-1950s, tens of thousands in the 1960s and hundreds of thousands by 1975. Because of the microprocessor it is now estimated that this number may exceed twenty million by the mid-1980s. It is interesting to remember that the designers of the first valve computers thought that the world's computing needs could be more or less satisfied by about four computers!

Now let us leave the technical development of semiconductor electronics in order to consider, briefly, the people and the companies who have created the new industry.

The growth of semiconductor electronics has taken place almost exclusively in the United States. Initially the large and well-established valve companies dominated the new industry too but fairly soon new companies specialising wholly in semiconductor electronics started to spring up. In 1954, Shockley left Bell to set up his own company in California and a few years later, in 1957, a group of engineers including Robert Noyce left Shockley's company to start their own company, Fairchild Semiconductor. This pattern has since been repeated many times and it is interesting to note how the three major technical innovations in semiconductor electronics have all been made by new firms—the silicon transistor by Texas Instruments in 1954, only five years after entering the electronics field, the planar process by Fairchild in 1959 and the microprocessor by Intel in 1971, each after the companies had only been in existence for two years. Mobility has become the norm amongst semiconductor engineers and the development of semiconductor electronics has been marked by the emergence of a new type of person—the technical entrepreneur. Two other factors have been important in this development. In the early stages the large American military market was important because of its demand for miniaturisation and its role as an early buyer of expensive new devices. Later, the growth of commercial and domestic markets was greatly helped by the availability of finance from banks who understood the special needs of the new industry.

Japan is the only country which has been able to challenge in any way the world dominance of the American semiconductor industry. However, the Japanese industry has developed very differently—via established valve firms and strong government involvement, with no military market, little research and development and practically no movement of individuals between companies.

The overwhelming impression left by the extremely rapid growth of the new microelectronics is one of ceaseless change which has proceeded almost spontaneously. It has been virtually impossible for anybody to plan for more than a short period because, time and again, the experts themselves have been taken by surprise by the direction of new developments and the relentless pace of change.

FUTURE DIRECTIONS FOR THE COMPUTER REVOLUTION

When the experts have so often been wrong, it may be foolish to attempt to look ahead. Yet some of the claims for the computer revolution are so sweeping that it is probably worth seeking guidelines for predicting what we may expect in the years ahead.

Let us first summarise the strengths and weaknesses of computers as identified in Chapter 6.7. Computers are capable of storing very large amounts of information and of processing that information quickly and accurately according to precisely defined rules. For tasks requiring these attributes, computers are much more efficient than human beings. It seems likely that the size of computer memories will continue to increase and that the cost of even very complex computer processing will continue to decline. For example, video discs are likely to enable even the smallest computer to have access to vast amounts of information. Techniques already exist whereby very large amounts of information—the equivalent of hundreds or even thousands of books—can be stored on a single video disc.

At the other end of the scale, very large computers can quickly scan and correlate data from a number of diverse data bases. These powerful machines could easily add considerably to the powers of central governments who could bring together all the diverse records for an individual—tax records, health records, employment records, criminal records (if any) and so on. Thus the privacy of individuals could be invaded without their knowledge. In most liberal countries the pressure of public opinion has led governments to issue guidelines or to legislate to try to preserve privacy. Yet these measures may not be enough to prevent the technically possible from happening. However, such possibilities are much more likely to become realities in totalitarian states which lack any effective public opinion. Perhaps it is fortunate that these states are, so far, not well versed in advanced computer technology.

The limitations of computers are related to their strengths. Because computers can handle enormous amounts of data, it is data input—both its quality and its speed—that is one of the crucial limitations. Data output is much less of a problem. Next, in the words of Lady Lovelace (see Chapter 6.7), a computer can still only 'do whatever we know how to order it to perform'. Therefore, it is necessary to devote much effort to breaking tasks down into simple steps and writing precise instructions for the computer to enable it to do those tasks. Now that computing power itself is so cheap, it could well be that the major bottleneck will be in writing efficient programs. Already many man-years of effort can be involved in writing even relatively simple programs and even more man-years in testing to ensure that they actually do what they are supposed to do. Already some programs are so large that it is virtually impossible for any individual to have a grasp of their totality. Also, by their very nature and flexibility, it is impossible to test all the possible ways in which a program or a computer language can be used. These properties of computer programs do not mean that such systems will be unusable since many of our more familiar information systems have similar properties (see Part 9). Yet it is likely to mean that the development of large programs will be more difficult and costly than is often claimed.

Finally, the most serious limitation of computers may well stem from some crucial differences between the human brain and the way that computers operate. In order to illustrate these differences let us briefly compare artificial computer languages with the many natural human languages.

These two types of language have some similarities, particularly if we consider high-level computer languages which may have many words in common with natural languages. However, there are major differences too. Computer languages have very precise logical structures which are designed and specified in detail by those devising the language. They deal with information which is centrally located in the computer's memory and has to be in a standard format. Precise rules define the overall structure and much of the content of any program which can be written and a single error of syntax or punctuation is often enough to destroy completely the meaning of a statement. Any self-correcting or checking procedures must be specifically designed by the programmer. So although a computer may be very versatile, it remains a machine which operates in a very precise and logical way.

By contrast let us consider natural languages. These have not been designed but have all evolved in an unplanned and spontaneous way and are so complex that no-one knows all their details. Natural languages are governed by some general grammatical rules but even today we have only a small understanding of the structure of language. Yet, despite this, we use our languages confidently—and without language virtually nothing of our culture would exist. And since they have successfully evolved, natural languages must incorporate self-correcting or error-eliminating mechanisms. Linguistic forms which do not lead to effective communication tend not to persist while more successful forms do and gradually evolve into more permanent features of the language. Natural languages also have the property that errors do not completely destroy the meanings of statements—partially incoherent or ungrammatical statements can transmit some information although, of course, they can also lead to misunderstandings. Finally, natural languages can use precise logical statements—for example, in formal logic, in mathematics and in computer languages—but these are just facets of their complex overall structure which in its totality has little similarity with the precise rules governing the operation of a machine.

These differences between natural and computer languages may help to explain the lack of success of computer translation projects—a development which was confidently predicted in the early days of valve computers. Yet despite much effort, computer translations are as far away as ever, although there seems to be a very good chance of using computer memories for the development of comprehensive dictionaries and other works of reference.

In conclusion, it is important that the major limitations as well as the advantages of electronic computers and microprocessors should be widely understood since the products of this very new science-based industry are certain to become much more common in the very near future.

REFERENCES

1   Bardeen, J., and Brattain, W. H., 'The Transistor, a semiconductor triode' in *Physical Review*, 15 July 1948, pp. 230–1.
2   Braun, E., and MacDonald, S., *Revolution in Miniature: The history and impact of semiconductor electronics*, (Cambridge: Cambridge University Press, 1978), pp. 49–50.
3   Ibid., p. 63.
4   Ibid., p. 112.

SUMMARY

1 Modern semiconductor electronics is a science-based industry which has its roots in the knowledge gained during the 1920s and 1930s by applying Schrödinger's wave mechanics to the behaviour of electrons in solids.

2 When transistors were discovered in 1947, they were seen primarily as replacements for thermionic valves. Since then many unexpected technical developments have led to integrated circuits being made which contain hundreds of thousands of transistors on a single silicon chip.

3 Silicon chips have been mass-produced both for specific purposes, as in calculators and digital watches, and for general application, such as computer memories and microprocessors which can carry out all the functions of the central processor of a computer.

4 Microprocessors are now so cheap that the number of digital computers in the world is likely to increase very rapidly; the actual costs of computing are now almost negligible and the main limitations on future applications are likely to be the input of reliable data and the costs of developing computer programs.

5 The computer revolution will undoubtedly change all our lives in ways which are likely to be as unplanned and unpredictable as was the technical development of silicon chips.

8.4 Some applications of biological knowledge

In this chapter we will briefly consider three important topics which involve the application of biological knowledge. These are the perennial problem of food production and two relatively new problems—pollution of the environment by man and the problems of biotechnology. Some of these involve applications of the new knowledge in genetics and molecular biology, which we described in Chapter 6.4. It is clearly not possible to discuss all aspects of these problems in one short chapter. All we can do is to indicate some of the important scientific and social factors which need to be considered in any public debate about these questions.

FOOD PRODUCTION

There have been two major revolutions in the way people live—the agricultural or neolithic revolution of 10,000 years ago, and the industrial revolution of the 18th century. In this book we have given most attention to the importance of science and technology in the industrial revolution. Yet in many ways the agricultural revolution is more fundamental. It was the discovery of agriculture and the domestication of animals which led to the first big expansion of the world's population and to the emergence of a settled way of life. Improvements in agriculture were also a vital factor in the industrial revolution which could never have developed without increased food supplies to feed the growing populations of the industrial towns. Food production is also vital in modern industrial societies, although it is easy to forget this in countries where only a few per cent of the population work in agriculture and food is plentiful.

In the early years of the 19th century, Thomas Malthus was greatly concerned about the possible inadequacy of future food supplies (see Chapter 4.2). His pessimistic predictions about food shortages have not so far come true in industrial societies. The growth of populations and of food supplies have kept roughly in step. In the 19th century this was primarily due to the cultivation of new land in many parts of the world. Examples include the United States, Canada, Argentina, the Ukraine, Australia and New Zealand. However, the growth of world food production in the 20th century has been primarily a result of very large increases in agricultural productivity. These increases are due to many factors. Some of the most important are the development of new varieties of plants and animals by selective breeding, the increased mechanisation of agriculture, the use of chemical fertilisers and pesticides, and the scientific evaluation of new agricultural techniques by statistical methods (see Chapter 6.7). All of these processes are in continual interaction and they all depend on the application of scientific knowledge at every stage. Perhaps the most underestimated feature of the growth in agricultural productivity is its dynamic nature. Nature is extremely adaptable, and there is a continual battle between man's attempts to raise yields with new varieties and new methods of pest control and nature's counter-attack with new types of pests or diseases. Such a dynamic process in the case of wheat has been described thus:

> The breeding of wheat for resistance to stem rust, a devastating disease, is a prime example. There are many kinds of stem rust. Pathologists, led by Elvin Stakman of the University of Minnesota, have devised ingenious methods of identifying them by inoculation of different hosts. The wheat breeder then develops a new variety which is resistant to the predominating races of stem rust. This is distributed to farmers and its acreage increases rapidly. But while the wheat breeder is hybridizing wheats, nature is hybridizing rusts. The reproductive stage of stem rusts occurs not on wheat but on an alternate host, the common barberry. On this plant new races of rust are constantly created. Although most of them probably die out, one that finds susceptible wheat varieties may multiply prodigiously and in a few years become the predominating race. The wheat breeder then searches the world for wheats resistant to the new hazard and again goes through all the stages of producing a new hybrid variety. The competition between man and the fungi for the wheat crop of the world is a biological 'cold war' which never ends.[1]

The dependence of improvements in agricultural productivity on the application of scientific knowledge and technological ingenuity is perhaps one reason why the liberal democratic countries of the first world are the most successful in producing agricultural surpluses. It has been estimated that the developed world with roughly 30 per cent of the world's population, produces around 60 per cent of world food supplies. It was the cereal producers of North America who sustained many of the developing countries during the years of bad harvests in 1965–6 and 1972–4. The European Economic Community, too, is well-known for its problems with agricultural surpluses—wine 'lakes' and butter or beef 'mountains'.

The centrally planned economies of the second world, with their emphasis on collective farms, have often not been able to grow enough food for their own populations. This was true in the Soviet Union in the 1930s when Stalin banned genetics, collectivised all the farms in spite of the opposition of the peasants, and generated the worst man-made famine in history in a country which had been a major food exporter in the early years of this century (see Chapter 7.4). And it is still true today, when the Soviet Union has to import large quantities of grain from North America and Argentina, that food shortages are endemic over most of Eastern Europe. It is generally recognised that the only really efficient agricultural units in the Soviet Union and Eastern Europe are the tiny private plots which are cultivated by the peasants for their own use.

It therefore seems likely that, despite the many problems involved in translating agricultural techniques from one region to another, it will be the continually evolving agricultural knowledge of the countries of the first world which, for the foreseeable future, will be most helpful in attempts to improve the world's food supplies.

POLLUTION OF THE ENVIRONMENT

Current problems of environmental pollution are, directly or indirectly, mainly a result of industrialisation and other consequences of the industrial revolution.

Pollution arises when substances are released to the environment, either intentionally or inadvertently, which can harm living things. Most attention is usually given to substances which can harm man, either directly or when concentrated in plants or animals eaten by man. However, pollution can harm other species too, and this has also given rise to concern.

In modern industrial societies, the number of possible pollutants and types of pollution is so enormous that it is clearly impossible to discuss them in any systematic way. Each case has to be examined separately and its assessment is likely to involve knowledge of many kinds—physical, chemical and biological—drawn from a wide range of sources. This can be seen most clearly if we briefly consider two well-known examples—air and river pollution in London and mercury pollution in Japan.

In the early 1950s the air in London and the water of the river Thames were both severely polluted. In the winter of 1952–3 air pollution by smoke and sulphur dioxide led to a severe fog or 'smog', following which about 4000 people died. The resulting publicity led to the Clean Air Acts of 1956, as a result of which smoke concentrations have been considerably reduced and London 'smogs' are a thing of the past. Better treatment of sewage and stricter controls on industrial emissions have also led to great improvements in the purity of the Thames. The river now has about 80 species of fish living in its lower reaches which in the 1950s had been deoxygenated and virtually lifeless.

This air and river pollution in London was caused by a large number of sources—both domestic and industrial. By contrast, the mercury pollution at Minamata in Japan was eventually found to have a single industrial source which could be eliminated once it was discovered. However, before this was done in 1960, 43 people had died and more than 60 others had been permanently disabled by a mysterious illness which first appeared in 1954. In 1956 the medical department at Kumamoto University were asked to investigate and they soon discovered that fish and shellfish from Minamata Bay contained up to 60 times the normal amounts of mercury. Fishing from the bay was banned and the source of the mercury was traced to a plastics factory, which eventually built a treatment plant to remove heavy metals. One important factor which contributed to the scale of the disaster was that some of the mercury in the effluent was converted into organic mercury compounds, which are much more poisonous. This disaster was one factor in the decision of the Japanese government to devote more resources from the 1960s onwards to research concerned with environmental pollution.

In these examples, two main factors were involved in limiting some of the dangers of pollution—the knowledge of physicists, chemists, biologists and epidemiologists in tracking down the sources of pollution, and the publicity which stimulated companies and governments into taking some action.

It seems that both these factors are necessary in limiting pollution. Scientific knowledge alone is not enough as perhaps can be seen by considering some cases of pollution in societies in which no effective public opinion exists. One major example here is the radioactive pollution of a large area in the Soviet Union in 1958, which we have already mentioned in Chapter 7.4. In this case scientists had warned of the dangers of the crude methods used to store radioactive waste but, in the absence of any effective public opinion, nothing had been done. Then when the accident did occur, the appropriate scientific expertise was not available (see Chapter 7.4). Also, the accident was never reported publicly and was not followed by any public enquiry.

This is not an isolated example of the harm which can arise when vital information is not publicly available. Indeed, in many Marxist states, the problem is not just that there is no effective public opinion. Rather it is that such states have comprehensive censorship organisations which often explicitly forbid the media even to mention cases of environmental pollution and other health hazards. For example, these are some of the instructions given to the censors in Poland:

> In school No. 80 in Gdansk a harmful substance emitted by the putty used to seal the windows was noticed. The school has been temporarily closed. No information whatsoever may be published about this subject.
>
> All global figures on labour hygiene and accidents in the sectors and branches of the economy are to be withheld.
>
> Information about the endangering of life and health of human beings caused through chemicals used in industry and farming have to be deleted.
>
> Banned is all information about food poisoning and epidemics affecting larger groups of people and especially in important plants, furthermore food poisoning in factory canteens, holiday centres, summer colonies.
>
> There should be no disclosure about the pollution of rivers flowing from Czechoslovakia, if the pollution is caused by economic activity on our territory.[2]

However, sometimes nationalism triumphs over censorship, since this last instruction continues thus:

> On the other hand, information about the pollution of these rivers can be released if this is caused by economic activity on Czechoslovak territory.[2]

The thoroughness of the censorship in Poland can be seen from an internal information bulletin which records 584 acts of censorship in a single fortnight, May 1–15 1974. The bulletin goes on:

> Considerable cuts were made in sixteen articles dealing with environmental problems. Information about the pollution of the air, earth and water, about the growing ecological damages, the wrong location of industrial plants which would accelerate the pollution process, the direct endangering of human health and the irreversibility of the damage, have been deleted.[3]

So it is perhaps not surprising that the short-lived rise of the Polish free trade union Solidarity has revealed information about 'some of the most savage industrial pollution of the twentieth century'.[4] An article in *New Scientist*, largely based on information from official sources, states that the historical city of Cracow

> ... is a virtual cauldron of destructive chemicals, which are rapidly destroying its unique architectural embellishments—the carved figures and flowers and leaves that make Gothic architecture Gothic. The faces of important stone statues at Wawel Castle have crumbled. Steeples fall off churches; balconies must be constantly replaced and repaired; the view down every narrow street is disrupted by the scaffolding of workers trying to hold the buildings together.
>
> Acid rain dissolved so much of the gold roof of the 16th century Sigismund Chapel of Wawel Cathedral that it recently had to be replaced. The chemical used to dissolve gold in the chemistry lab is aqua regia ('royal water' because it attacks 'noble' metals such as gold and platinum), a mixture of concentrated hydrochloric and nitric acids in a ratio of three or four to one. Something not unlike aqua regia falls daily in the Cracow rains, and converted the chapel's original gold roof into soluble chlorides.
>
> The industrial dust falling on Cracow is nine times the acceptable national limit. Annual dust emissions from the steelworks alone contain 7 tonnes of cadmium, 170 tonnes of lead, 470 tonnes of zinc and 18,000 tonnes of iron. . . .[5]

However, since martial law was imposed in Poland in December 1981, and Solidarity was dissolved, the flow of information about pollution in Poland has ceased.

GENETIC ENGINEERING AND BIOTECHNOLOGY

The use of biological techniques to mass-produce valuable products is not new. For thousands of years man has used naturally occurring yeasts to make wine, beer, cheese, yoghourt and, of course, bread. More recently fungi have been used in fermentation processes to mass-produce antibiotics like penicillin and tetracyclin.

What is new is the prospect of altering the genetic composition of naturally occurring micro-organisms so that they will make new products which are useful to man. The possibilities here are so encouraging that much research is now taking place into what has been christened biotechnology—a term which includes old-fashioned brewing and baking as well as many new and untried processes.

According to a recent report to the British government:

> Rapid advances in both cell and molecular biology have allowed more confident prediction that a given product can be produced by a biological process or organism at a reasonable cost. Genetic manipulation, involving the transfer of genetic material from one organism to another and giving the recipient some desired characteristic of the donor, has become a practical and quite general proposition. As a result, a rational rather than empirical approach to biotechnology is now feasible, with organisms tailored to specific needs and process conditions.[6]

The report claims that the new biological techniques

> ... will be relevant to a wide range of industries—including agriculture, food and feedstuffs, chemical, pharmaceutical, energy and water industries—and to such diverse products and services as bulk chemicals, antibiotics, vaccines, methane gas, metals, food additives, single cell protein, effluent treatment and waste recycling.[6]

One of the most fruitful applications of genetic engineering is likely to be the manufacture by cheap fermentation of large quantities of medically valuable proteins or peptides. For example, bacteria have been produced which will synthesise human insulin, although not yet in very large quantities. Anti-viral agents like interferon or vaccines for viral diseases are also likely to become available in the near future.

Other possibilities include the modification of the genetic constitution of bacteria so that they will produce useful enzymes which can then be used in large-scale industrial processes; the growth of single-cell proteins for food; and the production of cheap liquid and gaseous fuels from a variety of sources.

Apart from the technical and economic problems involved in genetic engineering, it is also necessary to consider the biological risks involved. In the early days of work on genetic manipulation in 1971–3, some biologists were concerned about the possibility of harmful effects from this research. Many biologists voluntarily agreed to stop working on such experiments until more was known about their possible effects. Safety procedures were then drawn up in consultation with government bodies such as the National Institutes of Health in the United States and the Genetic Manipulation Advisory Group in Britain. More recently these guidelines have been relaxed as knowledge about some of the possible risks has increased. However, these experiments must still be carried out under closely controlled conditions, and this is one reason why biotechnology is unlikely to grow as rapidly as microelectronics has done. Nevertheless, it is still likely to become a major new science-based industry in the relatively near future.

REFERENCES

1 Mangelsdorf, P. C., 'Wheat' in *Food: Readings from Scientific American*, (San Francisco: Freeman, 1973), p. 103. Copyright © 1973 by Scientific American, Inc. All rights reserved.
2 Labedz, L., 'How the censors build up a false picture of the world' in *The Times*, 26 September 1977, and also in Lendvai, P., *The Bureaucracy of Truth: how communist governments manage the news*, (London: Burnett, 1981), p. 116.
3 Lendvai, P., op. cit., p. 115.
4 Timberlake, L., 'Poland—the most polluted country in the world?' in *New Scientist*, 22 October 1981, p. 248. This first appeared in *New Scientist*, London, the weekly review of science and technology.
5 Ibid., p. 249.
6 *Biotechnology: Report of a Joint Working Party*, (London: HMSO, 1980), p. 16.

SUMMARY

1   Both the industrial revolution and life in modern industrial societies depend on an efficient agriculture to provide enough food for the vast majority of the population who live in towns.

2   Population growth and food supplies are keeping roughly in step in industrial societies primarily because of improvements in agricultural productivity due to the application of scientific knowledge and technological ingenuity.

3   Pollution of the environment is an inevitable consequence of large-scale industrialisation which can in many cases be reduced by the application of physical, chemical, biological and epidemiological knowledge.

4   Another vital factor in reducing pollution is publicity and the existence of an informed public opinion; consequently societies which censor all news of environmental pollution are likely to have severe pollution problems.

5   Genetic manipulation, involving the transfer of genes from one organism to another, is now a practical possibility and is likely to find application in many biological processes such as fermentation, the production of enzymes, and the growth of single-cell protein for food.

8.5 Malthus revisited? the population explosion and the limits to growth debate

In Chapter 4.2 we saw that demography or the study of the growth of human populations had its origins in the work of Thomas Malthus. Since that time the study of demography has become much more thorough and we now know much more about the many complex factors involved in population changes. In this chapter we will first look at some basic demographic facts. Then we will outline some of the main factors governing population changes, giving examples from both historical demography and studies of contemporary populations. Finally, we will briefly discuss the validity of some computerised predictions of future population changes such as those made in the well-known book *The Limits to Growth*.[1]

THE WORLD'S POPULATION

Let us look again at Fig. 6.1.8 (page 249) which shows the growth in world population from 400 BC to 1975. Note the rapid increase since 1800—by 80 per cent in the 19th century (900 million (M) to 1625 M) and by 140 per cent in the 20th century so far (1625 M to 3900 M). However, this diagram, since it gives a global average, conceals important differences in population growth between different parts of the world. Compare Fig. 8.5.1 showing population growth in Europe—a 120 per cent increase in the 19th century (from 180 M to 390 M) and a 60 per cent increase in the 20th (390 M to 635 M) with Fig. 8.5.2 for Asia—only a 50 per cent increase in the 19th century (625 M to 970 M) but an enormous 140 per cent increase in the 20th century so far (970 M to 2300 M). Between 1950 and 1975 the population of Asia increased by 850 M—a figure which is over 25 per cent more than the *whole* population of Europe in 1975 (635 M). And the potential future growth in Asia is even more rapid because so many of these populations are young. Look at Fig. 8.5.3 which shows age distributions which are typical of contemporary European and third world countries. On the right is the age distribution for Sweden in 1956. Note the high proportion of people over 60 which is characteristic of nearly all European countries; in recent years this proportion has increased further. On the left is the age distribution for Algeria in 1954 which has a very similar population structure to those of European countries in the 19th century. Note the high proportion, more than 43 per cent, under the age of 15 compared with only 24 per cent for Sweden. This means that there is great potential for further population growth in many third world countries. It has been estimated that the world's population could increase by more than 50 per cent in the next 30 years for this reason alone even if, as seems unlikely, each family had only two children.

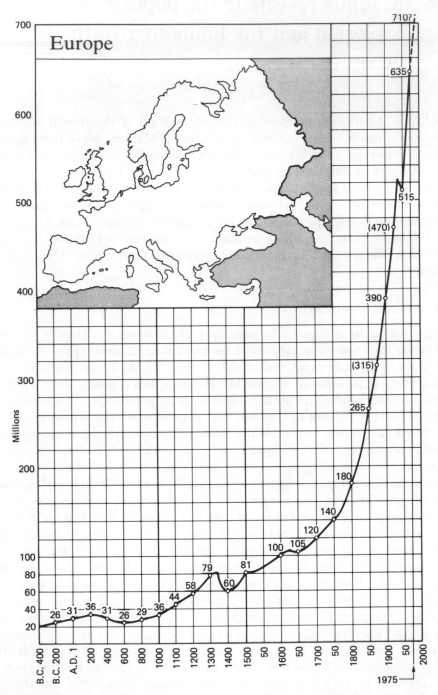

Fig. 8.5.1 Population of Europe, 400 BC to 1975. *Redrawn from McEvedy, C. and Jones, R., Atlas of World Population History, page 18, copyright © Colin McEvedy and Richard Jones, 1978, reprinted by permission of Penguin Books Ltd.*

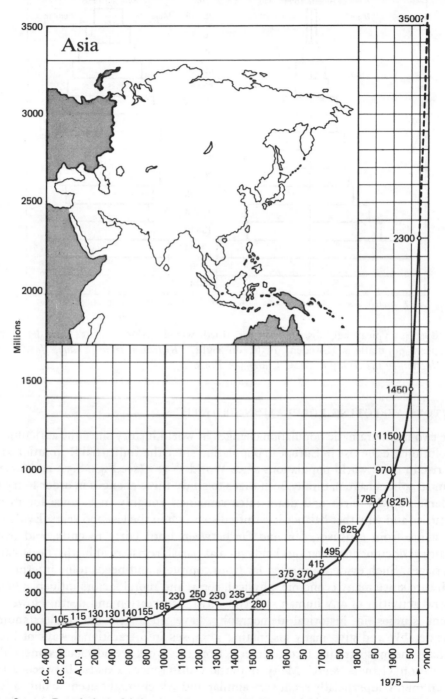

Fig. 8.5.2 Population of Asia, 400 BC to 1975. *Redrawn from McEvedy, C. and Jones, R., Atlas of World Population History, page 122, copyright © Colin McEvedy and Richard Jones, 1978, reprinted by permission of Penguin Books Ltd.*

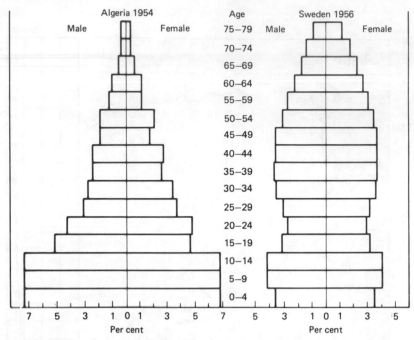

Fig. 8.5.3 Typical age distribution for third world (Algeria, 1954; on left) and European populations (Sweden, 1956; on right). *Redrawn from Young, J. Z., An Introduction to the Study of Man, Clarendon Press, 1971.*

FACTORS AFFECTING POPULATION GROWTH

The two most dramatic population changes in world history since the agricultural revolution are the rise in European population since the Industrial Revolution and the rise in third world populations since World War 2 (see Figs. 8.5.1 and 8.5.2). Demographers have naturally been eager to study these changes in order to try to understand why birth and death rates remained roughly in balance for many centuries and why that balance has shifted so much in the last 200 years. Evidence has come from censuses, few and far between until fairly recently, and from historical documents of various kinds such as parish registers of births, deaths and marriages. Such evidence is often far from complete and needs to be interpreted with care since many economic, physical, social and moral factors influence those central activities of any human society—birth, procreation and death. However, recent studies in historical demography have advanced our understanding considerably and give many fascinating glimpses into traditional ways of life.[2] Perhaps the most striking facts to emerge from this work are how local some of the population changes were. Many important differences existed between areas which might superficially seem very similar and which might even be only a few miles apart. So we should be wary of using overall national averages, like those given in Fig. 4.11.1 (page 213), since they may conceal large regional and local differences. However, some generalisations can be made about rates of mortality and fertility before and during the Industrial Revolution and in the modern world.

Changes in mortality usually preceded changes in fertility. So it is logical to consider first the decreasing death rates which underlay the rise in populations.

One major reason for the fall in death rates was a substantial reduction in crises of the type illustrated in Figs. 4.2.1 and 4.2.2 (pages 104 and 105) which were primarily brought about by harvest failures or epidemic diseases. Data for birth and death rates over long periods—from the 16th to the 19th century—show that in most years birth rates are much larger than death rates, often twice as large. So between the crisis years populations grew fairly quickly and were then cut back by sudden surges of mortality. Once these crisis years ceased or became less frequent, populations started to increase consistently. Amongst many reasons why these mortality peaks declined were improved transport—by road, canal, river and sea—to move food into areas with bad harvests; and the declining number of areas which depended on a single crop for their food supplies. Fig. 4.2.1 (page 104) shows that even in the 17th century some villages suffered much more in bad years than others. Aureil with its mixed agriculture was much less affected than Breteuil which depended on a single cereal crop.

Another major feature of the mortality statistics was the way in which age-specific death rates changed in western Europe in the 19th century. The figures for England and Wales shown in Fig. 8.5.4 were typical. Death rates fell much more among children, teenagers and young adults than amongst older adults. So the expectation of life grew rapidly (see Table 4.11.1, page 213). Fig. 8.5.4 also shows that these trends have continued throughout the 20th century

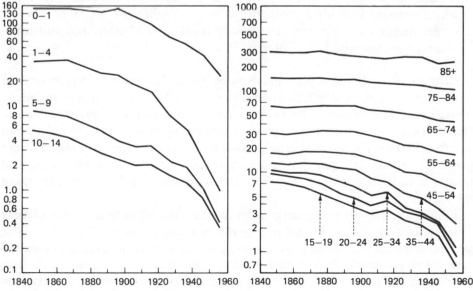

Fig. 8.5.4 Age specific death-rates per thousand for England and Wales, 1840–1960. *Redrawn from Wrigley, E. A., Population and History (London: Weidenfeld and Nicolson, 1969).*

and even extended to children in their first year of life whose very high death rates had remained virtually unchanged throughout the 19th century.

However, these average improvements in death rates concealed wide variations in different regions. Mortality was particularly high in the cities which grew so rapidly in the 19th century (see Table 4.10.1, page 196). In mid-century, expectation of life '... was only 24·4 years for men in Manchester according to the calculations of William Farr, at a time when it was 40·2 years in England as a whole, and already 51·0 in Surrey. ... In the worst parts of cities like Liverpool and Manchester expectation of life was well under 20 years.'[3]

According to Wrigley 'It is perhaps more accurate to say that high mortality was caused by urbanisation rather than by industrialisation. It was in the bigger cities that mortality was so very high.'[4]

Wrigley sums up these mortality changes thus:

During most of the nineteenth century improvements in health were a by-product of increases in wealth. The two went hand in hand. The age-specific death rates of the nineteenth century are amongst the most important and impressive testimonies to the economic benefits which came in the wake of the industrial revolution and which took place among the mass of the people in spite of the rapid growth of population. The long-standing connection between increasing pressure of numbers and declining real incomes per head which had marked so much of pre-industrial European history was finally broken by changes in the sources from which flowed the wealth of nations and of men. The productive powers of society could and did grow more quickly than populations so that increase of population was no longer difficult to reconcile with increasing individual prosperity. Neither fertility nor mortality was any longer density-dependent in the manner familiar to Malthus and still to be observed among wild life of all sorts.[5]

Changes in fertility took place much more slowly than changes in mortality. In most countries appreciable falls in fertility came half a century later than falls in mortality, leading to rapid rises in the size of families and in overall populations. But 'In time the control of conception within marriage became very widespread and fertility rates fell to levels in all probability lower than any to be found in any previous period. Any population which before the industrial revolution had displayed such low fertility rates would have disappeared quickly from the face of the earth'.[6]

France was the only European country in which the fall in fertility took place at roughly the same time as the fall in mortality and this fall in fertility took place in peasant communities as well as in the towns. Consequently the French population rose relatively slowly in the 19th century (see Fig. 4.11.1, page 213). It is interesting that this limitation in family size preceded the introduction of artificial methods of contraception and was not achieved by deferring the age of marriage. In fact once fertility in marriage began to be controlled in France 'couples married earlier and those areas in which marital fertility fell earliest and most were also those in which the proportions of men and women who were married in the young age-groups rose first and furthest.'[7]

Wrigley's conclusions on fertility are that:

When family size began to be deliberately limited in western Europe in the nineteenth century, the result was achieved largely by 'pre-industrial' methods, by *coitus interruptus* and by the procuring of abortions, both means which had been available to societies for centuries previously. . . .

. . . In the main, therefore, the tremendous growth in the use of contraception to limit family size from the late nineteenth century onwards was due to the much wider employment of means long known to European societies rather than to the opportunities afforded to them by the development of new techniques. The changes in marital fertility would have followed much the same path in all probability even if no new techniques of control had been invented.[8]

This somewhat surprising conclusion may be relevant to the rather different conditions in many parts of the world today. It certainly indicates that many complex social factors are involved in fertility changes in addition to the application of technical knowledge about contraception.

Now let us turn briefly to the rapid demographic changes which have taken place in many third world countries since World War 2. Again, changes in mortality led the way and were often much more rapid than mortality changes in Europe during the Industrial Revolution. For example, in a single year, 1946–7, the expectation of life in Sri Lanka rose from 42·2 to 51·8 years which is about as much as the rise in Europe over the whole of the 19th century. This was achieved by the almost complete eradication of malaria and the introduction of modern drugs for the treatment of other common diseases.

Crude death rates in the developing countries are now frequently lower than in European countries because a higher proportion of their populations is young, in age groups where the risk of death is slight.[9] (See Fig. 8.5.3.)

It is the combination of this decrease in mortality with continuing high fertility that has led to the growth of populations at about 2–3 per cent per year (see Fig. 8.5.2) and the preponderance of young people in third world populations (Fig. 8.5.3).

The long-term stabilisation of populations in the third world will obviously depend on reductions in fertility. So far there are indications that fertility rates are falling in some parts of the third world but, since the fall in mortality has been so rapid, a comparably rapid fall in fertility must not be too long delayed if populations of many countries are not to continue to grow even faster than those of Europe in the 19th century.

Is such a fall in fertility possible? And, if possible, is it likely to happen? It is often argued that fertility is unlikely to fall in many countries while a large family is regarded as a source of wealth and an insurance policy against hardship in old age. Similar arguments have been advanced by some third world governments who regard policies to restrict their populations as attempts to reduce their future power and influence. However, many other governments have actively tried to limit their populations as they saw all their efforts to increase food supplies nullified by parallel increases in population.

There are a number of reasons for thinking that fertility rates may fall significantly. First, fertility rates in a number of traditional societies are not always high. This implies that some form of fertility control already exists and may become more pervasive in the future. Secondly, the introduction of artificial contraceptive methods must have some effect in reducing fertility even if they are not completely effective or are not used by everybody. Thirdly, there are a number of countries where fertility rates have fallen rapidly in a very few years. In the six years from 1960 to 1966, the fertility rates fell by 25 per cent or more in urban Hong Kong and Singapore and in rural Taiwan. In Communist China little accurate data is available. However, fertility is reported to have fallen even more under the influence of social pressures for later marriage and measures such as the removal of child allowances for families with more than one child.

So, although predictions about future fertility trends are notoriously hazardous, rapid falls are certainly possible and could even happen fairly quickly. What does seem certain is that conditions across the enormous diversity of third world countries will differ even more than did conditions in the various parts of Europe during the Industrial Revolution. So the ways in which fertilities will change across the world are likely to be even more diverse than they were in Europe.

This fact makes it very unlikely that any sensible conclusions about world population growth can be drawn from attempts to treat the whole world as a single system, as was attempted in the Limits to Growth project. This project tried to construct a mathematical model which would simulate the future direction of world development. The model involved deriving equations linking average global estimates of five major variables—population, industrial production, food, pollution and the use of non-renewable resources—all of which were claimed to be growing exponentially. These equations were then solved by computer and the solutions purported to predict a major collapse in total world population within the next 50 to 100 years. The whole project has been severely criticised on a number of grounds, the most serious of which seems to be gross oversimplification, particularly in the attempt to use global averages for quantities like population trends which vary enormously both over time and across the world. That there are limits to the growth of world population is not in doubt. Yet any realistic estimate of these limits will need to take account of a multiplicity of factors. So it will need much more detailed and painstaking demographic research. Such research may not solve the world's population problems but it may help us to understand how to move towards their solution.

REFERENCES

1 Meadows, D. M., *et al.*, *The Limits to Growth*, (London: Pan, 1974).
2 See in particular Laslett, P., *The World we have Lost*, (London: Methuen, 1971), and Wrigley, E. A., *Population and History*, (London: Weidenfeld and Nicolson, 1969).

3 Wrigley, E. A., op. cit., p. 173.
4 Ibid., p. 174.
5 Ibid., p. 177.
6 Ibid., p. 180.
7 Ibid., p. 188.
8 Ibid., pp. 188–90.
9 Ibid., p. 204.

SUMMARY

1 Overall figures for world population growth conceal important differences between regions; for example, the population of Europe grew fastest in the 19th century, while that of Asia has grown so rapidly since 1950 that it has increased by 850 million—more than the whole current population of Europe.
2 During the Industrial Revolution overall death rates fell primarily because sudden surges in mortality due to harvest failures or epidemics became much less common; age-specific death rates also fell appreciably for children and young adults and life expectancies increased.
3 Fertility fell much more slowly than mortality and so European populations grew rapidly; the exception was France, where limitations on family size began to occur early in the 19th century.
4 Since World War 2, mortality rates in many third world countries have fallen much more rapidly than in Europe in the 19th century, due primarily to the spread of modern methods of disease prevention.
5 Fertility in most, but not all, third world countries has not fallen very much; therefore populations in most of the world are now growing rapidly and, since they contain many young people, are likely to continue to grow for some time yet.

8.6 The natural and the social sciences— similarities and differences

By the end of the 19th century the social sciences had become forces to be reckoned with—both in the ideas they developed and in the impact those ideas had on society (see Chapter 4.10). The writings of social scientists like Karl Marx and Max Weber were beginning to demonstrate the significance of man's new ideas about himself—that he is not the object of impersonal forces but that he can make and mould his own destiny.

For Karl Marx, man had seen the dream of a communist Utopia—and the working class of the world would unite to bring about the communist revolution. If we think about the political map of the world today, we can see how powerful Marx's ideas have been. In the space of some 60 years, over a third of the world's population has come to live under some kind of Marxist government (37 per cent), and these governments rule about 37 per cent of the world's land area.

For Max Weber, too, many of his ideas have become reality. For example, he wrote about bureaucracy, and we can see the growth of huge bureaucracies everywhere around us—from multi-national corporations like Shell or IBM, to bureaucracy in government and education. Some of Weber's fears about what bureaucracies do to people have also become reality—impersonal relationships, proliferation of committees, 'bureaucratic personalities' hidebound by red tape. Weber also foresaw how human life would become increasingly influenced by science and by rationality—and would lose a sense of mystery and wonder. People in modern industrial societies often feel like this—that, for better or worse, life has lost any sense of ultimate purpose or meaning. Weber also believed in the power of ideas; he disagreed with Marx's view that man's history is largely dependent on economic factors, as he believed that man is much more free than this to shape his own quality of life.

However, before we look in a little more detail at the scope of the human and social sciences in this century, we need to remind ourselves of some important ways in which they differ from the natural sciences.

DIFFERENCES BETWEEN THE NATURAL AND SOCIAL SCIENCES

These differences stem from the obvious but fundamental fact that the social sciences deal with human beings with minds of their own, who can think, feel react and respond. As Weber emphasised, this implies that understanding the meaning of what we observe can be just as important in the social sciences as the observations themselves (see Chapter 4.10).

The main differences are that:

(i)   the knowledge involved in human sciences is often more complex than in the natural sciences and there may seem to be more than one 'truth', or interpretation of events and data;

(ii)   human behaviour is harder to predict than some of the phenomena of the natural sciences; people are not molecules; they can react to the results of research and knowledge and therefore change the situation which is being studied;

(iii) moral issues and values are often more closely involved in social science research than in research in the natural sciences;

(iv)   the social sciences are often much more directly involved than the natural sciences in practical political questions; research in the social sciences is social and political policy in the making.

The point that 'truth' is not so clear cut as in the natural sciences was emphasised by a German sociologist, Karl Mannheim who, like Einstein, fled from Hitler's Germany. He wrote about how, in the social sciences, different systems of belief and competing versions of 'truth' can live side by side in ways which differ from natural science. For example, in a natural science such as medicine we believe, with good reason, that we can prevent smallpox by inoculation; or, in nuclear physics, that a given mass of Uranium 235 will explode by fission under given conditions. But when we turn to the social sciences we may find that people's versions of 'facts' are rather different. The Russian and American interpretations of the war in Vietnam may well be very different from each other. Or a Marxist view of the nature of mental illness may be very different from that of a Freudian, and there may be significant differences in the practice of psychiatry between western countries and the USSR. But, having said that, it is important to remember that there are issues on which data can be collected and on which there is general agreement. Examples include much of the demographic data in Chapter 8.5 or Durkheim's data on suicide (see Chapter 4.10).

Given these differences between the natural and the social sciences, the question arises as to whether the social sciences can be regarded as sciences of the same kind as the natural sciences. In Chapter 7.1 we discussed the criteria— logical coherence and the use of all available relevant evidence—by which claims to scientific knowledge are assessed, together with the associated values—open criticism, pluralism, tentativeness—involved in assessing claims to knowledge. It is clearly possible—and, I would argue, necessary—to employ these criteria and values in the social sciences too, and many social scientists attempt to do this. However, some of the most effective techniques used in the natural sciences for applying these criteria—the use of mathematics and of controlled experiments— are often much more difficult to use or may not be applicable at all in the social sciences. This, together with the complication caused by the complexity of many social situations and the extra dimensions introduced by the subjective nature of social meanings and by moral issues and values, means that the growth of knowledge in the social sciences is often slow and difficult. And there is another major problem in regarding the social sciences as similar in kind to the natural sciences—many social scientists deny that their subjects have any resemblance to the natural sciences. Some of these social scientists explicitly reject the methods of the natural sciences. They regard all social facts as suspect, particularly if they are quantitative, and they sometimes choose to work either on detailed case-studies which can be almost wholly subjective or within a fairly rigid ideological framework as is the case with some Marxist social scientists. So, as we noted in Chapter 4.10, there is a great deal of disagreement amongst social scientists even about the very nature of their subjects, and there is certainly no generally agreed body of knowledge on which we can confidently draw. Yet, despite this, it may be useful to try to indicate the range of subjects which the social and human sciences have covered in the 20th century and to discuss briefly a few topics in more detail.

A SKETCH MAP OF THE TERRAIN OF THE SOCIAL AND HUMAN SCIENCES

Fig. 8.6.1 shows some of the subjects making up the social and human sciences arranged on a sketch map with two dimensions or continua—from *micro* to *macro* studies and from those which primarily use *objective* data to those dealing more with *subjective* experiences and with experiences which are concerned with what philosophers of the social sciences call *intersubjective* aspects of life. These involve socially agreed meanings and expectations for individuals, for social groups or for whole societies; examples include the values and criteria by which the scientific community operates (see Chapter 7.1). At the micro end of the horizontal continuum are subjects which deal with individual people or, in some of the biological sciences, even with processes down to the cellular level. At the other end of this continuum the starting point is the study of whole societies in all their

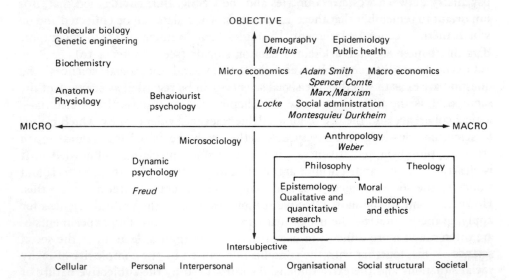

Fig. 8.6.1   Sketch map of the terrain of the human and social sciences and some of their founding fathers; this diagram should be studied in conjunction with the text. The inset dealing with philosophy/theology forms another dimension which is relevant to the whole diagram and not just to the bottom right-hand quadrant.

complexity. The vertical continuum ranges from those who deal primarily with relatively objective and quantifiable data—infant mortality rates or the distribution of wealth and poverty—to those who are concerned with the less tangible world of ideas and beliefs—the more qualitative aspects of human experience.

This sketch map of the human and social sciences also needs a third philosophical dimension (see box on Fig. 8.6.1) concerning both the methods used in these subjects and the moral, ethical and theological aspects of human experience. To deal with these topics adequately would require at least another whole book, but to leave them out altogether would be extremely misleading.

We obviously can't deal in this chapter with all the subjects covered on this sketch map. However, it may be useful to look at the various regions of the map in a little more detail in order to see the relationships between the work of different social scientists, including those whose work we discussed in Chapters 4.2 and 4.10. As we do this we will also briefly describe one or two 20th century developments in the social sciences.

At the top left of Fig. 8.6.1, in the 'micro/objective' quadrant, are to be found those natural sciences which are concerned with different aspects of human biology, together with behaviourist psychology—a school of psychology which deals primarily with objective data.

There are many different approaches to human psychology, but the behaviourists are amongst the most influential. The early behaviourists assumed that people are made in such a way that their behaviour can be effectively trained, controlled and predicted in any way we choose. Their theories are built on the concept of the conditioned reflex and response. In essence, they argue that our central nervous system can be trained by systematic rewards or punishments to react in programmed ways. The significance of the conditioned reflex and its possible use as a means of training was associated in the early days with the work of the Russian Pavlov, working in pre-revolutionary Russia. Pavlov experimented with dogs, showing how they could be trained to develop conditioned responses to certain stimuli. For example, he gave meat to hungry dogs at the same time as ringing a bell; the dogs salivated at the prospect of enjoying their meat; then, after a number of repetitions, they would automatically salivate when the bell rang, even if there was no meat. Pavlov always resisted those who wanted to apply his results to human psychology but a later worker in this field, an American called Watson, extrapolated from these ideas, believing that people could be trained in similar ways and that ultimately all human behaviour could be controlled using these principles. Watson had a vision of a 'universe unshackled by legendary folklore of happenings thousands of years ago; unhampered by disgraceful political history; free of foolish customs and conventions which have no significance in themselves, yet which hem the individual in like taut steel bands'.[1] He claimed that he was offering mankind a freedom which would change the world, because his methods could shape people into any kind of persons we choose: 'Give me the baby and my world to bring it up in and I'll make it crawl and walk . . . I'll make it a thief, a gunman or a dope fiend. The possibility of shaping it in any direction is almost endless'.[1]

Watson considered that 'anything that Nature or Nature's God can do, behaviourism . . . could do better'.[2] His object was to create a world in which people would develop, from birth onwards, in ways which would bring them automatically into conformity with group standards. This approach has subsequently been developed by psychologists like Skinner who concentrate on manipulating and predicting human behaviour in very precise experimental ways. In his work, Skinner has taken the white rat as a model as he claims that this is an excellent subject for experiment. He does not seem unduly concerned over the fact that rats are not human beings, saying 'the only differences I expect to see revealed between the behaviour of rat and Man (aside from enormous differences of complexity) lie in the field of verbal behaviour'.[3] We may feel that this is rather a large difference!

However, despite some of these grandiose claims, behaviourist psychology has been successfully used in specific areas—for example, in the work of Eysenck on the treatment of phobias and addiction.

Some people are so crippled with fear of certain situations—of enclosed spaces, claustrophobia; of open spaces, agoraphobia; or even of spiders, arachnophobia— that they cannot live normal lives. Behaviourist therapy is built on the concept of 'deconditioning'. It assumes that these people have developed their phobic reactions as a result of some unpleasant association with the fear-causing situation and so the treatment consists of trying to counteract this by introducing the patient to a graded contact with the object of his fear, in situations of support and comfort. Over time, the patient often becomes able to tolerate the thing he has been so afraid of and may be cured of his phobia.

The treatment of *addiction* uses negative stimuli rather than positive reinforcement and may therefore raise deeper moral questions. It is usually reserved for the treatment of conditions which are so damaging that the patient is willing to have the treatment despite its unpleasantness—such as people who are killing themselves by their addiction to hard drugs like heroin, or who are so addicted to alcohol that they are destroying their livers. It may also be used, perhaps more controversially, for some sexual conditions like paedophilia or some fetishisms—where there is a high probability that, if the person cannot be cured of them, he will have to remain incarcerated in some institution for the rest of his life for the sake of the safety of the general public. In essence, treatment consists of giving a drug which provokes severe symptoms such as nausea and vomiting. This is given at the same time as the patient is given the drink or the drug, or the sexual stimulus which is the cause of concern. After a number of occasions, he will tend, like Pavlov's dogs, to experience the physical reactions at the sight of the stimulus, even without the injection—and so, instead of yearning for his double gin or his shot of heroin, he will tend to feel repelled by the thought.

In summary, Eysenck claims that behaviour therapy is far more effective than psychotherapy or other forms of psychiatric treatment. He says: 'It is very difficult to look at the evidence as a whole (which is of course what a scientist should do!) and come to any other conclusion than that behaviour therapy not only has been shown to work but that it is the ONLY method of therapy for

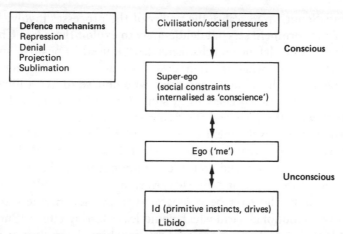

Fig. 8.6.2 Diagrammatic representation of Freud's theory of the structure of the psyche.

which this can be claimed. This seems to me to be the one conclusion which has been very firmly established. . . .'[4]

Other approaches to human psychology, such as that practised by those who follow in the tradition of Sigmund Freud, are to be found in the bottom left, or 'micro/subjective', quadrant of Fig. 8.6.1, along with microsociological approaches which concentrate on the details of interactions between individuals and small social groups.

Freud's work, like that of the behaviourists, has been extremely influential and yet it is very different from behaviourist psychology. One of the key elements of Freud's theory—his ideas about the structure of the human psyche—is illustrated in Fig. 8.6.2. Freud suggests that there are very strong tensions resulting from the clash between the primitive drives and instincts welling up from the Id and the repression of these which is necessary if people are to live in civilised society. The uneasy compromise or balance may be achieved by a number of 'defence mechanisms'. Examples include *repression*—where we bury uncomfortable urges or drives in the unconscious parts of our psyche; *denial*—where we try to make ourselves believe that they do not exist; *projection*—where we project on to other people guilt and frustration which we ourselves are feeling; and *sublimation*— where we channel the drives and instincts into socially acceptable forms.

Examples of projection and its deliberate manipulation can be found in some of the most horrific political movements of the 20th century. One form it has taken is what we call scapegoating and can be seen in the deliberate encouragement of anti-Semitism in countries such as Nazi Germany or the USSR. The key ingredients are: the identification and description of a group of people as 'outsiders' who are despicable and worthy of hatred and abuse; propaganda which fuels emotion and aggression towards these people; and the channelling of any feelings of frustration and aggression in the rest of society against this group in a way which unites everyone else. One need only to think of the six million Jews who died as a result of Hitler's anti-Semitism to realise the power of the processes involved and what can happen if a government sets out deliberately to exploit them (see Chapter 7.2).

On the other hand, defence mechanisms can be used in positive and humane ways—for example, in what is often described as the sublimation of sexual instincts by people such as monks and nuns who choose to live a life of celibacy and dedicate their emotional energies in some form of service, such as Mother Theresa's work with the poor and the dying in Calcutta.

Another significant development arising from the Freudian school of psychology is the theory of maternal deprivation which has highlighted the importance of providing a warm, loving and secure environment for children and the harm that may be done if this is lacking. Bowlby, working in the Freudian tradition, has shown how children who have been separated from their mothers and families may suffer severe emotional harm which could leave lasting effects. Such children may be unable to form satisfying personal relationships themselves and may tend to commit crimes of various kinds. One positive result of this work has been a revolution in the way we care for children who are orphaned or who have to be separated from their parents. Gone are the huge impersonal orphanages of 30 years ago; instead we try to foster or provide 'home-like' homes. Also gone are the days when parents were not allowed to visit children in hospital; now they are encouraged to visit as much as possible. The resulting psychological advantages are examples of the beneficial application of research in the human sciences in recent years.

Now let us turn to the top right of Fig. 8.6.1, the 'macro/objective' quadrant. It is here that we find the work of some of the founders of the social sciences—Montesquieu, Malthus and Locke on politics (see Chapter 4.2; Locke on psychology should come in the top left quadrant) and Comte, Spencer, Durkheim on suicide and the main thrust of Marx's writings with their emphasis on economic factors as the claimed objective determinants of social relationships (see Chapter 4.10). Here, too, we find demography (see Chapter 8.5), epidemiology, public health and social policy and administration—although, of course, these subjects have great bearing on individuals as well as on societies and so should perhaps extend some way into the 'micro/objective' quadrant, too. This is also true of economics, whether in the work of Adam Smith (see Chapter 4.2) or in more recent work, which includes different macroeconomic theories like those of Keynes and of monetarists like Friedman as well as microeconomic studies of individual firms or households.

We clearly can't discuss this enormous group of subjects in any detail, but one general point is important. Research in some of these subjects—for example, matters of social policy such as education or the provision of health care—can have a substantial impact on the policies adopted and hence on the lives and quality of life of millions of people. Here, in a very direct way, research in the social sciences can be social policy in the making.

Finally, let us look at the bottom right of Fig. 8.6.1, the 'macro/subjective' quadrant. Here we find subjects like anthropology and much of the work of two of sociology's founding fathers, Durkheim and Weber (see Chapter 4.10), with

Weber, in particular, emphasising the importance of beliefs in understanding how societies function. Weber was committed to a social science which would explain human behaviour both in terms of cause and of meaning. Both Weber and Durkheim wrote substantial works on religion—a topic which has been of major importance throughout human history and which we have put on our third dimension in Fig. 8.6.1. On religion, Weberians and Marxists are in profound disagreement. Marx predicted that religion was a result of man's alienation from his true relationship to other men and from his own true self and that this alienation was caused by the exploitative relationships of capitalism. When capitalism was overthrown and socialism took its place, man would no longer need religion as an 'opiate' and it would wither away. Weber had a rather different view: he believed that there is more to human life and experience than just the material and economic aspects, but he was concerned that the 20th-century growth of scientific and bureaucratic rationality would erode man's sense of mystery and wonder. Religious developments in this century have been interesting. Many western societies have seen some secularisation with a decrease in religious beliefs and institutions, although there are exceptions like the USA where there has been a general upward trend. In socialist societies, according to Marx's predictions, religion should be withering away—with some assistance in the form of maltreatment of religious believers ranging from outright persecution to more subtle forms of discrimination such as dismissal from jobs, difficulties in entering university or obtaining a house. However, against all the odds, religion is actually on the increase in some socialist societies in recent years. This seems to endorse Weber's view that people need some ultimate meaning or purpose to their lives.[5]

CONCLUSIONS

The social sciences, like the natural sciences, have great power and they influence people's treatment of each other. There are thus a few points which should be emphasised when we think about social science in the modern world.

First, because the reality they deal with is more complex than that with which natural sciences are concerned, they should be more tentative. Karl Popper emphasises that we can never 'know it all' and that we must avoid any dogmatic blueprints for a perfect society. If people allow themselves to think they know more than they do they run the danger of creating new totalitarian regimes—forcing individuals into moulds at the behest of their political leaders. We have seen how modern human sciences, whether the science of behavioural psychologists like Watson or of scientific materialists like the Marxists, can be used to try to make people conform to models derived from their theories.

Secondly, as social scientists, we should be committed to the main principles which underlie any science and particularly to the rule that we should try to take account of all available relevant evidence. Often, however, social scientists limit themselves to studying just one type of society. In Britain, this is seen in the vast numbers of books which are published criticising capitalist societies. Now it is clearly right and proper to be critical; but it is not right and proper to turn a blind eye to what is going on in other types of society. In other words, we need to see whether certain social problems are unique to one kind of society, or whether they occur in other types of society as well. We need to do this if we are to avoid the elementary fallacy of confusing correlation with cause—for example, by arguing that, because a particular problem occurs in capitalist society, it is *caused* by capitalism. Too many books on social science do just this—usually for ideological reasons—which reminds us of the point made earlier, that 'truth' may be more elusive in the social than in the natural sciences.[6]

Thirdly, we need to be aware in the social sciences of the intermeshing of beliefs and data. There is a much greater risk in the social sciences of data being distorted and interpretations being influenced by the beliefs and motivations of the people who are using them. It is much easier to deal in half-truths or to present selective data to support one's own wishes, because there is much less in the way of a concrete reality against which other people can test and refute our theories than is the case in the world of the natural sciences. Therefore, we need to beware of social science and to use it with caution. A well-known contemporary sociologist says that the ideal sociologist is one who 'thinks daringly and acts carefully'. Too often social scientists think daringly and act dogmatically. This is not science and needs to be resisted in the interests not only of science but also of all those whose lives may be affected adversely by that dogmatism. Yet there is a need for social theories and generalisations which help to make sense of our increasingly complex and interdependent world. So in the next chapter we will dare to 'think daringly' and, hopefully, 'act carefully' in putting forward such a social theory.

REFERENCES

1 Quoted in North, M., *The Mind Market*, (London: Allen and Unwin, 1975), p. 153.
2 Ibid., p. 153.
3 Ibid., p. 157.
4 Ibid., p. 168.
5 See, for example, Beeson, T., *Discretion and Valour*, (London: Collins/Fontana, 1974), or the journal *Religion in Communist Lands*, (Kestin College, Kent).
6 For further discussion see Cox, C. and Marks, J., 'What has Athens to do with Jerusalem?' in *The Right to Learn*, (London: Centre for Policy Studies, 1982), pp. 67–88 and Cox, C., *Sociology: An Introduction for Nurses, Midwives and Health Visitors*, (London: Butterworths, 1983).

SUMMARY

1 During the 20th century, the influence of the social sciences has increased—both in the ideas they developed and in the impact those ideas had on society.
2 The main differences between the natural and the social sciences stem from the obvious fact that the social sciences deal with human beings with minds of their own who can think, feel, react and respond.

3 Despite significant differences—greater complexity, the subjective element in social situations, the greater importance of moral issues—many of the same methods can be used in the natural and the social sciences.

4 The enormous range of subjects covered by the human and social sciences can be classified along two dimensions or continua—from macro-sociological studies of whole societies to micro-sociological studies dealing with individuals, and from studies which primarily use objective data to those concerned more with subjective experiences (see Fig. 8.6.1).

5 Because their subject matter is so complex and because their research can directly affect public policy, the social sciences should be suitably tentative in the conclusions they draw and in attempting to separate correlation from cause should be careful to compare similar phenomena in different kinds of society.

CONCLUSION

Science and the making of the modern world

In this final chapter I want to discuss some general ideas first put forward by the economist and social philosopher F. A. Hayek,[1] which concern phenomena that are 'the result of human action, but not the execution of any human design'.[2] Hayek first applied these ideas to the economic and legal framework of society but, as we shall see, they are just as useful in discussing the nature of science and its interaction with society.

However, before turning to these complex and difficult questions let us briefly remind ourselves of why it is important to try to understand them. As we saw in Chapter 8.5, the rapid growth in world population in this century is primarily due to the world-wide spread of industrialisation and the benefits brought by improvements in agriculture and by modern scientific medicine. Look again at Fig. 6.5.8 (page 305) which shows the dramatic fall in death rates from infectious diseases over the last 50 years. And remember, too, from Chapter 6.5, that smallpox has now been eradicated from the whole world. Because of modern science and technology and modern medicine, many hundreds of millions of people are alive today who would have died in infancy or childhood at any time in the past. And if modern science, technology and medicine were to vanish from the world or to be severely limited, many hundreds of millions would die. The death-toll arising from the loss of our sources of energy, from the loss of mechanised and scientific agriculture and from the abandonment of scientific medicine and preventative health care would completely overshadow that from any previous period in world history—whether it be the Black Death, the two world wars of this century, Hitler's concentration camps or Stalin's Gulag Archipelago. So, whatever the problems we face in the development of science and technology, we have no choice but to face them as best we can. Science has made itself indispensable for any future world society, no matter how it may evolve or be organised.

THE NATURE OF SCIENCE

First let us consider the nature of science and of the scientific community. For many people science is regarded as the major example of rationality in human affairs—and by rationality they usually mean that science works with definite fixed laws from which the detailed properties of particular systems can be worked out, in many cases with mathematical precision. This view of science is similar to that of Descartes who, as we saw in Chapter 3.3, emphasised the importance of deductive mathematics and took machines as his model for understanding all phenomena. Those who hold this view of scientific rationality often think of a scientist as being rather like the designer of a car or a building—as one who knows all about something and then sets out to specify in detail how it works or how it should be made. Let us use Hayek's terminology and call someone who holds this view a *constructive rationalist*.[3]

Of course, in many detailed areas, the constructive rationalist model does give a good picture of one important aspect of science. For example, Newton's laws of motion do enable us to calculate the details of many motions with great precision, and from Maxwell's equations (see Chapter 6.2) we can calculate many detailed properties of all kinds of electromagnetic radiations. However, even in these limited areas, constructive rationalism only tells part of the story. And when we come to consider the whole body of scientific knowledge which currently exists and the way in which that knowledge has developed, then a very different picture of science emerges. In Part 6 of this book we have discussed *some* key 20th-century developments in *some* of the major areas of science—modern physics, chemistry and biochemistry, genetics and molecular biology, medicine, the earth sciences and cosmology. What we have mentioned is only a tiny fraction of the scientific knowledge which now exists. This knowledge is made up of a vast network of theories, experiments and observations, much of which is recorded in numerous papers, monographs and textbooks but some of which is *not* published or even written down but is contained in the memories and accumulated experience of individual scientists. Look again at Fig. 6.1.3 (page 243) which shows the molecular structure of haemoglobin—the protein which as oxyhaemoglobin gives our blood its colour and makes our life possible by carrying oxygen to all parts of our bodies. To work out this structure required the skills and knowledge of hundreds of specialists in many different fields—medicine, biochemistry, chemistry, crystallography, physics, mathematics and computing—not to mention the engineering and technical skills involved in making the necessary scientific instruments. And, of course, knowledge of the structure of haemoglobin represents only one tiny part of the knowledge we now possess. The whole body of this knowledge is now so vast and complex that nobody can be aware of more than a fraction of it.

Yet it is not just a random assortment of separate and disconnected items. If it were, the sort of collaboration between specialists that led to the haemoglobin model of Fig. 6.1.3 would not have been possible. The parts of our scientific knowledge are interrelated in many complex ways and the whole has a structure and coherence of its own. Yet it is not a structure which has been designed or made or planned. As we have seen in our historical studies of the growth of science since the 17th century, the whole body of knowledge has *evolved* in many diverse, unplanned and unpredictable ways.

In addition we need to remember that our scientific knowledge would not exist if it were not for the scientific community. So in analysing the nature of science we need to consider the body of knowledge and the scientific community *together*. As we have seen, this community is made up of individual scientists, scientific societies (like the Royal Society), journals and institutions of higher education and research, all of which are in continual interaction. This whole structure is now very complex—remember, for example, that there are estimated to be about 30,000 scientific journals currently being published. As we have seen, this structure has grown and evolved over the years in a spontaneous and unplanned way from its rudimentary origins in the earliest journals and scientific societies of the 17th century—in parallel with the growth in scientific knowledge over the same period.

The primary purpose of the scientific community is to transmit reliable information throughout the whole complex structure. It does this by enabling effective and open public criticism to take place, by organising meetings and conferences, by publishing journals and monographs and by only admitting to full membership of the community those who adhere to two main criteria in judging claims to knowledge—Is this claim logical and coherent? and Does it take account of all the available relevant evidence? The whole community is thus a vast decentralised network for the transmission of information. Collectively it encourages the values and maintains the structures which experience has shown will make that information reliable. Yet the community does not collectively determine the *content* of what is published. Only the individual scientists—those who do the detailed work—can do that.

In this way, the whole scientific community can generate and use a vastly greater stock of collectively validated knowledge than any individual or centralised group could possibly do. Furthermore, the community's system of continual public criticism and evaluation has the great advantage of enabling faulty work to be eliminated.

It should now be clear that attempts to plan or direct the detailed course of future scientific development are likely to fail—not only because future discoveries, by their very nature, cannot be predicted, but also because no individual or small group can possess all the details of even our present knowledge; nor can they explicitly understand and control the detailed mechanisms and procedures by which scientific knowledge is acquired.

The growth of science and the scientific community does not proceed in any machine-like and predictable way, and so *constructive rationalism* does not give an adequate account of the development of science. Rather than resembling a machine, science and the scientific community make up a *spontaneous order* which has developed according to a process which Hayek calls *evolutionary rationalism*.[4]

The characteristic features of such a spontaneous order are that:

the order has evolved rather than been designed;
the order is so complex that no single person or group could know it in detail;
the order involves decentralised mechanisms for the transmission of information;
a framework of rules must be maintained which determines the structure but *not* the content of the order;
many self-correcting mechanisms occur naturally within the order;
the order can and does make use of constructive rationalism in limited areas but in its totality it bears little resemblance to a machine.

Constructive rationalism	Evolutionary rationalism
1 Designed or planned	1 Evolved *not* designed
2 Details known to designer/planner	2 So complex that no-one knows details
3 Information transmission from centre	3 Decentralised mechanisms for information transmission
4 Rules/laws determine structure *and* content	4 Rules which determine structure *not* content
5 Self-correcting mechanisms must be specifically designed	5 Self-correcting mechanisms occur naturally
6 Produces ordered structures like machines	6 Uses constructive rationalism in limited areas but in its totality bears little resemblance to a machine
Designed or made order	Spontaneous order

Fig. 9.1   Main features of evolutionary rationalism and spontaneous orders contrasted with constructive rationalism and designed orders.

Fig. 9.1 contrasts the characteristic features of evolutionary rationalism, which lead to the formation of spontaneous orders, with the features of constructive rationalism, which lead to orders or structures which are designed or made.

I would argue that scientific knowledge and the scientific community taken together form a complex spontaneous order which has all the features outlined above and that this model of science is the only realistic one we can use in any discussion of its future development. It is inconceivable that anyone could have designed the scientific order which, over the last three centuries, has created the most significant and universal body of interlocking theoretical and empirical human knowledge yet achieved.

HOW SCIENCE INTERACTS WITH SOCIETY

Now let us move on to the second main topic of this chapter—the nature of the interactions between science and society. In doing this, we are going to use some of the same ideas which we have just outlined in our analysis of the nature of science. The differences between constructive and evolutionary rationalism—and between designed orders and spontaneous orders—are just as useful in analysing societies as they are in discussing the nature of science. Indeed Hayek first developed them while discussing the nature of social and economic institutions.

First let us look briefly at the societies in which science has flourished over its long history. When we look at the historical development of science, we can identify different societies at different times which were the main centres for scientific activity at those times.[5] Modern science had its origins in ancient Greece and in particular in the period of the independent city-states of the 5th and 4th centuries BC. It then went into decline in the Dark Ages, apart from some important work in the Arab world, and next really prospered in Italy after the Renaissance—around the time of Galileo in the late 16th and early 17th centuries. After Galileo's trial and condemnation by the Church, the centre of scientific activity shifted in the later 17th century to the England of Newton and the newly-founded Royal Society. Then in the 18th century the enthusiasm for science shown by the men of the Enlightenment made France the scientific centre of the world—and this became particularly evident in the new scientific institutions created after the French Revolution. In the 19th century, Germany, in trying to compete with France, created many new universities and gradually became the most respected scientific nation; this pre-eminence lasted right up until Hitler came to power in the 1930s. Then the stream of distinguished refugees from Hitler's Germany—Einstein among them—gave a great stimulus to science in the United States, and since World War 2 the USA has become the world's leading scientific and technological power.

Can we learn anything about the conditions science needs in order to flourish from the way scientific pre-eminence has shifted over time? Is there anything that the Greek city-states have in common with Renaissance Italy, the England of Newton, pre- and post-revolutionary France, 19th-century Germany, and the United States today? With the possible exception of 18th-century France, different though they were, one common factor can be found. They were all societies in which different centres of power existed and which therefore were potentially able to be more tolerant of people or groups with differing opinions. This tolerance did not necessarily extend to all members of these societies but, so long as it enabled scientists to work with some degree of independence—and to move elsewhere if necessary—then the spontaneous order of science had a chance to develop. As we have seen, such independence is a necessary condition for science to flourish, but it is not, of course, a sufficient condition.

However, when we come to the 20th century we are not just concerned with whether or not science will flourish. The key questions are much more concerned with whether we can apply science effectively, whether we can continue to develop science-based industries and whether we can achieve the benefits of science while minimising some of its adverse consequences. Amongst these adverse consequences, as well as physical problems such as environmental pollution and the using up of natural resources, we should also include the problems involved in applying or misapplying the social sciences (see Chapter 8.6).

So, as well as looking at the nature of science, we need to analyse different kinds of society. In this analysis it will be most useful to concentrate mainly on the major industrialised societies—that is on the liberal capitalist societies of the first world and on the Marxist socialist societies of the second world. In doing this we may well learn something which will be helpful in analysing the problems of third world countries, many of which are in the early stages of industrialisation (see Chapter 8.1).

LIBERAL CAPITALIST AND MARXIST SOCIALIST SOCIETIES

In analysing these societies we are going to use some of the same ideas as we did in analysing science—that is, the differences between constructive and evolutionary rationalism and between designed and spontaneous orders.

According to constructive rationalism, rational actions are those which are determined entirely by known and demonstrable truths, and rational social institutions are those which are deliberately designed to achieve specific, defined purposes. The very success of constructive rationalism in limited areas—such as the design of cars or silicon chips or buildings—may lead to the presumption that all social institutions are, and ought to be, the product of deliberate design.

Yet this is neither feasible, nor does it actually occur. It is not possible for any individual or small group to know all the relevant facts needed to design complex social institutions—to think that this is possible is to suffer from what Hayek calls the *synoptic delusion*.[6] Most social institutions in modern industrial societies have not been consciously designed. Many of these institutions have evolved gradually and man has often been successful because he followed accepted practices and observed rules which he does not fully understand.

As we saw in Chapter 8.3, natural languages are particularly good examples of spontaneous orders which have evolved rather than been designed. Even today we have only a small understanding of the structure of natural languages. Yet we use these languages confidently and, without them, virtually nothing of our culture would exist.

Therefore evolutionary rationalism is as important in understanding man and society as is biological evolution in understanding man as a species. This is particularly true of the complex industrial societies of Europe and the United States where many significant developments have not been planned in advance but have evolved in a spontaneous way. Think, for example, of the expansion of the railways in 19th-century Europe, the replacement of steam engines by electric motors, or the massive rise in the use of oil as a major fuel and raw material since 1950. Perhaps the most recent example is the rapid progress in the micro-miniaturisation of electronic circuits during the late 1960s and early 1970s which has led to the term 'silicon chip' becoming part of everybody's vocabulary (see Chapter 8.3). This arose almost by chance as a result of a number of very specialised developments and it was so unexpected that it even took the electronics experts by surprise.

Developments like these, unplanned as they are, could not take place in a society without some order and structure. However, in liberal societies, this order is not a designed or made order but a spontaneous order, similar to that of science and the scientific community—and it has many of the features listed in Fig. 9.1.

Probably the most important feature is the third one—the vital role played by the transmission of information amongst the many separate centres of power and decision in liberal societies. Without freedom of expression, communication and access to information, such a pluralistic society could not function, and it is one of the major duties of governments in such societies to maintain these freedoms. As we saw in Chapters 8.2 and 8.4, it is the free-flow of information and the possibility of public criticism, in the media and elsewhere, that enables such societies to check abuses of power and to devise correctives to some of the harmful effects of scientific and technological developments—that is, to allow the natural emergence of self-correcting mechanisms (see Fig. 9.1, point 5).

It is because the fundamental structures of liberal societies are so compatible with those of science and the scientific community that such societies have been so successful both in developing science and in applying it. Another important aspect of this compatibility can be seen if we consider the fundamental economic structure of liberal capitalist societies—the free market whose principles of operation were first spelt out by Adam Smith in the 18th century (see Chapter 4.2). The market, too, is a spontaneous order whose details are not known to anybody and which functions effectively because it has a decentralised process of information transmission. This is provided by the market prices obtaining under free competition. These prices transmit information which enables each part of the system to respond to the rest and to plan its own detailed activities. One of the main functions of governments in liberal societies is to maintain the conditions which allow the market order to function since it is the market order on which the overall prosperity of the economy depends. As Hayek puts it:

> . . . if anything is certain it is that no person who was not already familiar with the market could have designed the economic order which is capable of maintaining the present numbers of mankind.[7]

So liberal capitalist societies depend primarily on evolutionary rationalism and spontaneous orders for many of their fundamental structures. However, to recognise this is not to deny the importance of planning and constructive rationalism *in limited areas*. In almost all real situations *both* kinds of rationalism are involved. So, if we appreciate this, we ought to be able to estimate the potential benefits and limitations of conscious design and to distinguish situations where the constructive rationalist model will be most fruitful.

Now, by contrast, let us consider the Marxist socialist societies of the second world. These societies take constructive rationalism and designed orders for their model. They are governed by a single party, the Communist Party, which attempts to control and centrally plan all aspects of life—politics, the economy, education and all religious, cultural and even recreational activities. Freedom of expression and of access to information are expressly prevented, and anything which threatens the central role of the party is condemned and suppressed.

On our analysis this means that the values and structures of such a society are the very opposite of those of the scientific community. So one might expect the sort of conflicts between the government and the scientific community—the Lysenko affair, the persecution of scientists like Vavilov and Sakharov—which we described in Chapter 7.4.

We would also expect that, in a society which bases itself on constructive rationalism and a centrally planned economy, a great gulf would exist between what such societies promise and what they achieve—for they are denying the central importance of evolutionary rationalism in the development of modern industrial societies. In addition, by suppressing freedom of expression and public criticism, they are preventing the development of any feedback mechanisms which might correct the errors of the central planners. So we might expect the relative failure of applied science in such societies described by Sakharov, together with events like the disastrous environmental pollution in the Urals nuclear accident reported by Zhores Medvedev (see Chapter 7.4).

Much of this analysis also applies to the similarly totalitarian National Socialist government in Hitler's Germany (see Chapter 7.2). That government also tried to control all aspects of life and expressly prevented freedom of expression and public criticism. The main difference was that the economy in Hitler's Germany was not centrally planned to anything like the same extent as it is in Marxist states. But Hitler did realise that what he was doing was completely opposed to the spirit of science and the scientific community. Remember he said: 'Our national policies will not be revoked or modified, even for scientists. If [this] means the annihilation of contemporary German science, then we shall do without science for a few years'.[8]

However, Marxist societies do not share this contempt for science. Their claim has always been that they are organising their societies according to the principles of scientific socialism. On our analysis, this claim is based on a total misunderstanding of the nature of science and is a clear example of scientism— the attempt to apply what are thought to be scientific principles in areas where they are not applicable and to justify this by appealing to the prestige of science.

Instead of this arrogant and dogmatic attitude, what we need is a better appreciation of the real nature of science and scientific activity, and a large dose of the tolerance and humility which is central to the political and scientific philosophy of men like Hayek and Popper.

It is the open liberal societies of the world which—by trial and error, and by the processes of public criticism and attempts at piecemeal improvement—have established most current technologies and defined the acceptable limits within which they operate.

I am convinced that, if the benefits of the scientific and the industrial revolutions are to be successfully transmitted to the less industrialised countries of the world, then these countries and the world—will have to devise institutions and forms of government which operate on similar principles to those of the liberal societies which created those revolutions. I hope that the facts and the arguments I have presented in this book will go some way towards convincing you of this.

REFERENCES

1 von Hayek, F. A., *Law, Legislation and Liberty*, Vol. 1, *Rules and Order*, Vol. 2, *The Mirage of Social Justice*, Vol. 3, *The Political Order of a Free People*, (London: Routledge and Kegan Paul; Chicago: Chicago University Press, 1973, 1976, 1979).

2 Ferguson, A., *An Essay on the History of Civil Society*, (London: 1767, p. 187); quoted in von Hayek, F. A., op. cit., Vol. 1, p. 20.

3 von Hayek, F. A., op. cit., Vol. 1, p. 5; von Hayek used the term constructivist or, more recently, constructivistic, but we will use constructive rationalism throughout.

4 Ibid., Vol. 1, p. 5, p. 37.

5 Ben-David, J., *The Scientist's Role in Society*, (New Jersey: Prentice Hall, 1971).

6 von Hayek, F. A., op. cit., Vol. 1, p. 14.

7 Ibid., Vol. III, p. 164.

8 See Chapter 7.2, reference 6 (page 377).

SUMMARY

1   If modern science, technology and medicine were to vanish from the world, many hundreds of millions of people would die; therefore science is now indispensable for any future world society no matter how it may evolve or be organised.

2   Science and the scientific community have grown and evolved in a spontaneous and unplanned way from their rudimentary origins in the 17th century; consequently they are not well described by constructive rationalism which emphasises deductive mathematics and takes machines as models for understanding all phenomena.

3   Like natural languages, science and the scientific community are spontaneous orders, which have evolved rather than been designed, and are so complex that no individual or group can know them in detail; they all involve decentralised mechanisms for the transmisssion of information and are products of evolutionary rather than constructive rationalism.

4   The basically decentralised structures of liberal capitalist societies, which are kept functioning by the transmission of information amongst many separate centres of power and decision, are best described as spontaneous orders governed by evolutionary rationalism; by contrast, Marxist socialist societies, which attempt to control and centrally plan all aspects of life, take constructive rationalism and designed orders for their model.

5   Because their fundamental structures are so compatible with those of science and the scientific community, liberal capitalist societies are likely to continue to be much more successful than Marxist socialist societies in developing and applying science.

Guide to further reading

So many sources have been used in writing this book that a complete list would be far too long to be of very much use to students. Some indication of the main sources can be obtained from the references given at the end of each chapter and these references will, of course, usually contain details of other sources of information.

In this short guide to further reading I shall confine myself to mentioning a few books and articles which I have found particularly stimulating and useful in the ten years since I first began systematically to explore the issues discussed in this book. One important criterion for inclusion is whether or not a book has been comparably stimulating on a second reading as it was when first encountered.

Two books which first aroused my interest in the continuities of the historical development of science are *Origins of Modern Science: 1300–1800*[1] by Herbert Butterfield and *The Edge of Objectivity*[2] by C. C. Gillispie. Although some of their interpretations have, inevitably, had to be revised in the light of later scholarship they both provide fascinating and readable introductions to many important aspects of the history of science. Professor Butterfield's book is particularly noteworthy as it was a pioneering attempt by a distinguished historian to evaluate the place of science in the development of the modern world.

Much detailed information about the history of science can be found in the multi-volume *Dictionary of Scientific Biography*[3]. At a more popular level *The Ascent of Man*[4] by J. Bronowski and *The Force of Knowledge*[5] by J. Ziman are useful introductions to the development of science and its interaction with society from the earliest times to the present day; both books assume very little prior knowledge of science and are well illustrated. Another useful popular book is the *Hutchinson History of the World*[6] by J. M. Roberts which gives more space to the influence of science than is sometimes done in such general histories.

The Scientist's Role in Society[7] by J. Ben-David provides a good sociological introduction by a professional sociologist to the interaction between science and society in many different societies ranging from ancient Greece to 20th-century America. By contrast, *Little Science, Big Science*[8] by D. de Solla Price, who incidentally is not a sociologist, provides much fascinating information, some of it quantitative, about the scientific community and how it has changed. Both these books are more accessible to the general reader, and more in tune with the spirit of science and the scientific community, than many more recent writings by sociologists of science.

It is particularly difficult to identify accessible and concise sources of information for the chapters in Section 7 which deal with the development of science and technology in different societies and cultures. However, *Soviet Science*[9] by Zh. Medvedev and *Scientists under Hitler: Politics and the Physics Community in the Third Reich*[10] by A. D. Beyerchen, provide readable and accurate surveys of the fate of science under the two major totalitarian governments of the 20th century.

At the heart of this book lies a deep respect for two major co-operative human achievements—the acquisition of the enormous body of scientific knowledge we now possess and the development of those open liberal democratic societies which have enabled scientific freedom—and many other freedoms—to flourish. Two major thinkers, Karl Popper and Friedrich Hayek, have written clearly and profoundly about the connections between the philosophical principles underlying such societies and those shaping the development of science. During the dark days of World War 2 both men wrote books—Popper's *The Open Society and its Enemies*[11] and Hayek's *The Road to Serfdom*[12]—intended for a wide readership, in which they sought to expound and to justify the principles of open liberal societies. They have also written other more specialised works which repay careful study. These include Popper's *The Logic of Scientific Discovery*[13] and *Conjectures and Refutations*[14] and Hayek's *Law, Legislation and Liberty*[15] and *The Counter-Revolution of Science*[16]. Both men have also recently set out some of their key ideas in more popular form—Popper in his intellectual autobiography *Unended Quest*[17] and Hayek in his Hobhouse Lecture *The Three Sources of Human Values*, given at the London School of Economics in 1978 and reprinted as an Epilogue in Volume 3 of *Law, Legislation and Liberty*.

Two other writers have significantly influenced my treatment of the status of scientific knowledge and its relation to other forms of knowledge. The first is M. Polanyi, whose short book *The Tacit Dimension*[18] deals clearly and briefly with some of the more personal aspects of scientific discovery. The second is M. Hesse who has been an invaluable guide through her clear and authoritative expositions of the philosophy of science and, in particular, of her own network model of science. The introduction *The Task of a Logic of Science* to her *Structure of Scientific Inference*[19], together with the first two chapters *Theory and Observation* and *A Network Model of Universals* have been especially useful. In addition her *Revolutions and Reconstructions in the Philosophy of Science*[20] gives a valuable and balanced introduction to many recent debates and developments in the philosophy, history and sociology of science, while her concluding essay in that book, *Criteria of Truth in Science and Theology*, extends the debate about the nature of knowledge in fruitful and relatively uncharted directions.

REFERENCES

1 Butterfield, H., *Origins of Modern Science: 1300–1800*, (London: Bell, 1951).
2 Gillispie, C. C., *The Edge of Objectivity*, (Princeton: Princeton University Press, 1960).
3 Gillispie, C. C. (ed.), *Dictionary of Scientific Biography*, 14 vols, (New York: Charles Scribner, 1972–6).
4 Bronowski, J., *The Ascent of Man*, (London: BBC Publications, 1974).
5 Ziman, J., *The Force of Knowledge*, (Cambridge: Cambridge University Press, 1976).
6 Roberts, J. M., *The Hutchinson History of the World*, (London: Hutchinson, 1975).

7   Ben-David, J., *The Scientist's Role in Society*, (New Jersey: Prentice Hall, 1971).
8   De Solla Price, D., *Little Science, Big Science*, (Columbia: Columbia University Press, 1963).
9   Medvedev, Zh., *Soviet Science*, (Oxford: Oxford University Press, 1979).
10 Beyerchen, A. D., *Scientists under Hitler: Politics and the Physics Community in the Third Reich*, (New Haven: Yale University Press, 1977).
11 Popper, K. R., *The Open Society and its Enemies*, (London: Routledge and Kegan Paul, 1966).
12 von Hayek, F. A., *The Road to Serfdom*, (London: Routledge and Kegan Paul, 1944).
13 Popper, K. R., *The Logic of Scientific Discovery*, (London: Hutchinson, 1972).
14 Popper, K. R., *Conjectures and Refutations*, (London: Routledge and Kegan Paul, 1974).
15 von Hayek, F. A., *Law, Legislation and Liberty*, 3 vols, (London: Routledge and Kegan Paul, 1973, 1976, 1979).
16 von Hayek, F. A., *The Counter-Revolution of Science*, (Indianapolis: Liberty Press, 1979).
17 Popper, K. R., *Unended Quest*, (London: Fontana, 1976).
18 Polanyi, M., *The Tacit Dimension*, (New York: Doubleday, 1966).
19 Hesse, M., *Structure of Scientific Inference*, (London: Macmillan, 1974).
20 Hesse, M., *Revolutions and Reconstructions in the Philosophy of Science*, (Brighton: Harvester, 1980).

INDEX

Academie des Sciences, 83, 120, 125, 126

Aeroplanes, 242

Agriculture, 1, 4, 6, 97, 113, 129, 157, 212–13, 224, 329–30, 383, 393, 394–6, 397, 415, 423, 432, 434, 464–5

Air pumps, 50

Alchemy, 87–9

Alexander (the Great), 10

Alexandria, 10

Algae, 6

Algebra, 60, 231

Alienation, 204–5

Almagest, 18, 19

American Declaration of Independence, 113–14, 117, 119

American Revolution, 1, 113, 117, 119–20, 127

Amino-acids, 283–92

Ampere, A. M., 252

Anaesthetics, 301–3

Anatomy, 53, 56, 161, 296

Aniline dyes, 192

Animalcules, 48

Annales de Chimie, 150

Anthrax, 301

Antibiotics, 242, 305

Antisemitism, 373, 485

Antisepsis, 301, 303

Apes, 5, 53, 183

Apollonius, 14–16, 76

Aqueducts, 135

Archimedes, 41

Aristarchus, 11

Aristotle, 10, 12, 13, 18, 36, 37, 53, 149

Arithmetic, 231

Arkwright, Richard, 130, 132, 141

Astbury, 284

Astrology, 11

Astrolabe, 19–20

Astronomia Nova, 28, 29

Astronomy, 6, 9, 13–16, 19, 30, 35, 44, 46, 53, 90–1, 231, 251, 325, 338–9, 344

Athens, 9, 10

Atomic bomb, 241. *See also* Nuclear bombs

Atomic theory, 152–4, 264, 311, 315–20, 340, 361, 453

Babbage, Charles, 331–4, 458

Babylon, 4

Bacon, Francis, 80–1, 143

Bacteria, 6, 50, 301–2

Bacteriophage, 284

Balances, 145, 147, 165

Balloon flights, 149

Barometers, 45, 49–51, 135

Batteries, 154, 172, 193, 252

Beagle, 175–6

Becquerel, A. C., 315

Behrens, C. B. A., 117–18

Benzene, 192

Bernouilli, J., 325

Berthollet, C. L., 126

Biblical criticism, 88. *See also* Religious attitudes

Big Bang Theory, 6, 343, 345–6

Binomial theorem, 67, 231

Biochemistry, 242

Biology, 174–83, 276–92

Biometry, 329

Black, Joseph, 145–6, 157, 165

Blast furnaces, 97, 100, 132, 197

Blood circulation, 47, 56–7, 297

Bohr, Niels, 273, 315–20

Boulton, Matthew, 139, 141, 142, 157, 191

Boyle, Robert, 51, 82

Brahe, Tycho, 25–7, 65, 89, 325, 345

Brewing, 438

Bridges, 134, 188, 234–5

Broglie, Louis de, 320

Bronowski, J., 334, 371, 376

Bronze age, 4

Brownian motion, 264

Brunel, Isambard Kingdom, 188–90

Buffon, 349

Bunsen, R. W., 217, 316

Butterfield, Prof. H., 4

Calculators, 330–1, 458–9

Calculus, 67, 76

Calendar reform, 19, 125–6

Cambrian period, 6

Canals, 135, 136, 160–1, 235–6

Cancers, 306

Carboniferous period, 6

Carding, 130

Carnot, Sadi, 127, 142, 165–8

Cast iron, 132

Cave paintings, 4

Censorship, 467–8

Censuses, 325–6

Charles II, 46, 66, 82

Chemistry, 144–5, 192, 311–23

Chetverikov, S. S., 282, 383, 394, 396

Child labour, 198–9
China, 44, 222–39, 418–27, 430
Chinese Revolutions, 418–19, 421–7
Chinese Science, 222–39, 418, 420
Cholera, 200, 208–9, 301
Chromosomes, 282
Churches. *See* Religious attitudes; Roman Catholic Church
Clapeyron, 168
Clavius, 36
Clocks, 45, 52, 231–2
Coal, 6, 189–92, 214, 244
Coal gas, 190, 191
Coal tar, 191, 192
Coalbrookdale, 132, 135, 138
Coke, 132, 190
Colonisation, 217–18, 225, 433
Colour, 67, 68–9
Comets, 27, 46, 75, 231
Communications (scientific), 78–85, 378–9, 410, 491
Communist Manifesto, 203–4, 248
Compass (magnetic), 44, 231
Computers/computing, 324–37, 397, 453–62
Comte, Auguste, 107, 201–2, 204, 486
Condorcet, Marquis de, 94, 117, 125, 126, 201
Confucius, 222, 224, 419
Constellations, 11, 35
Continental drift, 352–6
Copernican Revolution, 18
Copernicus, Nicolaus, 4, 9, 11, 18–30, 36, 37, 61, 65, 75, 219, 338
Cornish pumping engine, 186–7
Cosmology, 6, 13, 28, 265, 343
Cotton industry, 1, 129–32, 198–9, 438
Coulomb, C. A., 252
Crick, Francis, 245, 284, 285, 292
Crop rotation, 113. *See also* Agriculture
Crystallography, 284, 336, 369
Cult of Reason, 125, 126
Cultural Revolution, 424–5
Cuvier, Georges, 161–2

D'Alembert, Jean, 92, 96, 99
Dalton, John, 152–4, 168, 311, 361
Dante, 2, 338
Darby, Abraham, 132, 430
Dark Ages, 18, 296
Darwin, Charles, 3, 113, 162, 175–84, 248, 276, 327–8, 349
Darwin, Erasmus, 142, 157
Das Kapital, 203
Davy Humphry, 154–5, 170
Death rates, 475–6, 477, 490. *See also* Demography
De Humani Corporis Fabrica, 53

Delbrück, Max, 284
Democracy, 90
Demography, 103, 177–8, 196–211, 212, 217, 222–4, 248–9, 276, 282, 305, 325, 383, 395, 397, 412, 441, 464–5, 471–8
De Motu Cordis, 57
De Revolutionibus, 21, 22–5
Descartes, René, 46, 47, 58–63, 65, 73, 75, 79, 361
Descent of Man, 3, 182
Dialogue . . . . Chief World Systems, 38
Diderot, Denis, 96–101, 117
Diffraction, 285, 320–3
Digges, Thomas, 24
Dinosaurs, 5
Discourse on Method, 61
Diseases, 103–4, 196, 199–200, 208–9, 292, 300–9, 490
Dissection, 53
Divine Comedy, 2, 3, 18–19, 358
Division of labour, 109–10, 204, 206, 437
DNA, 242, 383, 394
Dobzhansky, T., 282, 383, 394
Donne, John, 25
Doppler effect, 343
Drosophila, 280–3, 383, 396
Drugs, 298, 304
Durkheim, Emile, 201, 206–7, 486
Dyes, 192
Dynamics, 9, 41, 42, 58, 71
Dynamo, 172, 192–4

Earth's age, 349–56
Eclipses, 265
Ecliptic, 11, 231
École polytechnique, 126, 166
Economics, 103, 108–11, 205–6, 429–39, 486. *See also* Government, Politics
Edison, T. A., 193
Egyptian civilisations, 4
Einstein, Albert, 220, 238, 251, 256–62, 263–4, 346, 360, 368, 443
Electric motor, 170, 192
Electricity, 144, 170–3, 192–5, 252
Electrolysis, 154
Electromagnetic waves, 253
Electromagnetism, 171, 251–6
Electron microscope, 245
Electronics, 241, 322–3, 334–7, 453–4
Electrons, 241, 260, 315–23, 453–4
Empires, 225
Encyclopedia, 96–101
Energy, 441–52
Engels, Friedrich, 203
Engineering, 126–7, 186–9, 235, 409
Enlightenment, 94, 96, 115, 117–27

Epicyclic model, 15
Epidemics, 103–4, 200, 298
Essay . . . . . . Human Understanding, 106–7
Ethnography, 294
Euclid, 10, 11, 40, 41, 42, 60, 76
Euler, 99
Evolution (biological), 177–84, 203, 283, 327
Evolution (earth), 349–56
Evolution (stars), 347–9
Eysenck, H. J., 484

Fabricius, 297
Factories, 1, 130, 196–9, 204–5
Factory Acts, 198
Faraday, Michael, 170–3, 192, 252
Fascism, 373–5
Fermat, P. de, 79, 325
Fermi, Enrico, 299, 270
Ferranti, 334–5
Fertilisers, 113
Fertility (human), 475–8
Feynmann, Richard, 254
Fisher, R. A., 282, 329–30, 383
Fluxions. *See* Calculus
Food supplies, 111–13, 177, 395, 415, 423, 434
Fossils, 6, 158–62, 180
Foucault pendulum, 24
Franklin, Benjamin, 120
Fraunhofer lines, 340–1
French Revolution, 1, 101, 113–15, 118, 120–7, 145
Fresnel, 127
Freud, Sigmund, 485
Friction, 442

Galapagos Is., 175
Galaxies, 3, 6, 35, 338–49
Galen, 53–4, 296
Galileo, 9, 30, 32–42, 46, 47, 52, 58, 65, 75, 76, 79, 103, 219, 325
Galton, Francis, 328–9
Gaslighting, 191
Gauss, Karl F., 252, 325, 326, 328
Genes, 282, 291, 396
Genetics/code/engineering, 242, 276–92, 330, 383–4, 394–7, 468–9
Geology, 144, 147, 156–63, 175, 349–55
Geometry, 10, 11, 58–9, 99, 231
Geothermal power, 445
German industrialisation, 216–17, 368
German (Nazi) Science, 368–76
Germanium, 313, 456–7
Government, 106, 107–8, 108–9, 110, 120–7, 202, 226, 246, 267, 309, 363–6, 368–76, 378–83, 389–406, 409–17, 417–27, 438, 460–1, 465, 486
Gravitation, 67, 71, 75, 265

Great Quadrant, 25, 26
Greece, Ancient, 4, 9–17, 18, 42, 44, 89, 294
Green Revolution, 433
Gunpowder, 228, 229

Haber, Fritz, 371
Haemoglobin, 284, 336
Haldane, J. B. S., 282, 383
Halley, Edmund, 71, 82, 231, 325
Hargreaves, J., 130
Harmonice Mundi, 30, 89
Harmonics, 11, 30, 89
Harvey, William, 44, 56–7, 80, 297
Hayek, F. A., 490
Heat/Heat engines, 165–70
Heisenberg, W., 273
Hellmont, van, 79
Heredity, 276–83
Herschel, William, 251
Hertz, Heinrich, 242, 254
Hippocrates, 10, 53, 294
Hippocratic oath, 295
Hiroshima, 271, 414
Hitler, Adolf, 370–4, 376
Hobbes, T., 79
Hofmann, A. W., 192
Homer, 10
Homo sapiens, 5, 6, 175
Hooke, Robert, 46, 70, 82, 158, 160
Hospitals, 298, 306, 437
Housing, 197, 208, 305
Hubble, Edwin, 343
Hutton, James, 157
Huyghens, Christian, 52, 65, 325
Hydraulic engineering, 235
Hydrogen, 149, 445
Hydroelectric power, 195, 445

Inert gases, 315
Inquisition, 39, 103
Instauratio Magna, 80
Insulin, 283
Integrated circuits, 456–60
Inverse-square law, 75, 252
Iron age, 4
Iron manufacture, 1, 97, 100, 129, 132–5, 190–1, 197, 214–15, 234, 413
Irrigation, 235

Japanese science, 408–17, 430
Jefferson, Thomas, 120, 127
Jenner, Edward, 300
Joule, James, 168–70, 442
Jute, 433, 438

Kay's shuttle, 130
Kekulé, 311
Kelvin, Lord, 183, 350

Kepler, Johannes, 27–30, 36, 46, 65–6, 339
Kepler's laws, 30, 75
Kirchhoff, G., 316
Kuhn, Thomas, 18, 24, 63, 75, 360

La Dioptrique, 46, 61
Labour camps, 391, 393, 398, 422
Laennec, R. T. H., 300
La Géometrie, 60, 61
Lagrange, J. L., 91, 126
Laplace, P. S., 91, 97, 126, 325
Lascaux paintings, 4
Lasers, 307
Lavoisier, Antoine, 125, 147–52, 298
Laws of Motion, 73, 94
Leeuwenhoek, Antony van, 47–9, 301
Leibniz, G. W., 331
Lenard, P., 267, 368
Lenses, 46
Les Météors, 61
L'Esprit des Lois, 107
Lettres Philosophiques, 107
Liberalism, 106, 110, 373, 378–83, 405
Lice, 48
Liebig, J. van, 217
Light, 68–70, 193, 256, 264. *See also* Optics
Linnacan Society, 175, 178
Linnaeus, C., 174–5
Lister, Joseph, 303
Locke, John, 106–7, 378, 486
Locomotives, 187. *See also* Railways
Logarithms, 330
Logical Positivists, 358–60
Looms, 133, 331, 438
Louis XIV, 117–18
Louis XVI, 122, 123
Lunar Society, 142, 157
Luther, Martin, 21
Lyell, Charles, 162–3, 349
Lysenko, T. D., 396–7

Mach, Ernst, 241
Magdeburg hemispheres, 52
Magnetism, 144, 170–3, 228, 229, 230, 252, 353. *See also* Electromagnetism
Malpighi, 47
Malthus, Thomas, 111–13, 177, 178, 202, 212–13, 248, 464, 471, 486
Manhattan project, 270, 386
Mannheim, Karl, 481
Maps, 160, 353. *See also* Demography
Marconi, 242
Marx, Karl, 109, 184, 203–5, 206, 248, 480, 486
Mass, 72, 258, 443
Mathematics, 10, 11, 14–16, 22–4, 32, 58, 66, 67, 72, 76, 90, 101, 325, 327, 490
Mauve (discovery), 192

Maxwell, James C., 251, 252–6
Mechanical philosophy, 60, 63, 91, 330
Mechanics, 90, 91
Medical knowledge/profession, 10, 52, 53–7, 200, 294–309. *See also* Diseases, Hospitals
Medici, Cosmo di, 34
Medvedev, Zhores, 389, 396–7, 399–401, 403–5
Mendel, Abbé Gregor, 242, 276–81, 327
Mendeleev, Dmitry, 311–15, 347, 361
Mercury/poisoning, 50, 466
Mersenne, Marin, 79–80
Mesons, 414
Meteorology, 61
Meteors, 231
Method of Chemical Nomenclature, 149
Metric system, 122
Micrographia, 46, 50, 70
Microprocessors, 456–62. *See also* Computers
Microscopes, 45, 46–9, 245
Milky Way, 35
Mines, 135, 189, 199, 434. *See also* Coal
Molecular biology, 276, 283–92, 383–4, 394–6, 398, 468
Montesquieu, Baron de, 107–8, 127, 486
Moon, 11, 27, 34, 35, 338
Momentum/motion, 13, 14, 41, 66, 72–3, 75, 172, 258, 260, 442
Morgan, T. H., 279–81, 383
Müller, H. J., 217, 282
Mussolini, Benito, 373–4
Mutations, 282

Napier, John, 330
Napoleon, 127
National Socialism (Nazism), 267, 270, 368–76, 393
Natural gas, 444
Natural selection, 179, 182, 283, 330
Neanderthal man, 5
Needham, Joseph, 227, 234
Neolithic revolution, 4, 6
New System of Chemical Philosophy, 153
Newcomen, 136–8
Newton, Isaac, 4, 9, 42, 58, 63, 65–76, 87–102, 105–6, 219, 251, 339, 360, 442
Niagara Falls, 195
Nova, 25, 231, 345
Nuclear accidents, 385, 399–401
Nuclear bombs, 272, 384–6, 398, 399, 424, 425
Nuclear energy, 384–6, 398, 441–52
Nuclear fusion, 445
Nuclear hazards, 385, 399, 446–9, 451
Nuclear power, 441, 444–5, 446–51
Nuclear reactions, 260–1, 268–72, 350, 443

Observatories, 25
Oersted, H. C., 170, 252

Oil deposits, 6, 444
Oldenburg, Henry, 82
Olympic Games (Greek), 9 294
Oppenheimer, J. R., 378, 386–7
Opticks, 90, 94–6, 105, 219
Optics, 61, 68–70
Origin of Species, 175, 177–80, 202
Orrery, 92–3
Ostwald, W., 241
Oxygen, 147

Parabolic paths, 41, 42
Paracelsus, 144, 296–7
Paradiso, 2, 3
Paré, A., 298
Pascal, B., 50, 79, 231, 325, 331
Pasteur, Louis, 301–2
Pauling, Linus, 285
Pavlov, I. P., 483
Pearson, Karl, 329
Peking Man, 5
Pendulum clock, 45, 52
Periodic table, 241, 311–15, 347, 361
Perkin, William, 192
Perutz, Max, 284
Philosopher's table, 93–4
Philosophy of science, 357–66, 493–7
Photoelectric effect, 273
Physics, 144, 165–73, 220, 241, 256, 315
Physiocrats, 108–9
Physiology, 53
Pinmaking, 97, 100, 109
Plague, 67
Planck, Max, 264, 267, 371
Planetary motion, 30, 46, 65, 67, 71, 73–5, 90–1. *See also* Astronomy, Cosmology
Planets, 11–16, 23, 27–30, 34–6, 46, 66, 251–2, 338–9
Plantations, 434
Plato, 10, 11
Poland, 467–8
Poliomyelitis, 305
Politics and science, 125, 267, 270, 363–6, 368–76, 389–406, 408–17, 418, 420, 493–7
Pollution, 296–7, 466–8
Popes. *See* Religious attitudes
Popper, Karl, 115, 265–6, 358–9
Population study. *See* Demography
Power stations, 194, 444
Priestley, Joseph, 142, 145–7, 252
Principia Mathematica, 242
Principia Naturalis, 65, 70–6, 90, 105, 219
Principia Philosophica, 61–3, 75
Principles of Geology, 162, 177
Printing, 44, 228, 229
Prisms, 68, 69
Projectiles, 41, 42, 66, 75, 90

Proteins, 283–4
Protons, 347–8
Psychiatry, 307–8
Psychology, 103, 106, 201, 307–8, 483–8
Ptolemy, 15–16, 18, 23, 36, 37, 53
Public health, 209, 299, 304, 466
Punched cards, 332
Pythagoras, 9, 10, 11, 89

Quantum theory, 264, 273, 292, 322

Racialism, 373, 434, 485
Radar, 344, 454
Radioactive dating, 6, 350–2
Radioactivity, 315–20, 350, 399–400, 443, 458, 467
Radio astronomy, 6, 339, 344–7, 458
Radiotherapy, 307
Radio waves, 242, 254
Railways, 1, 186–9, 192, 193, 216, 409
Ramsay, 315
Rationalism, 490–3, 494–7
Red Guards, 424–5
Reflection, 95–6
Reformation (religion), 21
Refraction, 68, 95–6
Relativity, 251, 256–62, 264–5, 346
Religious attitudes, 3, 18, 21, 25, 92, 94, 103, 104, 110, 117, 122, 125, 145, 182–3, 207, 219, 247, 296, 371–6, 394, 408, 487
Renaissance, 44, 296
Respiration, 298
Retrograde motion, 14
Robespierre, 126
Rocketry, 247, 424. *See also* Cosmology
Rocks, 159–63
Roman Catholic Church, 32, 34, 36–9, 122, 125. *See also* Religious attitudes
Roman Empire, 10, 18, 224
Romantic movement, 101
Röntgen, W. K., 306
Royal Institution of London, 154
Royal Society of London, 46, 71, 82, 87
Russell, Bertrand, 242
Russian Revolution, 1, 247, 389
Russian Science, 389–406
Rutherford, Ernest, 268, 315–16

Sakharov, Andrei, 398, 403–4
Saliva, 49
Sanger, Frederick, 283
Sanitation, 196, 199–200, 305
Satellites, 35, 46, 90, 338–9
Savery, Thomas, 135
Sceptics, 79
Schools, 126, 375–6
Schrödinger, Erwin, 273, 292, 320–3, 453
Scientific revolutions, 4, 360

Scientific societies, 46, 71, 82
Scientism, 247–8
Seismographs, 234
Semiconductors, 322, 453–60
Shockley, William, 454–6
Silicon chips, 334–6, 453–60
Simplicio, 39
Slide rules, 53, 330
Smallpox, 234, 300
Smith, Adam, 108–11, 157, 201, 202, 204, 484
Smith, William, 159–61
Social Darwinism, 184, 248
Society and science. *See* Government, Politics
Social science/Sociology, 103–15, 184, 196–211, 422, 480–8
Socrates, 10
Solar power, 445
Solubility, 152
Solzhenitsyn, A., 398
Spacecraft, 90, 242, 244, 458
Sparta, 9, 10
Spectra, 68, 317, 340–3
Spectroscopy, 245, 316–17
Spencer, Herbert, 178, 184, 202–3, 204, 486
Spinning, 1, 130, 433
Spinning Jenny, 130
Stalin, Josef, 393–6, 403–4
Stark, Johannes, 267, 368
Starry Messenger, 33–4, 36, 338
Statistics, 324–37
Steam engines, 1, 129, 135–41, 165, 186–7
Steam ships, 188
Steel production, 198, 214, 215, 234, 413, 423
Stellar parallax, 24, 27
Stephenson, Robert, 187
Stethoscope, 300
Stone Age, 4
Stoney, G. J., 241
Student's test, 329
Sub-luminary sphere, 13
Submarine cables, 192
Suicide, 207, 481
Sulphonamides, 305
Summa Theologica, 18
Sun, 11, 73, 261, 350. *See also* Solar power
Sunspots, 36
Surgery, 298, 307
Szilard, L., 269
Systema Naturae, 174

Technology choice (third world), 429–40
Telegraph, 192
Telescopes, 3, 25, 33, 35, 45–6, 68, 231, 338, 344. *See also* Astronomy
Thermionic valve, 454
Thermodynamics, 168–70, 183, 350, 442
Thermometers, 45, 52, 165

Thermoscope, 52
Third world, 430–40, 471
Thomas Aquinas, 18
Thompson, J. J., 241, 315
Thought reform (China), 422
Tides, 75
Torricelli, 50–1, 135
Trajectories, 41
Transistors, 454–8
Trevithick, Richard, 187
Tuberculosis, 300, 304–6
Tull, Jethro, 97, 99
Two New Sciences, 40, 41, 42, 49, 58
Two-sphere universe, 11–18

Universities, 127, 217, 296, 368, 405, 408, 410, 424–5

Vaccination, 234, 300
Vacuum pumps, 45
Vavilov, N. I., 394, 396, 401–2
Vesalius, 44, 53–6
Vibration, 11
Vitriol, 89
Volcanoes, 163
Volta, A., 154, 172, 252
Voltaire, 91–4, 107, 117, 122–3
Von Guericke, 51, 135
Vulcanists, 157

Wallace, Alfred, 113, 178
Warfare thesis, 36
Wars, 123, 127, 246–7, 270, 376, 379, 384, 392, 398, 413–14, 418–19, 481. *See also* Government, Politics
Water pumps, 49
Water supplies, 130, 234
Waterwheels, 130, 234
Watson, James, 245, 284, 285, 292
Watt, James, 1, 138–41, 142, 157, 191
Wave mechanics, 273, 320–3, 453
Wealth of Nations, 108
Weaving, 1, 130, 194, 331, 433
Weber, Max, 201, 205–6, 480
Wedgwood, Josiah, 142
Werner, Abraham, 156, 349
Wheat production, 464
Whewell, William, 219
Whitehead, A. N., 242
Wilkins, Maurice, 285, 292
Wilkinson, John, 139, 141, 145, 198
Wren, Christopher, 71, 82
Wright, Sewall, 282, 383

X-rays, 242, 245–6, 284, 285, 306, 315, 336, 369